Princes, Brokers, and Bureaucrats

Princes, Brokers, and Bureaucrats

Oil and the State in Saudi Arabia

Steffen Hertog

Cornell University Press
Ithaca and London

This publication has been produced with the assistance of the European Union. The contents of this publication are the sole responsibility of Steffen Hertog and can in no way be taken to reflect the views of the European Union.

Cornell University Press gratefully acknowledges support from the Fondation Nationale des Sciences Politiques, Paris, which aided in the publication of this book.

Copyright © 2010 by Cornell University

All rights reserved. Except for brief quotations in a review, this book, or parts thereof, must not be reproduced in any form without permission in writing from the publisher. For information, address Cornell University Press, Sage House, 512 East State Street, Ithaca, New York 14850.

First published 2010 by Cornell University Press

Printed in the United States of America

Library of Congress Cataloging-in-Publication Data

Hertog, Steffen.
   Princes, brokers, and bureaucrats : oil and the state in Saudi Arabia / Steffen Hertog.
     p. cm.
   Includes bibliographical references and index.
   ISBN 978-0-8014-4781-5 (cloth : alk. paper)
   1. Petroleum industry and trade—Government policy—Saudi Arabia. 2. Economic development—Political aspects—Saudi Arabia. 3. Bureaucracy—Saudi Arabia. 4. Patron and client—Saudi Arabia. 5. Saudi Arabia—Economic policy. 6. Saudi Arabia—Politics and government. I. Title.
   HD9576.S33H47   2010
   330.9538'05—dc22   2009032730

Cornell University Press strives to use environmentally responsible suppliers and materials to the fullest extent possible in the publishing of its books. Such materials include vegetable-based, low-VOC inks and acid-free papers that are recycled, totally chlorine-free, or partly composed of nonwood fibers. For further information, visit our website at www.cornellpress.cornell.edu.

Cloth printing   10 9 8 7 6 5 4 3 2 1

Contents

Preface	vii
List of Acronyms	ix
Dramatis Personae	xi
Introduction	1
1. Unpacking the Saudi State: Oil Fiefdoms and Their Clients	9
Part I: Oil and History	37
2. Oil Fiefdoms in Flux: The New Saudi State in the 1950s	41
3. The Emerging Bureaucratic Order under Faisal	61
4. The 1970s Boom: Bloating the State and Clientelizing Society	84
Part II: Policy-Making in Segmented Clientelism	137
5. The Foreign Investment Act: Lost between Fiefdoms	143
6. Eluding the "Saudization" of Labor Markets	185
7. The Fragmented Domestic Negotiations over WTO Adaptation	223
8. Comparing the Case Studies, Comparing Saudi Arabia	246
References	277
Index	289

Preface

The journey that a book project takes is almost always longer than planned. Mine brought me back to Saudi Arabia half a dozen more times after I first thought I had collected enough material for a manuscript. The first time, I was a bit reluctant to come back. By now I have developed an attachment to the Kingdom and its people which is probably for life, and Riyadh has become a kind of second home. It is not only curiosity, but also true friendships that have brought me back time and again. I have discovered that discussing Saudi politics with even the best Western experts does not compete with a relaxed chat with curious and critical local friends. I thank all these friends, most of whom will remain unnamed, for their time, friendship, and generosity.

I also extend my gratitude to the Daimler-Benz Foundation, who made possible the research on which the initial draft of this book was based, the Princeton Environmental Institute for supporting my follow-up research in 2006–07, and the Kuwait Foundation for the Advancement of Sciences for funding the Kuwait Program at Sciences Po that supports my current position, and has provided me with the space and resources to finish the book in a world-class intellectual environment—and not just anywhere, but in Paris! This book also owes much to its editor, Roger Haydon at Cornell University Press, and the great copyediting team there.

A number of individuals should be thanked by name: Philip Robins for allowing me to work at my own pace while helping me to navigate Oxford University's institutional intricacies; my mentor, friend, and colleague Giacomo Luciani for teaching me the virtues of patience, and much else; Abdulaziz Al-Sager and his pioneering Gulf Research Center for their repeated crucial assistance and trust in me. My deep gratitude for, among other things, the numerous discussions and insights that helped to shape this book also goes to Abdulaziz Al-Fahad, Abdulaziz Al-Bsaily, Suliman Al-Mandeel, Abdulaziz Al-Uwaisheq, Greg Gause, Bernard Haykel, John Sfakianakis, and Bob Vitalis.

I am particularly beholden to my young colleagues and friends Stephane Lacroix, Thomas Hegghammer, and Christopher Boucek—it is rare that scholarship and friendship go together so neatly. I am also much indebted to my old friend (and non-Middle East expert) Sebastian Karcher for patient and perceptive comments on many half-baked versions of the manuscript that contained a lot more tedious detail than the current one.

Most of all, I thank my wife, Ekaterina, who was the one person to help me through the really rough patches during the last seven years, and she knows how many there were. Engaged in parallel career tracks, academics understand each other all too well, and they too often are forgiving of the histrionics and privations of the profession. I would like, on this occasion, to pledge betterment!

Finally, my deep gratitude goes to my parents, for supporting me when going off to London a decade ago to study the Middle East seemed a flight of fancy, and for their unending support ever since. I dedicate this book to them.

I gratefully acknowledge permission to reprint parts of the following articles: "Shaping the Saudi state: human agency's shifting role in rentier-state formation," *International Journal of Middle East Studies* Vol. 39, No. 4 (November 2007); and "Two-level negotiations in a fragmented system: Saudi Arabia's WTO accession," *Review of International Political Economy,* Vol. 15, No. 4 (October 2008). The latter journal is available online on http://www.informaworld.com.

Acronyms

CCI	Chamber of Commerce and Industry
CDS	Central Department of Statistics
CR	Commercial Registration
CSCCI	Council of Saudi Chambers of Commerce and Industry
FIA	Foreign Investment Act
GCC	Gulf Cooperation Council
GOTEVOT	General Organization for Technical and Vocational Training
IPA	Institute of Public Administration
KAEC	King Abdullah Economic City
MoC	Ministry of Commerce (also cited as MoCI, i.e. Ministry of Commerce and Industry, the title it acquired in 2003)
MoDA	Ministry of Defense and Aviation
MoF	Ministry of Finance
MoFA	Ministry of Foreign Affairs
MoI	Ministry of Interior
MoIE	Ministry of Industry and Electricity
MoJ	Ministry of Justice
MoL	Ministry of Labor (also cited as MoLSA, i.e. Ministry of Labor and Social Affairs, the title it had until 2004)
MoMRA	Ministry of Municipal and Rural Affairs
NG	National Guard
PCA	Presidency of Civil Aviation
RCJY	Royal Commission for Jubail and Yanbu

SABIC	Saudi Basic Industries Corporation
SAGIA	Saudi Arabian General Investment Authority
SAMA	Saudi Arabian Monetary Agency
SCT	Supreme Commission for Tourism
SEC	Supreme Economic Council
SIDF	Saudi Industrial Development Fund

Figures in the book are often cited in Saudi Riyals (SR). The Riyal has been pegged to the U.S. dollar at a rate of 3.75 SR/US$ since 1986.

Dramatis Personae

(only the functions appearing in the narrative are given)

Abdallah bin Abdulaziz: head of National Guard, 1962–, Crown Prince, 1982–2005, King, 2005–

Abdallah bin Faisal bin Turki: Governor of the Saudi Arabian General Investment Authority (SAGIA), 2000–04

Abdulaziz bin Abdulrahman ("Ibn Saud"): King, 1932–53

Alamy, Fawaz: Deputy Minister of Commerce, 1996–2008

Ali, Anwar: Pakistani IMF expert and Governor of the Saudi Arabian Monetary Agency, 1957–74

Dabbagh, Amr: Governor of SAGIA, 2004–

Fahd bin Abdulaziz: Minister of Education, 1953–60, Minister of Interior, 1962–75, Crown Prince, 1975–82, King, 1982–2005

Faisal bin Abdulaziz: Crown Prince, 1953–64, King, 1964–75

Faqih, Usama: Minister of Commerce, 1995–2003

Al-Gosaibi, Ghazi: Minister of Industry and Electricity, 1976–83, Minister of Health, 1983–85, Minister of Labor, 2004–

Khalid bin Abdulaziz: Crown Prince, 1965–75, King, 1975–82

Mish'al bin Abdulaziz: Minister of Defense, 1951–56

Mit'eb bin Abdulaziz: Deputy Minister of Defense, 1951–56, Minister of Public Works and Housing, 1975–2003, Minister of Municipal and Rural Affairs, 2003–

Naif bin Abdulaziz: Minister of Interior, 1975–

Al-Namlah, Ali: Minister of Labor and Social Affairs, 1999–2004

Salman bin Abdulaziz: Governor of Riyadh, 1962–

Saud bin Abdulaziz: King, 1953–64

Sulaiman, Abdallah: Saudi Minister of Finance, 1932–54

Sultan bin Abdulaziz: Minister of Agriculture, 1953–55, Minister of Communications, 1955–60, Minister of Defense and Aviation, 1962–, Crown Prince, 2005–

Sultan bin Salman: Secretary General of the Supreme Commission for Tourism, 2000–

Yamani, Hashim: Minister of Commerce and Industry, 2003–08

Princes, Brokers, and Bureaucrats

Introduction

The Puzzle: In the Telex Room

When I entered the telex room of a notable Saudi ministry in the summer of 2003, it dawned on me that something was wrong with communications between the different parts of the Saudi government. A Sudanese expatriate employee with a massive turban sat before several flickering twelve-inch amber computer monitors dating from the 1970s. Occasionally, an old matrix printer in a corner, which at first appeared to be decommissioned, would come noisily back to life and hammer out a message on its endless reel of paper.

This was the way Saudi ministries communicated with each other in 2003, after billions had been poured into modernizing the Kingdom's administration, after three consecutive years of massive oil revenues.

It was an interesting case of bureaucratic archaeology in its own right, but the telex room was also a piece of the broader empirical puzzle I had been investigating for some time: during the previous decade, practically all of the Saudi political elite had converged on a raft of economic reforms, including privatization, Saudization of labor markets, capital markets reform, liberalization of foreign investment rules, and—to the extent that there were clear views at all—WTO accession. State elites had become quite willing to redefine the role of the administration, and the private sector, having grown capital-rich over the years, was now willing to assume new developmental tasks.

But although policy consensus appeared widespread, the record of Saudi economic reform since 1999 had been decidedly uneven. This intrigued me, as it fit neither the official narrative of wisely guided development nor the Western clichés of inescapable Saudi corruption. Why did the regulation of the Saudi labor market prove so elusive, while more specific projects, such as the privatization of Saudi Telecom in 2003, were conducted successfully? Why was it so much harder to improve the regulatory environment for foreign investors in the 2000s than building a world-class infrastructure had been a quarter of a century earlier? Why did some policies get implemented swiftly while others foundered in bureaucratic limbo?

Saudi Arabia's mixed record did not square easily with the accepted ways of explaining the Kingdom's political economy. In particular, theories of the "rentier state," for which the Kingdom has always served as a primary example, painted with too broad a brush. While they did provide a useful way to think about some generic problems of oil-based development, they were less useful in explaining degrees of success and failure—arguably the most interesting puzzle in a complex system like Saudi Arabia, by no means a developmental failure. Rentier theories predict that oil income will allow states to act independently of demands in society, and that it will empower the state to the detriment of social forces, albeit at the cost of weakened regulatory capacity of state institutions and pervasive rent-seeking. All true of Saudi Arabia—in some regards. Yet none of these predictions capture the complexities of Saudi policy-making or the variation in outcomes.

The Sociology of Sharing the Wealth

Eventually, I realized that the Saudi story pointed up a crucial weakness of rentier theory: although the literature predicts that resource-rich states and economies will exhibit specific features—and is often right in these prognostications—the accounts of *how* these outcomes come about, where they exist, are usually brief and general. Much of the rentier state debate lacks empirical analysis of the causal mechanisms on any but the most general level.

Attention to specific causal mechanisms, however, is what we need to address the Saudi puzzle of highly heterogeneous reform outcomes, in which rentier effects seem to obtain on some occasions but not on others. This need for specificity is the guiding motif of this book. I argue that oil has mattered a great deal in shaping the Saudi state, its power structures, and its policy-making patterns but not always in ways we might have expected, and through avenues we can understand only if we grasp the specific ways that oil has affected the power structures within the Kingdom over time.

To tease out the channels through which oil income has influenced Saudi politics, I chose a strategy that required more than a year of field research and archival work on three continents: tracing the *sociology* of rent distribution in the kingdom. If oil income matters, how so? What exactly is done with it? What do people do when they build a rentier state and negotiate policies within it? Who gets access to state resources and how? What kinds of power relations are established in the process? To tackle these issues, I had to look at concrete social networks within and around the Saudi state. Rentier theories work mostly with macro-aggregates such as "state," "society," or "business." My research supplies meso- and micro-foundations to rentier processes, explaining the impact of rent distribution on specific institutions, societal groups, and social networks. It is probably the first to do so systematically.[1]

Grasping the concrete impact of oil money on power relations has allowed me to delimit—and, on occasion, contradict—some of the bigger claims from the rentier state debate and to improve the causal underpinning of others. Starting with the onset of oil-based state building in the 1950s, this book traces the causal path from external rent incomes, via the interaction of rent with local social structures, to policy outcomes in the 2000s as conditioned by these structures.

A first major finding is that elite decisions are of enormous import in shaping the state, especially at the early stage of state building. There is no automatic mechanism that produces corruption, rent-seeking, and a weak bureaucracy. While Saudi royals have on many occasions used their fiscal authority to build personal fiefdoms or to employ veritable armies of idle bureaucratic clients, on others they have used their resources to build efficient administrative bodies by purchasing international expertise and offering attractive career paths to ambitious nationals. If anything, large increments of oil income have *increased* the menu of institutional options available to the elite, resulting in a state apparatus with highly varied components.

The Saudi state apparatus has played a crushingly dominant role in national politics. Saudi politics has been highly centralized around the regime elite, whose patronage and largesse undermined the autonomy of social groups. Yet the disjointed nature of state growth and the hub-and-spoke structure of a system almost entirely centered around the royal family have led to great heterogeneity and indeed fragmentation of social groups and, crucially, state institutions at the lower levels of the polity.

And while politics has been decidedly top-down and dominated by vertical relationships, the regime quickly built up large-scale fiscal obligations toward its various clients in society, be they groups or individuals. Over time, this paternal largesse has proved difficult to reverse, much reducing the leadership's autonomy in freely disposing of its oil money, a dynamic that highlights the great importance

---

1. John Davis (1987) has looked at local tribal politics in the Libyan rentier system in the 1980s, but with mostly ethnographic methods and employing the rentier concept only as a very loose framework.

of distributional decisions made at earlier historical junctures. The large cascades of rentier clients accrued over time have been useful in pacifying society on the political level, but their immovable presence in and around the bureaucracy makes reform and day-to-day administrative control more difficult.

After a dash of rapid development and change through expansion, the Saudi state has thus emerged as a surprisingly fragmented, immobile behemoth—albeit one with some very efficient and capable parts. To explain why in today's Saudi Arabia some reforms work while others do not requires that we "unpack" the state. We need to understand its structure and its relations to society on the meso-level of specific organizations and social groups as well as the micro-level of individual clients.

The meso-level fragmentation of the Saudi state means that the capability to coordinate and integrate policies between different institutions and networks is low—witness the failing amber screens in the telex room, where messages arrived as if sent from another continent. Conversely, reform policies that involve only a few organizational players or are coordinated through powerful international actors are managed more successfully. Even then, however, the success of implementation is conditioned by how many micro-level actors need to be activated and steered in the process, be they administrators in the clientelist bureaucracy or the individuals in society whom they are supposed to regulate. While Saudi society is weak in terms of collective organization, small-scale social networks in the bureaucracy can often scupper policy. Meso-level "width" and micro-level "depth" of a policy are crucial determinants for its success or failure. The differentiation of rentier politics on meso- and micro-levels allows for a more nuanced understanding of the "state capacity" of rentier bureaucracies, a factor that can vary strongly within a system depending on the institutional context.

A New Take on Received Wisdom

In-depth analysis of the paradigmatic Saudi case allows us to give the rentier state debate stronger foundations. It adds nuance to the assertion that rentier states are autonomous from society. Initially, in fact, oil income did give regime elites large fiscal leeway. This autonomy can strongly decline over time, however, as the state gets tied down in society as it incurs micro-level distributional obligations that are difficult to reverse.

The Saudi case confirms the theoretical expectation that rents empower the state vis-à-vis society. The mechanism underlying this outcome is that dependence on rents tends to make social groups subordinate to the regime and to undermine their internal cohesion. Yet large oil surpluses also seem to tempt political elites to add ever more organizations to the state apparatus to address both political and administrative problems, resulting in the fragmentation of the state itself. This is a heretofore undocumented outcome of rent income that seems to obtain in other rentier cases, too.

As important an insight is that "flabby" bureaucracies with limited regulatory powers, generally attributed to rent income, are more contingent than the literature would have us expect. Depending on how the leadership decides to use its newfound income in the state-building process, they may appear or may not. If they do, then the underlying causal mechanism is the use of bureaucratic employment as a patronage resource that saps individual-level bureaucratic incentives. In other cases, handpicked technocratic clients of the Al Saud have built up lean and well-managed "islands of efficiency" in the Saudi state apparatus, with separate recruitment and incentive systems and, in some cases, explicit mandates to bypass the rest of the bureaucracy. Unlike what has been argued before, I find that the presence or absence of national-level taxation has little to do with the state's regulatory powers.

The rent-seeking and corruption that the literature predicts do occur in Saudi Arabia, not least as some state institutions double as the private fiefdoms of senior players in the regime. The extent of corruption varies greatly from institution to institution, however, and it usually is not the main cause for unsatisfactory policy results. Those have more to do with state fragmentation and clientelist inertia in the bureaucracy.

What I call "segmented clientelism"—a heterogeneous system of formal and informal, rent-based clientelism in which vertical links dominate—allows for a finer-grained analysis than do rentier state theories per se. It allows us to understand where rentier-derived phenomena kick in and where not as well as which concrete power relations underlie them.

The main ambition of this book is to provide a comprehensive, revisionist account of the Saudi political economy—a primary exhibit for the claims of rentier theory and a very important case in its own right. As it is theoretically informed, the book has obvious comparative implications, and evidence from other rentier states suggests that bureaucratic fragmentation, islands of bureaucratic excellence, and declining regime autonomy through the "locking in" of clientelist fiscal obligations over time are indeed more general phenomena.

Although oil creates a propensity toward these outcomes, however, it does not produce them automatically, and rents are not their only possible cause. In its historical-institutional approach, this book is a call for relative specificity. Oil income can influence politics through specific mechanisms, but when these mechanisms are activated and how they combine is difficult to predict ex ante, as both pre-oil social structures and elite decisions along the road of oil state-building influence how rents are used, which after all are merely a passive resource.

By making more modest and contextualized claims, I hope to work toward making the rentier debate a genuine social scientific research program. After we have established that something distinct can be said about the effects of rent income on politics, we will have gone through the grand and abstract claims typical of a young subdiscipline. It is time to "normalize" the debate by becoming more modest, doing the hard slog of empirical research, attending to specific causal mechanisms, and acknowledging the importance of contexts and boundary conditions.

Map of the Book

The structure of the book follows the argument I have adumbrated above. After a general framework chapter that situates the Saudi case in the literature and explains my theoretical and conceptual argument in more detail, chapters 2 through 4 analyze the historical roots of the modern Saudi bureaucracy and its relationship to society.

Chapter 2 narrates how the sudden surplus of oil in the early 1950s led to the rapid expansion of a state that was created above society and that early on was divided into different "fiefdoms," which were often used for material patronage. Chapter 3 analyzes how this system, initially in a state of flux, was consolidated into a more stable and formalized bureaucracy under Faisal, while retaining core features of excessive centralization, parallelism of administrative units, and clientelistic employment. The last of the history chapters outlines how "segmented clientelism" expanded to encompass all of Saudi society in the 1970s as the state accelerated its growth, expanding laterally through the addition of new institutions and creating networks of clientage and intermediation throughout all social strata.

Chapters 5 to 7 address three recent policy cases, scrutinizing in detail how the historically accumulated and increasingly rigid arrangement of Saudi bureaucratic institutions has affected policy-making and regulatory reform since the late 1990s. They cover attempts at foreign investment reform, policies of labor market Saudization, and attempts to adapt Saudi regulations to WTO requirements, respectively. Though outcomes differ from case to case, all of them show a great heterogeneity in the quality of institutions, regular communications failures, fragmented policies, the diffuse blocking power of mid-level bureaucrats, networks of brokers around administrative institutions, and games of patronage and informal clientelism—but also some surprising, if local, successes.

The concluding chapter situates the case in its historical context, summarizes the comparative findings, and sketches its implications to other policy areas not studied in detail. Then I return to my broader comparative themes, relating the Saudi case to the rentier debate as well as the debate on state-society relations in developing countries in general.

The Data

Joel Migdal has called for an "anthropology of the state" in order to overcome idealized assumptions of state coherence and a clear state-society divide.[2] I started my field research in Saudi Arabia as a "participant observer," working as a consultant to Saudi public organizations for more than a year. I have not used any *specific* data from that work, for two reasons. First, I signed a nondisclosure agreement, as almost any consultant anywhere would do. Second, even when it comes to "softer" stories of mundane bureaucracy, I would consider it a breach of my Saudi friends' and colleagues' confidence to relate such anecdotes explicitly. Everyone I dealt with knew that I was studying Saudi economic policies, but I suspect many of them would not want to end up as protagonists in stories about their daily work.

Nonetheless, my work gave me an acute sense of Saudi bureaucratic life in general, and much of my *generic* analysis is inspired not only by interviews and written material collected in the Kingdom but by my personal experience as a temporary Saudi bureaucrat as well. In all cases, I made sure that whatever I observed was generalizable to other agencies through targeted data collection outside of my home agency.

During that year and in five shorter follow-up visits, I conducted more than 120 interviews with Saudi bureaucrats, ex-bureaucrats, businessmen, bankers, expatriate advisors, and diplomats. My primary documents are various consultancy reports, official reports, laws and decrees of the Saudi government and its agencies, documents of international agencies, and Saudi, regional, and international newspapers and trade journals.

For the historical narrative, I have used archival resources on three different continents: U.S. and British diplomatic documents up to 1975, the extensive records of a Ford Foundation mission on administrative reform dating from the 1960s and 1970s, and other official documents held at the Institute of Public Administration in Riyadh, the Philby archive at St. Antony's College in Oxford, the

2. Migdal 1994: 15.

Mulligan Papers at Georgetown University (an extensive collection of Aramco documents), the Middle East Documentation Unit at the University of Durham, and the private paper collection of Lady Caroline Montagu at Mapperton (Beaminster, Dorset) on Saudi economic policies in the 1980s and 1990s. Complementing the archival record, I conducted further interviews in New York, London, and Riyadh with Saudi and Western "old hands" in business, administration and, diplomacy. Finally, I draw on many Ph.D. dissertations written by Saudis in the West about Saudi administrative development issues as well as contemporary country studies and reports in 1950s and 1960s trade journals.

Chapter 1

Unpacking the Saudi State
Oil Fiefdoms and Their Clients

After almost two decades of economic drift and stagnation, the Saudi government started an ambitious program of reforms in the late 1990s, putting not only the Kingdom's long-term economic survival but also standard theories of Saudi politics and economics to the test.

Contrary to both academic pessimists and regime propagandists, there has been great variety in policy outcomes—some quite successful, others not. On the one hand, after many half-hearted attempts at trade reform, the Kingdom witnessed swift regulatory changes when the government pushed for accession to the WTO in 2005. The Saudi regime also engineered the successful privatizations of Saudi Telecom in 2003 and the national insurance company in 2004–05 more smoothly and transparently than is customary for privatizations in most developing countries. On the other hand, the government has been struggling to streamline large parts of the bureaucracy that deals with foreign investors, despite high expectations of FDI reform and a dynamic new body in charge of revamped foreign investment policies since 2000. Similarly, it has been unable to pursue coherent labor market policies, although this is arguably the most pressing item on the regime's laundry list of reforms.

Curiously, the policies that faltered were the very ones on which there probably was the widest agreement among state elites. How to explain that? Received ways of thinking about Saudi political and economic development are of little help in solving the puzzle. Both academic and folklore explanations of the Kingdom's political economy make much of its character as "rentier state," a somewhat artificial oil-dependent entity with some characteristic deficiencies.[1] In the fields of administration and economic policy, these deficiencies include pervasive administrative corruption and "rent-seeking" behavior as well as low regulatory capacity of the state. As it does not have to levy domestic taxes, it is politically autonomous from society, but it also lacks a powerful bureaucracy that would be able to implement economic rules consistently.[2]

Something is to be said for all these ideas, but they are of little use in explaining the mixed record of reforms since 1999: after all, rents are supposed to have a uniformly negative impact. Puzzlingly, moreover, levels of corruption and rent-seeking have had little to do with reform success or failure. The new body in charge of foreign investment reform was by most accounts clean and staffed with well-intentioned administrators, but outcomes in terms of the prevailing bureaucratic atmosphere were mediocre. Conversely, labor policies improved at least somewhat when their implementation was delegated to a particularly weak and ill-reputed agency, the Ministry of Labor, in 2004.

System-wide explanations of rent-seeking or regulatory failure clearly will not do. The argument of this book is that in order to explain the striking heterogeneity of outcomes, we instead have to do something that rentier theorists have been reluctant to do: unpack the state. Oil has undeniably had a profound impact on the Saudi administrative apparatus throughout its modern history. This impact, however, has been far from uniform across the state, neither on the level of organizations nor on the level of individual bureaucratic behavior.

The Argument

The modern Saudi state was created rapidly between the 1950s and the early 1980s through the decisions of a few Saudi royals. These decisions were taken in a top-down fashion but often resulted in the oil-funded recruitment of many clients into the growing state apparatus, not least to co-opt and control society—essentially exchanging jobs and state services for political quiescence. The fiscal autonomy of the regime elite allowed them to expand the state into different directions, resulting in the parallel existence of state agencies of different quality and composition. In this hierarchical, vertically divided hub-and-spoke system, the only common denominator has been the central role of the Al Saud ruling

---

1. The discussion of rentier theories here is cursory. I size up their specific claims against evidence from the empirical chapters more systematically in chapter 8.
2. Chaudhry 1997; Vandewalle 1999; Beblawi 1987; Isham et al. 2005; Mick Moore 2004.

family as patrons and controllers of the purse and the only macro-level political force in the kingdom. Consequently, on the "meso-level" of politics—that of relations between different organizations—the state has witnessed little horizontal cooperation. Communication has been largely vertical, with the patrons on top of the system.

This pattern is reproduced on a smaller scale within state agencies, on the "micro-level" of individual organizational units and bureaucrats. Bureaucrats in large parts of the oversized state apparatus tend to be de facto distributive clients of the state, with job entitlements that sap individual motivation. This results in a penchant for referring matters upwards while giving superiors little control over the day-to-day behavior of low-level administrators. The two factors combine with the concentration of power in the Al Saud's hands to reinforce centralization and the dominance of vertical exchange.

The idiosyncrasies of policy-making in Saudi Arabia are best explained by these state structures and how they interact with society. The distinction of two levels of analysis—meso and micro—is crucial here: while vertical communication dominates on both levels within the state and can stymie policy coordination, state-society relations play out differently on each level. The fragmented state apparatus dominates politics and policy-making on the meso- or organizational level, as rent-based state-building has left no space for the emergence of powerful organized groups in society that could throw around their weight in the policy-making process (business being only a partial exception). Instead, society has been turned into a congeries of fragmented clienteles.

On the smaller, micro-scale of day-to-day bureaucracy, hierarchies are also steep, but centralized control is much less effective. Outside of the Kingdom's insulated islands of administrative excellence, clientelist bureaucratic employment undermines performance incentives and monitoring mechanisms, which leads to lax enforcement of rules and the pursuit of individual-level social interests. Small-scale personalized networks between state and society can make the large resources of the fragmented and hierarchical Saudi state available to society, thereby softening the administration's rigidity but also potentially decreasing the consistency with which policies are implemented.

The regime's use of oil rents has shaped state-society relations differently on different levels: the regime dominates on the organizational level, and it is usually politics *within* the state that shapes policy-making processes. Once policies are to be implemented, however, they are often subverted by society on the micro-level, where the state's many individual clients are difficult to police. It is the interplay of meso- and micro-politics in the Saudi distributive system that determines policy outcomes. In the fragmented Saudi context, it makes little sense to analyze "state" and "society" as coherent (macro) aggregates, as rentier theories often implicitly do.[3] Oil-based distribution has fragmented society on the meso-level but empowered it on the micro-level.

Saudi politics are shaped by the rent-based structures of "segmented clientelism": "Clientelism" denotes unequal, exclusive, diffuse, and relatively stable relationships of exchange within and around the state apparatus, on both meso- and micro-levels. It is "segmented" because of the parallel and often strictly separate existence of institutions and clienteles in the Saudi distributional system, only partially balanced through cross-cutting small-scale networks.

This system is indeed built on rents and distribution. Unlike the standard accounts however, my explanation does not take either category as an abstract aggregate but analyzes the actual power structures and social relations that mediate them over time. It therefore allows for heterogeneous state structures and historical contingencies. The timing of elite decisions has a strong impact on the state's meso-level shape, as the state's malleability has declined as it has grown larger. Client groups in society, moreover, are not static but can develop their expectations and capacities, as is the case with the Saudi private sector.

How Special Is the Saudi Case?

The underlying premise of the account—the dominance of vertical linkages in a fragmented state based on clientelist power relations—is simple, yet its historical details are necessarily complex. Both dimensions of analysis beg the question of how special the Saudi case is.

All states are heterogeneous systems. Entities as different as Mubarak's Egyptian bureaucracy and the U.S. federal administration consist of large fiefdoms that are often hard to bring in line.[4] In this sense, some aspects of the Saudi story might not be that exceptional. Yet they point to problems of economic reform and policy-making that are seldom scrutinized systematically, as most political scientists still implicitly treat the state as a unitary actor—certainly in the case of authoritarian regimes. In this sense, my account of veto players within the fragmented Saudi state apparatus opens avenues for new comparative research on politics *within* the state that could cross-cut authoritarian and democratic systems.

Beyond proposing broad-brush comparisons, however, I would argue that the heterogeneity of the Saudi state is exceptional even relative to other fragmented

---

3. As mentioned, the Al Saud are the only macro-level player, present throughout the whole system. But they are themselves a very diverse group that often lacks coherence on policy matters that do not directly impinge on regime survival.

4. Bianchi 1989; Moe and Caldwell 1994; Weaver and Rockman 1993.

systems. This has to do with how rents and royal decisions combined to shape the Saudi state over time. Oil surpluses have enabled the building of veritable states within the Saudi state; The Ministries of Defense and Interior, the National Guard, and the religious bureaucracy have reached a level of internal autonomy that is almost unrivalled among modern states. The parallelism of infrastructures within the state and the low level of interagency coordination are striking even by the messy standards of policy-making in Cairo or Washington. The related dominance of vertical over horizontal links in Saudi politics is highly unusual: many states have strong hierarchies, but it is rare for one to have so few countervailing forces that can integrate politics. The Kingdom knows no ruling party, no parliament, and no organized pressure groups that could force a stronger horizontal integration of the system.

Macro-historians might argue that such a system is not so new. Human history has seen many large, hierarchical, and heterogeneous clientelist systems, even if most predated the twentieth century.[5] Yet in Saudi Arabia, large parts of the central state play important, direct, and simultaneous roles in the lives of its citizenry, giving special salience to institutional fragmentation. This makes the Kingdom different from more conventional clientelist systems in, say, nineteenth-century Latin America, where the reach and resources of the central state were much more limited and clientelism was more mediated through local social elites and not usually formalized through state employment and other services. Rent-based clientelism in Saudi Arabia is characterized by the overwhelming role of a state that is also fragmented in itself.

The concluding chapter will say more about how the concepts developed here shed light on other cases. Suffice it to say that no other place looks exactly like Saudi Arabia, yet many of the individual causal processes described here make—yet under-researched—cameos elsewhere. They never do so in precisely the same overall constellation, but teasing out how similar social mechanisms recombine in different ways is the essence of historically oriented social science.

Oil and History: Institutionalizing a Hub-and-Spoke System

A Segmented State

Oil has transformed Saudi Arabia beyond recognition—that much is a truism. But although rent income confronted the Saudi elites with unconventional choices when they built the Saudi administration, rents did not predetermine

---

5. Eisenstadt and Roniger 1984; Eisenstadt 1973.

these choices or their immediate outcomes.[6] If anything constrained the shape the state took, it was pre-oil socioeconomic and political conditions, not the oil money with which the state was built. Different parts of the Saudi state demonstrate that elites can use oil income for markedly different purposes; oil surpluses make unconventional institutional designs possible rather than predetermine one specific trajectory. At the same time, the degree of freedom the Saudi rentier elites had in shaping their state has not been constant over time, and specific patterns of organization have been "locked in" surprisingly quickly. In some cases those elites created massive rent-seeking networks, in others large but passive bureaucracies, in yet others islands of remarkable efficiency. What has remained constant over time and across the state apparatus are the strict hierarchies centered around its elite and, just as important, the deep cleavages between different institutions.

Several parallel spheres have emerged over time within the Saudi state. They include the agglomeration of large and often overstaffed ministries run by commoners, the bureaucratized religious establishment, the vast and impenetrable bureaucracy of the Ministry of Defense and Civil Aviation under Prince Sultan, the National Guard under (now) King Abdallah, the large and omnipresent Ministry of Interior under Prince Naif, a few specialized and often quite efficient administrative bodies outside of the regular bureaucracy, the autonomously managed oil giant Saudi Aramco, and the Presidency of Youth Welfare run by a few younger princes. In many cases, these spheres do not communicate much with each other—on the rare occasions that they do, it is through their small elites or through certain senior princes hovering above the bureaucracy. Different parts of the state tend to operate on different rules and report to different principals. The religious institutions, including the judiciary, even operate on different assumptions about the very legal basis of the Saudi state, often refusing to recognize the "secular" regulations of other bodies.

This segmentation of spheres, which is sometimes reproduced on a smaller scale within organizations, has a history in which contingent decisions and structural conditions combined in a way that might only be conceivable in a young and rapidly growing oil state. We will first address the political logic of this expansion and then the mechanisms that have reinforced the growth and development trajectories of specific institutions.

There are underlying regularities in what appear to be haphazard and fragmented developments when viewed through the lens of standard approaches to

---

6. As Pete Moore (2004) has argued in his account of state-business relations in Kuwait and Jordan, states economically dependent on one sector can look quite different politically. Institutional outcomes on a meso-level need to be analyzed to understand policy output. For a standard sectoral-determinist work, see Shafer 1994.

rentier states and state-building. Historical institutionalism, a social science approach that is more open to context and contingency, offers us tools to make systematic sense of how oil and history interacted to create specific institutions. The idea of path dependency allows us to fathom the impact of early decisions on institutional development further down the road. The concepts of layering and sequencing illuminate how the timing of decisions has been crucial in shaping the state's components and point us to the declining leeway Saudi elites had in this process.[7]

The different pillars and sub-pillars of the Saudi state were created from the top down within a few decades, in a rapid fashion belying the self-ascribed conservatism of the Saudi leadership. Saudi Arabia, as Kiren Chaudhry has aptly put it, has not seen a "national moment" of political struggle or integration; rather, it is to a large degree an elite imposition.[8]

The Al Saud indeed enjoyed an extraordinary degree of autonomy from society when it came to designing their state during the 1950s and 1960s, a situation that orthodox rentier state theory predicts: in the absence of taxation, it holds, society cannot dictate regime policies. It was mostly intra-elite conflicts and top-down decisions that determined distributive decisions and institutional outcomes. I differ from Chaudhry's analysis of Saudi state formation in that I see no larger social group constraining the decisions of the Al Saud and their sparse followers during this era: neither merchants, nor tribes, nor *ulama*. Post–World War II Saudi history never saw class alliances or other horizontally integrated, collective actors contesting the state and its resources. Instead, the elite were pretty much free in their use of the sudden and massive influx of external income, which dwarfed any previous resources and gave the Al Saud enormous fiscal leeway to disburse to groups and institutions as they saw fit. As Michael Herb has demonstrated for a number of Gulf states, the state apparatus was in many cases a resource that could be apportioned in order to settle conflicts within the ruling family.[9]

Autonomy to set up new institutions did not mean that the character of these institutions would not be influenced by preexisting social structures. Largely illiterate Saudi society offered few potential "Weberian" bureaucrats to recruit, and the state depended heavily on Arab and Western expatriate advisors. The absence of a bureaucratic-regulatory tradition in the Saudi realm meant that the new formal administration would be met with distrust in society. The fiefdom character

---

7. Historical institutionalism has been loosely employed before to analyze historical phases of rentier system evolution and to point to the importance of pre-oil institutions (Crystal 1995; Vandewalle 1999: 6), but never to explain variation *within* rentier systems. Its tools are arguably more suitable to do the latter.
8. Chaudhry 1997: 192.
9. As Herb (1999: 11) underlines, political agency mediates the impact of oil.

of various ministries was the outcome of isolated distributive decisions, but those decisions were also embedded in established patrimonial traditions in the elite, clientelist expectations of new bureaucratic recruits and (closely related) the absence of an established distinction between public and private.[10]

Still, the elite had tremendous leeway in deciding on the shape of individual institutions as they operated above society at large, which in the early years had little to do with the fledgling administration. Ministries and security agencies went through a flurry of reorganizations and expansions in the 1950s, according to the elite whims and scuffles of the moment.

In this sprawling and often unplanned process, the state was seldom forced to act in a particularly integrated fashion. Instead, rising oil income allowed for lateral, sprawling growth of bureaucracies, leading to the evolution of small "statelets within the state," each of which looked quite different from one another. As long as the pie kept expanding, the creation of new institutions seemed the easiest way both to manage the balance of power in the expanding elite and to address new developmental problems. The state and the growing social groups attached to it evolved as a highly heterogeneous hub-and-spoke system centered around the royal family, from which all distribution originated in the final instance and which became the center of both state and, gradually, society.[11] Many of the strong vertical links that dominate today's Saudi state originated in the early phase of state-building.

How the Timing of Decisions Has Shaped the State

As the process of state-building was so top-heavy, elite conflicts and royal decisions at an early stage had tremendous impact on the composition of this segmented state. Saudi institutional formation is a powerful example of what historical institutionalists call path dependency, the idea that an event has much more important ramifications if it occurs early in a process than if it does so later.[12]

Conflicts and accommodations between senior princes in the 1950s and early 1960s had a much more fundamental impact on the setup of the Saudi state than they would in the 1980s or 1990s. Institutions were much more pliable tokens of political games at a moment when a formal state barely existed and whole ministries could be set up by fiat simply to balance forces within the Al Saud. Many

---

10. As Thelen (2003) points out, every account of path dependency needs to involve some analysis of constraining structures at the "front end" of the path-dependent process.

11. Fandy 1999.

12. Thelen 1999. Path dependency is also used by Terry Lynn Karl (1997) in her account of the Venezuelan rentier state, but in a much more deterministic fashion, leading to uniform outcomes of inefficiency and waste.

of these highly personalized early deals and accommodations left a permanent imprint on the state, turning specific ministries into fiefdoms and segmenting the Saudi system at the very top.

The great leeway that elite-controlled state growth gave for institutional design at an early stage also resulted in the creation of several efficient bureaucratic islands. The trust that King Abdulaziz put into an American advisor in the early 1950s led to the creation of what has until today been probably the most professional central bank in the Middle East, an enclave of efficiency shielded from day-to-day politics and patronage. The boom of the 1970s witnessed similar patterns when various specialized administrative bodies were created outside of the regular bureaucracy and were granted operational autonomy under young technocrats to run industrial zones or heavy industry operations. Oil, in other words, created great leeway to accommodate elites in various arenas but also permitted the creation—and funding—of important islands of bureaucratic efficiency, vertically linked to the Al Saud and no one else.

Much of Saudi state-building has consisted of the building of patron-client relations and the concomitant cooptation of individuals into the distributive system.[13] Initially, cooptation occurred mostly within the state elite through the granting of institutional fiefdoms or posts to princes or, on a smaller scale, to senior commoners. Increasingly, however, the regime used state institutions and resources to co-opt and reshape significant parts of society, granting the more pliable parts of the religious establishment its own educational and judicial bureaucracy, employing members of various tribes in the National Guard, and employing townspeople in the rapidly growing and consistently overstaffed bureaucracy.[14] In a more diffuse fashion, merchants were individually co-opted through state procurement and commercial licenses, and the state greatly increased its reach into society through the provision of free health and educational services, frequently taking over tasks that had been supplied privately before. Specific groups, such as the Western-educated technocrats in the state's assortment of efficient administrative islands, were virtually created from scratch as client elites through the provision of new career paths and extensive scholarship programs.[15] The Saudi political elite rapidly imposed a large state on its subjects, thoroughly refashioning social structures.

---

13. Even top-level technocrats that are selected for performance tend to become personal clients of senior princes over time.
14. Crystal 1995: 11. An informal welfare system of keeping people unproductively employed is used in other MENA states, too; cf. Cunningham and Sarayrah 1993: 176; Waterbury 1983; Wickham 2002; Waldner 1999: 39. The relation of public sector to private sector employees among nationals is uniquely high in Gulf rentier systems however; cf. ILO 1994: 23.
15. In Ross (2001), the "group formation" effect is one hypothesis to explain lower democracy scores in rentier states.

Unlike other Middle Eastern cases of top-down state-building, this imposition was not particularly violent, however.[16] It was instead mostly based on the nurturing of a plethora of clientelist relationships, made possible by a steadily growing supply of external rent.

The Decreasing Leeway to Reengineer the State

The great leeway the Al Saud enjoyed in shaping the growing state through the early 1980s stands in striking contrast to their restricted maneuverability in the 1990s and 2000s. Yet both are outcomes of the same system of "additive," fragmented, cooptation-based pattern of state-building. Since the end of the oil boom in the mid-1980s, the expansion of the Saudi system has ended and with it, by and large, institutional change. Whereas numerous new administrative bodies could still be added to the system in the 1970s, the option of changing the state's shape through growth has much narrowed since then. For their part, existing institutions often turned out to be change-resistant too.

This decreasing autonomy to change institutions is not the result of macro- or meso-politics such as the emergence of organized interest groups and political contestation on the national or institutional level. In fact, if anything, Saudi society grew more politically fragmented in the course of state-building in the 1960s and 1970s. King Faisal suppressed political opposition, did not tolerate independent organizations in society, and made social elites increasingly dependent on the state. Vertical links of clientage between state and society predominated.

However, while society grew more dependent, it also progressively grew into the state. As bureaucratic employment and service provision grew, more and more Saudis acquired immediate stakes in the bureaucracy and its distributional policies. The various tokens of distribution—employment, social services, contracts, handouts of various types—helped to keep Saudi society quiet and fragmented but also created claims on state resources. The clientelist deal, based on highly unequal material and organizational resources, basically was a trade of material distribution for quiescence. Contrary to the prevailing static view of rentier states, however, it also served to constrain the state over time. Having built a system based on cooptation, the Saudi regime found it hard to retract its commitments.

Individuals, once employed, proved almost impossible to fire, and subsidized state services, once deployed, were difficult to scale down. Maintenance of both became an immovable and growing fiscal item. Entitlements tended to be sticky: Saudis increasingly relied on the state, adapted their expectations of what they

---

16. Cf. Heydemann 1999; Tripp 2002.

were entitled to, and often became tied to specific institutions through thick social bonds, while the ruling family legitimized itself with rhetoric of paternal, inclusive care.[17] The national payroll has increased every single year for decades; even in years of large deficits, bureaucratic job guarantees were never cut back. Some bureaucratic institutions have lingered on decades after losing their function. Institutionalization and payroll commitments have served to "lock in" the organizational landscape of the Saudi state, greatly increasing the path dependency of early decisions.

In the less efficient segments of the state, rapid bureaucratic growth also prevented effective administrative control of mid- and low-level bureaucrats, who could draw a salary and remain idle or establish their own informal networks of brokerage through which state resources would be locally distributed. The Al Saud perpetuated the patriarchal centralization of the administrative system to limit these effects but were unable and probably unwilling to eradicate them.

The large army of Saudi bureaucrats can hardly act as a coherent force. However, as the case studies will show, bureaucratic sluggishness or self-interest on a micro-scale can aggregate to something like a diffuse veto over policies. Bureaucrats can exert locally circumscribed control over state resources. As historical institutionalist Kathleen Thelen points out, new institutions can open political opportunities for otherwise marginal groups. They certainly empowered the numerous petty bourgeois townspeople in the Saudi bureaucracy—on a small scale, perhaps, but pervasively.[18]

The distributional system remained highly hierarchical, but its various commitments came to bog it down. Although Saudi society remains weak on the meso-level, it seeped into large parts of the state on the micro-level, making institutions inflexible and difficult to control and change. The state has been able to change mostly through the accretion of new units. Changing existing institutions has proved much harder, perpetuating administrative fragmentation. Few institutional innovations or reforms have been made since the end of the oil boom; regime autonomy regarding administrative change and reform has decreased dramatically. There is evidence that other rentier states such as Kuwait or Oman—however different they are in other regards—have seen similar processes of early elite-driven institutional fluctuation and later clientelism-based stasis.[19]

This means that state autonomy in the specific but important field of institutional engineering is not constant. In rentier states with weak pre-oil state

---

17. In Jordan, the customs department is seen as a "second tribe" of social networks to which loyalty is owed (Cunningham and Sarayra 1993: 57); allegiances in Saudi Arabia are similar.
18. Thelen 2003: 216.
19. Cf. Crystal 1995; Allen and Rigsbee 2000: 34–64.

traditions, it instead moves from very high to pretty low. More specifically, early autonomous decisions on distribution seem to lay the groundwork for later immobility.[20] Limits on the regime's autonomy are not imposed by powerful, collective social forces but are rather created by the regime itself through building clienteles and incurring micro-level distributive obligations. A regime can lose autonomy without encountering organized opposition. In the case of Saudi state-building, the emphasis has shifted decisively from "structure" to "agency."[21]

The stickiness of institutions and their jurisdictions after the early 1960s highlights the importance of "sequencing," that is, when an institutional decision is taken.[22] Who came first has been crucial. The new efficient bureaucratic enclaves of the 1970s were already restricted in their reach by the presence of preexisting ministries, taking care only of specific portfolios (industrial zones, ports, urban planning for Riyadh, and so on). Similarly, the Ministry of Planning, which came of age in the 1970s, could never evolve into the all-integrating agency its creators claimed it to be.[23] When the General Investment Authority was created in 2000 as another "special administrative body" to take care of foreign investors (see chapter 5), the regulatory turf was already so clearly divided among the Ministries of Commerce, Industry, Interior, and so on that the new organization found it hard to find anything to regulate at all. Frequently unwilling to change existing ministerial structures, the regime has tended to add additional "layers" of specialized institutions (and often also regulations), with varying degrees of success.[24]

Oil has in many ways enabled the emergence of institutional heterogeneity and strong vertical divisions in the Saudi state, which have become defining features of the polity at large. Starting in the early 1950s, oil opened the door for institutional sprawl in the Kingdom, with the only common denominator being the central patronage role of the Al Saud. The administrative system was never tightly integrated, and—despite bureaucratic professionalization—it never had to be. The adoption and creation of various parallel groups of clients as well as the "granting" of specific segments of the state to various elite players have led to a costly process that is possible only on the basis of large external rent income. Over time, this vertically segmented and often redundant hub-and-spoke system has come at

---

20. Saudi state autonomy is a prime case of an independent variable in state-society relations, which in turn is influenced by the effect it spawns; cf. Migdal 2001: 24.

21. No matter how generalizable, this nuanced finding is a clear answer to the criticism that historical institutionalism cannot distinguish the causal roles of individual intentions and institutional constraints; cf. Koelble 1995: 239.

22. Cf. Pierson and Skocpol 2002.

23. Unlike, for example, the legendary Economic Planning Board in South Korea, which integrated administrative action through its overriding powers, or the MITI lead agency in Japan; cf. Evans 1995; Chibber 2004.

24. On layering, see Thelen 2003: 226–27; Streeck and Thelen 2005.

the price of decreasing flexibility and, as we shall see, functional inconsistency of various parts of the state.

So far, we have focused mostly on explaining the emergence of the broader institutional (or "meso-level") landscape through the Al Saud's use of fiscal power. The following section will show more systematically how patterns of vertical segmentation also explain individual behavior in institutions in and around the distributive Saudi state. The two together will help us to explain the peculiar patterns of policy-making and, in particular, policy implementation addressed in the case studies. Many of the relationships that characterize the organizational landscape at large are also found on the micro-level—there, however, social networks can countervail the bureaucracy's fragmentation and overcentralization.

Bureaucrats and Brokers in a Segmented State

The chapter so far has employed the term *clientelistic* in a general sense to describe the nature of Saudi state institutions as they emerged after World War II. The next section is devoted to individual behavior within state institutions and between them and actors in society.[25] In this context, the concept of clientelism needs to be defined more precisely.

According to Luis Roniger, "Clientelism involves asymmetric but mutually beneficial, open-ended transactions based on the differential control by individuals or groups over the access and flow of resources in stratified societies."[26] This, I argue, encompasses most of the power relations in the Saudi state and between state and society in the Kingdom. As indicated, the predominant exchange pattern is one of material benefits and access to the state and its networks in return for loyalty or, in the less personalized cases of large-scale bureaucratic distribution, little more than passive quiescence. This exchange happens on a scale that would be inconceivable without the availability of large rents.

More specific definitions of patron-client relations indicate that they are entered into voluntarily, at least in principle, and involve an element of unconditionality or "long-range credit." This means that acts of assistance do not have to be reciprocated immediately. They are also "particularistic" in that they involve local, clearly delimited sets of actors, while excluding others. All of this applies to the Kingdom.

---

25. Much of what follows is based on my participant observation as advisor to the Saudi government for one year.
26. Roniger 2002: 1.

More strictly anthropological definitions of clientelism also include the personal and informal nature of patron-client relations and the fact that it transcends simple kinship relations. Special stress is put on interpersonal obligations and concepts of honor.[27] Many politically and administratively relevant unequal relationships in the Saudi distributional system indeed have all of these characteristics: senior princes tend to have individual commoner ministers or businessmen as clients; within ministries, mid-level individuals can greatly profit from personal patronage of ministers and their deputies. Smaller businessmen can profit from patronage of bureaucrats. There are elective patron-client relationships within the large royal family. Princes often maintain an extended household of clients, which can include the descendants of manumitted slaves, resembling the *manumissio*-based clientelism in ancient Rome.[28] Personal patronage among elites is so pervasive that even most Saudi intellectuals have a princely patron.[29] All these relationships tend to be quite long-term; perceptions of obligation and honor are strong. Kinship often plays a role in constructing these relationships, but not a determinant one.

Ultimately, however, this book is about processes of economic policy-making that involve large institutions and groups. In this context, Saudi Arabia is too complex a system to be run only on personalized relations. The scale of bureaucratic employment and other forms of distribution means that something more than the deployment of favors for personal followers has been at play in the growth of the Saudi system. Bureaucratic and service-based cooptation was a large-scale process, often formalized and anonymous in nature—but nevertheless an unequal process of exchange based on great asymmetry in resources and delimited sets of actors. To capture these processes, I prefer to use Roniger's more open definition of clientelism cited above, with additional emphasis on its voluntary, unconditional, and exclusive nature. This understanding allows us to analyze aggregates of persons as clients and to include both formal and informal exchanges.[30] It is best understood as "radial" adaptation of the more precise anthropological definitions.[31] It is still specific enough to give it quite particular political ramifications.

---

27. Eisenstadt and Roniger 1984: 48–49.
28. Ibid.: 52.
29. Client intellectuals tend to praise the wisdom and foresight of their patrons; the more daring ones are wont to denounce royal family factions to which they are not affiliated.
30. The underlying logic of exchange in formal and informal clientelism is fundamentally the same: making state assets and networks available to clients in exchange for loyalty or quiescence. Formal clientelism does not only exist in Saudi Arabia, of course. In Brazil, for example, "clientelism is becoming more and more bureaucratized and impersonal and tends to involve entire categories of persons in the role of both patrons and clients." Roniger 2002: 6.
31. Collier and Levitsky 1997.

Sinecures and Hierarchies in the Bureaucracy

The incredibly rapid growth of the Saudi bureaucracy has led to the clientelistic employment of many mid- and low-level bureaucrats with limited qualifications and limited motivation. It did not take long for jobs to be seen as an entitlement. These entitlements have been tied to the broad patronage of individual princes such as Sultan in the vast Ministry of Defense or Abdallah in the National Guard, but they also exist in a more diffuse way in other parts of the bureaucracy where they are usually the result of anonymous state employment policies.[32] Clientelist public employment exists in many developing countries, but seldom has it been deployed in such pure form and as rapidly as in the Saudi case.

It would have been difficult to create a cohesive bureaucratic corps and meritocratic orientation within a few decades under the best of circumstances. Due to the growing distributional welfare function of public employment in the Kingdom, it became almost impossible in many—though not all—parts of the administration.[33] The regime's rapid top-down state-building has afforded its elites full control over which institutions to create. But despite rigid hierarchies, these elites do not fully control most organizations once they are called into being.

In the clientelist bureaucracy, senior bureaucrats command few positive or negative incentives to encourage productive outcomes, so control of subordinates often focuses on the prevention of abuses.[34] A strategy of minimal effort is usually safe and comfortable for low-level employees.[35] Outcome-oriented activism can be risky, as it entails taking more responsibility and becoming visible, often with little prospect of reward. In the event that something goes wrong and responsibility can be traced to a specific unit, administrators and their superiors are put at risk in a strictly hierarchical system. Inactivity, by contrast, is difficult to punish, as it is so prevalent; usually it is difficult to replace an inactive individual with someone better.[36] Inactivity of client bureaucrats is more pervasive—and more relevant as a bureaucratic problem—than the corruption and rent-seeking rentier theories lead one to expect.

---

32. The size of individual followings confers status within the royal family, giving incentives to expand one's patronage; Field 1984: 104.

33. That many citizens are at the same time bureaucrats has created a vicious cycle of stalled reforms in Libya too; Vandewalle 1999: 159.

34. On bureaucrats' lack of available incentives, see Al Saud 1996: 155. As senior administrators admit, mid-level bureaucrats are very difficult to control. Interview with a former deputy minister, Riyadh, December 2005, and various other discussions in Riyadh.

35. As several mid-level bureaucrats have told me, taking initiatives only leads to trouble; keeping one's head low is the most convenient strategy. Middle management was criticized for its passivity even during the boom years, when the chances of mobility were much higher; cf. Madi 1975: 165–66.

36. Discussions with Saudi bureaucrats, Riyadh, 2003 to 2005.

These structures help to explain why vertical exchange also predominates on a smaller scale. Exerting limited control over mundane bureaucratic behavior, principals in the administration are reluctant to delegate much authority. Conversely, effort- and risk-minimizing mid- and low-level bureaucrats frequently refer decisions upwards. Ministers often have to decide on even banal matters such as small travel allowances or individual license applications; some business licensing issues are even decided on the cabinet level. Communication with other agencies is particularly centralized; there are cases of assistant deputy ministers having to ask for permission from their minister to meet representatives of other institutions even for noncommittal discussions.[37] Problems of bureaucratic control combine with a polity in which administrators are, in the final analysis, solely accountable to the very top, to the Al Saud as meta-patrons and distributors. Pervasive overcentralization and lack of horizontal communication are the result.

Horizontal communication and coordination is deficient not only between different agencies but often also within individual organizations.[38] Superiors trying to control large apparatuses of employees rather discourage horizontal cooperation. Many hoard information and play different groups of subordinates against each other,[39] for example by assigning the same task to different working groups operating separately.[40] The duplicate structures and vertical divisions the state exhibits on the meso-level are reproduced on a smaller scale within agencies. Transaction costs are high, structures and efforts redundant, and results often unsatisfying. In a group of 180 countries assessed for "government effectiveness" by the World Bank in 2005, Saudi Arabia finished 96th, a position far below its peers with comparable GDP per capita.[41]

Informal Networks between Bureaucracy and Society

With the bureaucracy's internal communications so hamstrung, formal interaction with actors outside of the state apparatus can be even more dysfunctional. The sudden imposition of large, overcentralized, and inward-looking bureaucracies, often staffed by poorly motivated administrators with little capacity to make

---

37. Al Saud 1996: 155.
38. Ibid.: 131.
39. See the discussion of "vertical" vs. "horizontal" trust in Breton and Wintrobe 1986.
40. This semblance of competition allows at least for a comparison and hence for partial control of different results. Gambetta (1988: 220) points out that employing other agents' mutual distrust can help to constrain them.
41. The Kingdom ranked 40th for GDP per capita (data from World Bank Governance Indicators and World Development Indicators). For nonrentier states, GDP per capita and government effectiveness are correlated closely, making Saudi Arabia (as well as other rentiers) an outlier case. The rankings for other governance indicators yield similar results.

independent decisions, has made for a patchy relationship with society. Formal trust between bureaucracy and society—that is, trust that does not rely on specific personal relations—has been at least as low as between institutions within the state.[42] Individuals have little leverage over the unaccountable, hierarchical bureaucracy, which is often difficult to reach through formal channels. Applications might not move from one desk to another, license decisions are postponed, documents are rejected on the basis of obscure or obsolete formalisms.

Conversely, although the state reaches widely into society by distributing jobs and other resources, society has never been formally penetrated in terms of compliance with bureaucratic rules. In many ways, distribution has been a one-way and top-down process that has not created specific levers of control over individual behavior—whether over bureaucrats or over Saudi nationals outside of the state apparatus. Businesses as well as individuals are reluctant to disclose information to the state or stick to its rules in various fields, be it building standards, insurance duties, or labor regulations.[43] Until today, efforts at gathering information about private economic activities in the Kingdom have been half-hearted, undermined by bureaucratic passivity and widespread distrust of the state. In a vicious cycle, such distrust is fed by the administration's inconsistent regulatory track record, which undermines expectations of fairness and leads to evasion.

Although many developing countries have limited regulatory capacities, few have such large and well-endowed bureaucracies as Saudi Arabia with such a far-reaching presence in individuals' lives. Limited capacity to gather information and influence individual behavior often leads the administration to heavy-handed interventions in business life as a last resort. These can do much collateral damage.

Given the story so far, the Saudi system appears severely dysfunctional, if not due to rent-seeking, then due to its hierarchical rigidities and limited predictability. However, on the micro-level—and only there—it offers countervailing mechanisms that can make it surprisingly supple and can mitigate bureaucratic inaccessibility and fragmentation, if at a cost.

The more difficult a bureaucracy is to deal with, the higher the demand for assistance. As bureaucratic principals have imperfect control over their subordinates, the latter can often offer supplicants assistance by dint of their personal, small-scale, informal networks. In the face of general opacity, Saudis frequently get things done through friendship or kinship links to bureaucrats, which can be used to circumvent or negotiate hierarchies, bend rules, and make things happen.

---

42. Trust can be more or less personal or institutionalized; Koniordos 2004. Anonymous formal trust is different from "focalized trust" in specific persons and networks as described by Roniger (1990: 16).

43. In an early 1990s survey with 702 Saudi respondents, low trust toward the bureaucracy and low civic responsibility, were prevalent; cf. Al-Mizjaji 1992.

Corruption is another means to getting one's way if formal mechanisms don't work. The politically fragmented Saudi society has little way to exert collective political pressure on the Saudi state, but it can access and influence it on a small scale—an unintended outcome of the rigidity and centralization of state structures.

However, direct personal contacts are often not sufficient to cut through a large and complex system, and impersonal "market" corruption is still not pervasive.[44] There is another, more widespread mechanism to access the bureaucracy: resorting to "brokers," individuals between the state and a claimant who can make the hierarchical and impenetrable bureaucracy accessible. They usually do this through some kind of privileged linkage, which is often informal in nature. In an environment in which formal trust is scarce and access to the state uneven and often discriminatory, they peddle their positions as intermediaries. Due to the large role of the state in Saudi business and society and the large resources still at its disposal, need for such assistance is pervasive.

Beyond their function of intermediation, the brokers around the Saudi state have little in common. They can be well-connected senior relatives of supplicants in specific agencies. They can be well-networked Saudi joint venture partners who help foreign investors acquire licenses, import tariff exemptions, or labor import allowances. They can be so-called *mu'aqqibs,* usually lower-class Saudis who have established networks in specific government agencies, which help them to expedite paperwork for less patient or well-connected customers.[45] They can be followers of individual princes or ministers who relay crucial information on state contracts or arrange for access to decision-makers (needless to say, on bigger projects, the princes or ministers themselves might be brokers). It is said that when the king passes through one of Saudi Arabia's important provinces, one can arrange a meeting with him for a fee ranging from 500,000 to 1 million Saudi riyals through the local governor's entourage.[46] Often, individuals are brokers only in a specific context and in an occasional way, by virtue of a connection they happen to have.

The more impenetrable the system and the larger the resources involved, the higher the value of specific, discriminatory access.[47] This equation explains the

---

44. On this concept, see Leff 1964.
45. I have encountered many Saudis making a living as double-brokers, registering a company in their name that is actually owned and run by expatriates who pay them a fee (see chapter 5) and using their actual working time as "paper chaser" in the local administration.
46. Discussion with Saudi social activist, Riyadh, February 2007.
47. Access is universal in theory but uneven in practice (Roniger 1990: 5). "It is the combination of potentially open access to the markets with continuous semi-institutionalized attempts to limit free access that is the crux of the clientelist model" (Eisenstadt and Roniger 1981: 280). What is true about the market is also true about bureaucracy.

inordinate expansion of various types of brokers in the course of Saudi state growth—an expansion that the state has sometimes encouraged and formally licensed, as the history chapters will show. The amount of brokerage witnessed in Saudi Arabia might be unique to rich rentier systems like the Kingdom.

Brokers and their networks can bridge public and private in a complex, hierarchical environment of low formal trust.[48] Access to brokers itself is often still personalized and frequently described as *wasta,* an Arabic term that indicates personal intermediation. In many areas, however, it has become a more monetized and anonymous phenomenon, based on specific exchange and contract-like agreements: through local law firms, muʿaqqib offices around ministries, "free visa" peddlers who resell imported manpower, and so on (see chapters 5 and 6).[49] Such interaction is pervasive in the case studies in this book, and they often define state-business interaction and patterns of policy implementation, even in areas in which conventional rent-seeking plays no significant role. The value of specific access also explains the subtle interest of many mid-level bureaucrats in maintaining networks of intermediaries and in keeping the bureaucracy relatively inaccessible. Locally specific deals of brokerage can undermine the consistent implementation of policy and the clear formation of collective interests.

Although informal social networks can provide some element of horizontal access, they themselves are frequently tied up in hierarchies and patron-client relations. Mu'aqqibs, for example, can have princely patrons who work as advisors or on the deputy ministerial level in an organization.[50] Well-connected contracting firms that subcontract to smaller client firms often rely on clientelist links of their own to senior royals. Until 2004, individuals informally trading in labor permissions frequently enjoyed patronage of senior figures in the Ministry of Interior. In a multi-layered system, brokers can be "patron-brokers" providing access to sets of less privileged clients.[51] Often, they are themselves clients.[52]

---

48. Heydemann 2004: 29n8. Lively examples of (personalized) intercession on a municipal level can be found in Al-Salem 1996.

49. Similar developments have been observed in Jordan; cf. Cunningham and Sarayrah 1993: 14, 86.

50. One taxi driver who took me from Riyadh airport to the city in 2004, for example, earned some extra income by chasing documents in the notoriously opaque Ministry of Interior; he did this by virtue of a prince's patronage, about whose generosity he seemed genuinely enthusiastic.

51. Roniger 1990: 3.

52. With brokerage enmeshed in specific hierarchical relations, their function of enabling horizontal communication and cooperation is often locally circumscribed. This is not to say that all Saudis are trapped in dyadic patron-client relationships. In addition to the option of monetized brokerage, it is fairly common to have multiple patrons or friendly brokers in different (though seldom completely separate) contexts. Indeed, it is often necessary for doing business with a segmented state. Although most interactions remain specific and local, Saudi Arabia is more fluid and complex than traditional clientelist systems. Against this background, Madawi Al-Rasheed's (2005) account of separate circles of power and

Islands of Efficiency in the State

Although the flabbiness of the state and the size of its resources have allowed the social penetration of many fields of administration, it is more pronounced in some areas than in others. As indicated, the reasons for this are often historical. Princely networks of (re)distribution have not been allowed free rein everywhere. Informal trade in labor allowances has been more pervasive than trade in most other types of licenses. Certain types of brokerage have become much less pervasive with increasing professionalization of the domestic private sector, for example subcontracting (and sub-subcontracting) of government contracts through shell companies.[53]

Just as important, the hub-and-spoke nature of the Saudi state has allowed for the emergence of institutional islands that have been largely separated from games of brokerage. This is true, at least in regional comparison, of the religiously dominated judiciary but also of a few efficient state agencies—the Central Bank SAMA, industrial giant Saudi Basic Industries Corporation (SABIC), the oil company Saudi Aramco, the Saudi Ports Authority—that are beholden only to senior members of the Al Saud.

Due to scarcely restrained control over the additional increments of oil income accruing to the Kingdom year after year, Faisal and, after him, Fahd have been able to separate significant chunks of the growing administration from the regular civil service and its stifling logic of cooptation and clientelist entitlement. While the leeway for doing so has shifted over time, the social foundations that insulate any given institution have been stable. In the 1970s, the deployment of a new generation of Saudi graduates with Western degrees and expensive expatriate advisors helped to protect new agencies from the clientelist syndrome and allowed for relatively swift and transparent administration. Compared to the regular bureaucracy, the leadership gave much more discretion in hiring and firing and allowed for higher salary levels and more flexible incentives. Recruitment has been highly competitive, drawing the most motivated and educated nationals.[54]

To be sure, most leaders of insulated agencies have been disposable technocratic clients of the royal family without any power base of their own, just like other high-level bureaucrats. But the vertically divided Saudi system allowed for great variation between different forms of technocratic clientelism. With no broader civil service traditions to speak of, Saudi administrative units could be designed

---

patronage around a number of senior princes seems too "dyadic," as in fact many senior nonroyal players, especially those with a longer history, do have access to several patrons.

53. Interview with former deputy minister, Riyadh, November 2005.

54. Aramco and SABIC are by far the preferred employers among graduates of the small and elite King Fahad University for Petroleum and Minerals; see the survey in Gulftalent 2005.

from scratch and in different ways. Royal principals could give their technocratic agents in specific, insulated areas clear incentives to produce efficient outcomes and the authority to bypass the rest of bureaucracy.

Barbara Geddes describes how various insulated institutions were swamped with patronage politics in Brazil when the political circumstances changed and the leadership was in need of new political allies.[55] In Saudi Arabia's more static and top-down polity, different segments of the state have by and large remained clearly divided, with little horizontal pressure from interest groups in society or the broader state that would lead to patronage appointments or the emergence of rent-seeking networks. Insofar as there was such pressure, it tended to come from royal family players—who were usually prevented by senior royals from preying on historically insulated agencies and who thus concentrated on established fields of rent-seeking that offered easier prey.

As has been pointed out in the Latin American context, bureaucratic insulation is a contingent and political process about which it is hard to generalize.[56] Yet in the Saudi case there has been an underlying logic: a segmented and hierarchical system of parallel institutions that has provided stable and protected spaces for efficient bureaucracies accountable only to the top.[57]

Patterns of Policy-Making under Segmented Clientelism

What does this all mean for policy-making and policy outcomes, our initial puzzle? The case studies will demonstrate in detail how the hierarchical and segmented Saudi system copes with different policy challenges. The following paragraphs anticipate some underlying commonalities.

Negotiations over economic reforms in developing countries are often analyzed in terms of "coalitions" between the government and various forces in society.[58] The above account of state formation indicates that this is not the most useful way of looking at policy processes in Saudi Arabia, as there are few coherent allies in society the regime could conduct negotiations with. Clientelism tends to be diffuse and offers few clear mechanisms of interest aggregation. There are no unions or syndicates representing worker interests or professional groups, no consumer or environmental groups. Formal civil society is fragmented or nonexistent, as the

---

55. Geddes 1990, 1994.
56. Schneider 1991: 229.
57. The literature on the developmental state occasionally mentions "pockets of efficiency" (Geddes 1990; Schneider 1991; Evans 1995; Hout 2007), but they remain undertheorized. Most research continues to be focused on the aggregate qualities of the state instead of its components.
58. Waterbury 1989.

state dominates the formal organizational field, the meso-level of politics.[59] This means that policy-making on labor or trade issues, for example, is much more state-dominated in Saudi Arabia than in most other systems (I discuss the partial exception of business lobbying below). Different interests in society tend to be at best notionally represented through the sectoral responsibilities and interests of different ministries. As policy is mostly negotiated *within* the state, the process most of all reflects the internal structure of the state.

The general attitudes of individual technocrats are often liberal and pro-reform. However, the hub-and-spoke structure of the state leads them to defend the interests of their own institutions on specific issues. The primacy of organizational interests and hierarchies trumps individual attitudes, as it is difficult to engineer cross-cutting, horizontal reform coalitions that might enable broader policy deals. Secrecy and compartmentalization of information are prevalent, as information, if shared at all, is made available to superiors rather than peers.[60]

Individual agencies tasked with a given policy often fight isolated struggles. Other agencies tend to ignore what their peer institutions are doing, waiting until specific royal guidelines force them to act on an issue. Most major decisions are not made in the cabinet but among royals and personal advisors. Commoner ministers therefore are oriented upwards rather than toward the cabinet as a collegiate body.[61] Due to the presence of different princely patrons and the steep hierarchies at play, the senior technocratic level is often factionalized. Senior princes, moreover, are not necessarily versed or interested in technical issues of policy and are severely overtaxed due to the system's strong tendency to refer matters to the royal hub of decision-making.[62] They are also surrounded by "gatekeepers"—personal advisors, secretaries, and so on—who act as access brokers and filter information.[63] When royal attention is drawn to a problem, this often results in a sudden rush of activity among their bureaucratic clients, which however tends to be short-lived due to limits of the royal attention span and the bureaucracy's general penchant for inactivity.

Just as individual agencies are struggling for themselves, different agencies often struggle in parallel, without being willing or able to coordinate their approaches on the same matter.[64] More often than not, this results in contradictory regulations, incompatible procedures, and duplicate efforts. In policy areas like

---

59. Hertog 2006.
60. Vandewalle (1999: xv) mentions compartmentalization of information as a prevalent characteristic also of the Libyan bureaucracy too.
61. Phone interview with former British ambassador to Saudi Arabia, June 2005.
62. Interview with former British diplomat, London, June 2005.
63. Interview with former deputy minister, Riyadh, November 2005.
64. Cf. Al-Sabban 1982.

health, education, or infrastructure, the rapid and parallel sprawl of numerous organizations in the boom years has led to a fragmentation of jurisdictions, which exacerbates this problem. Agencies led by senior princes are specifically resilient to horizontal coordination, as they have evolved as separate fiefdoms exclusively beholden to their respective patron ministers.

In this setting, locally delimited policies such as privatization or capital market reform are significantly easier to implement than policies that involve more institutions and hence require slow and cumbersome communication funneled through the center. As the case study on labor market Saudization will show, it can take a painful redefinition of jurisdictions to create a minimum level of coherence on more complex matters.

This all points to the conclusion that it is not in the main nepotism and corruption that impede coherent policy-making on a senior bureaucratic level. In many cases, high-level civil servants are rather "Weberian" figures—well-educated, reasonably "clean," and committed to producing results in the context of their institution. However, the organizational structure of the state and its agencies—its meso-level structure—often prevents these characteristics from translating into coherent policies.[65]

A fragmented state governed by the royal family's consensual policy style is bound to see slow decision-making. And even when a policy finds broad consensus on the cabinet level, individual agencies often dilute its implementation, and cast smaller vetoes on bylaws or other specific procedures. The Saudi system, although highly centralized and hierarchical, arguably contains numerous veto players[66]—a concept not yet applied to authoritarian systems, as they are often misunderstood as unitary actors.[67]

On a senior level, the powerful Ministry of Interior, whether intentionally or not, has often scuppered or modified significant aspects of new policies. While normal line agencies are sometimes beaten into submission, fiefdoms with more historical depth, such as Interior, Defense, or the religiously controlled judiciary, seem to have almost unlimited staying power.[68]

---

65. For a similar argument on India, see Chibber 2002.
66. Tsebelis 1995.
67. In fact, U.S. politics—the system that inspired the veto player model—offers striking similarities to the policy-making process in Saudi Arabia. The separation of powers requires that many players are paid off to push through a policy; there is no central conflict-resolution mechanism; and policy-making is fragmented; Weaver and Rockman 1993: 436. Presidents tend to add new bureaucratic structures instead of reshaping what exists; Moe and Caldwell 1994: 176. Parliamentary systems tend to be more cohesive.
68. A line agency is a ministry with specific sectoral responsibility.

Beyond the deeper fiefdoms, there are ample opportunities for what might be called *low-level vetoes:* Although at the end of the day every senior technocrat is at the mercy of the Al Saud and will follow their specific orders, instructions from the top are often quite general in nature, and many policies get implemented without senior princes paying particular attention. As long as there is no overriding pressure, ministries can often act as veto points on cross-cutting policies. Indeed, in some ways this is positively encouraged by overcentralization and vertical communication. Agencies such as the Al-Riyadh Development Authority or the Ministries of Labor and Commerce are not decision-makers on large policy questions, but they are tasked with fleshing out the details. Not able to introduce new policies by themselves or to force other agencies to cooperate, they still often have the "power to say no."[69]

Significantly, internal efficiency of an organization is of little help if the cooperation of other players is required for policy implementation. Efficient "insulated agencies" are most useful to manage sector-specific projects over which they have full ownership, such as the infrastructure projects of the 1970s and 1980s. The newfangled General Investment Authority in the early 2000s, by contrast, was powerless to change the set ways of the bureaucracy that foreign investors encountered in other ministries.

The exertion of veto power does not require coordination, whereas pursuing a proactive policy usually does. The more fragmented a system is, the more vetoes and the less successful coordination can be expected. Vetoes can be effectively overruled through reassignment of jurisdictions (see chapter 6) or through external pressure (see chapter 7), but this does not happen frequently or easily. The regime's autonomy in changing institutions is constrained, and external pressure rare.

As elucidated above, the segmented hierarchies of the Saudi state also exist on lower levels, and when it comes to executing policies, low-level bureaucrats and their networks can also cast individual vetoes, even if they are otherwise incapable of proactive behavior.[70] Micro-level actors such as administrators and "brokers" below the technocratic elite do not usually influence policy-making, but they have great impact on implementation.

At the height of the reform drive of the 2000s, senior princes commissioned confidential studies about the non-implementation of policy in Saudi Arabia, reflecting both their concern and their relative powerlessness over the issue.[71] Even

---

69. Phone interview with former British ambassador, June 2005.
70. Typically, bureaucrats are more capable of preventing a file from moving within the bureaucracy than of accelerating its movement; cf. Bardhan 1997: 1324.
71. Interview with advisor to the Saudi government, Riyadh, April 2004.

when royal elites manage to push decisions through the senior echelons, they often enjoy limited control over execution.

Joel Migdal reminds us that "implementors are often far from the sight of state leaders—often even from the sight of the top personnel in their agencies."[72] The Saudi state is large, and top personnel are much tied up in cabinet and representative duties (one senior Saudi bureaucrat once told me that his minister "knows perhaps five people" in his own ministry). Control over the clientelist low-level bureaucracy is limited, which gives it much scope for delaying decisions.

Facing distrust and overcentralization, low-level administrators are, if anything, even more reluctant and less able to coordinate with their peers in other agencies than their bosses. When policies change, new rules are sometimes ignored or interpreted in counterproductive ways.[73] Insufficient qualification of mid-level bureaucrats is part of the story, but a larger part is the clientelist mixture of entitlement thinking, risk avoidance, overcentralization, and lack of effective supervision.

Frequently as resilient as mid-level bureaucrats, informal social networks and brokers also tend to adjust to new rules in unintended ways, and in many cases new regulations create new mechanisms of brokerage or new ways "around" the rules. This is true especially when demanding regulatory policies combine with low supervisory capacity of the bureaucracy. Low trust between the fragmented and hierarchical bureaucracy and its subjects often makes brokerage the more reliable alternative to following rules by the letter. The Saudi regime has been quite reluctant to clamp down on networks of brokers in large areas such as contracting and the informal market for labor permits—probably because brokerage can help to make the state's opaquely administered goods and services accessible to wider circles, but also because brokers can make new (sometimes unimplementable) policies bearable, even if in a selective and costly way.

In sum, the organizational fragmentation of the state makes decision-making and implementation more difficult, while micro-level bureaucratic clientelism and brokerage tend to undermine implementation even further. The more agencies are involved and the more micro-structures need to be mobilized in a given policy area, the harder it is to produce results. State capacity, a term much bandied about in the debates about rentier and developmental states, is not uniform but is influenced by *width* and *depth* of policy: by how many organizations are involved, and by how many individuals need to be mobilized, both in state and society. This explains why some reform policies have been remarkably successful, but others much less so: While industrialization in the 1970s or capital markets reform in the 2000s involved limited sets of actors, foreign investment and labor market

---

72. Migdal 2001: 84.
73. SAGIA 2004: 63.

reform were more diffuse policies stumbling over numerous high- and low-level veto players.

In the absence of larger organized groups in society, the state still dominates policy-making. There appears to be one increasingly salient exception to the atomization of social interests: Saudi business, organized through the Chambers of Commerce and Industry. Unlike the state, with its constant obligation to disburse largesse, business has been strengthened through decades of rent accumulation and increasingly caters to private as opposed to state demand. It has increased its economic independence from the regime as state contracts dried up in the 1980s, private capital formation has increased, and productive and managerial capacities have grown substantially.[74] A significant share of Saudi business has moved beyond the rentier clientelism of the boom years, showing once again that a static appraisal of rentier systems is misleading.

Large parts of Saudi business have grown exasperated with the bureaucratic rigidities of the state. At the same time, the regime increasingly needs private capitalists as drivers of economic diversification and job creation. Especially in recent years, therefore, the regime has increasingly integrated Saudi business into formal economic policy-making through various new bodies and corporatist mechanisms of consultation.[75] Unlike other groups, business has been allowed to act as a collective lobby group, and the Chambers, although partially state-directed, control sizeable funds and an organizational infrastructure that has been built up over decades.

Business as a collective actor should be expected to work against the state-dominated and disjointed nature of Saudi policy-making. Yet state-business relations themselves reflect both the historical dominance and the fragmentation of the state. State-business consultation itself is often fragmented, as diverse state agencies and political patrons get involved in it. Moreover, many businessmen are still tied up in individual networks of clientage, even if such links do not immediately relate to rent distribution anymore. They can be used to informally convey collective positions, but they can also be used to pursue individual interests.[76] Trust and reciprocity between bureaucracy and business are generally low.[77] Collective negotiations over policy therefore are often perfunctory affairs, as the state cannot be trusted to implement its own rules and businesses have ample

---

74. Luciani 2005.

75. For more details, see Hertog 2006b.

76. The degree of personal clientelism varies by sector. It is still rather strong in contracting and still somewhat more present in trading than in the industrial sector. Interview with Michael Field, Gulf business consultant, London, July 2005.

77. Cf. Schneider and Maxfield 1997.

individual opportunities to avoid or bend formal regulations through patrons, brokers, and other mechanisms.

Despite its economic maturity and channels of collective action, business operates within a deeply ingrained tradition of state paternalism in which the state takes the initiative on all major changes. On more complex policy issues, business remains "policy taker" and tends to give in to state demands once royal wishes have been clearly articulated. Policy consultations, albeit increasingly institutionalized, are still "granted" by the state, and last-minute interventions by the regime can still reverse policies.

Insofar as it acts coherently, business often attempts to cast vetoes over government policies rather than delivering proactive policy input—making it yet another veto player in a complex system shot through with such actors. As in the state apparatus, defense of established stakes trumps other interests. There is no particularly clear sectoral differentiation of business interests, and despite solid organization, the research capacities of the Chambers are limited.[78] Today's easy exit option of moving capital abroad tends to undermine corporate coherence still further.[79] In the clientelist Saudi system, individual maturity of businesses has not yet translated into collective maturity. Business, like other groups in state and society, remains part of the system of segmented clientelism.

---

78. "Encompassing," cross-cutting business associations are generally seen as an advantage for coherent policy-making (Schneider 2004). The Saudi chambers often lack internal coherence, however.

79. Hirschman 1970.

PART I

OIL AND HISTORY

The first half of the book explains where the institutions that underpin the Saudi Kingdom's modern political economy come from and how oil income has allowed them to take the idiosyncratic shape they have. It will try to bring individual agency and personality back into the "structuralist" political economy accounts of Gulf states and rentier systems in general.[1] More specifically, it will address how the scope of individual agency has shrunk over time and how early contingencies and choices have determined long-term institutional outcomes.

Unlike the general expectation of rentier literature—and unlike standard political science accounts of Saudi state-building—the pages to come will show that large flows of oil income into the early, underdeveloped Saudi state did not necessarily lead to institutional decay or an all-encompassing expansion of rent-seeking networks.[2] To be sure, government agencies frequently evolved into personal fiefdoms with ill-defined administrative structures and limited regulatory powers. But while princes, bureaucrats, and their cronies siphoned off much rent in the process of state-building, the state's rapidly growing resources also allowed for the emergence of royally protected islands of efficiency with impressive administrative capacities, such as SAMA, the Central Bank, or the various institutions in charge of the Kingdom's rapid industrialization in the 1970s.

Saudi elites used the unprecedented autonomy that oil gave them at an early stage of state formation for different purposes of institution-building in different

---
1. Cf. Chaudhry 1997; Delacroix 1980; Luciani 1990; Karl 1997; Vandewalle 1999.
2. Chaudhry 1997.

areas of the state. Thanks to growing oil surpluses, different parts of the state were able to develop independently, creating both enormous redundancies and deep cleavages within the bureaucracy, which fragmented the state organizationally. This oil-enabled, meso-level splintering of the state apparatus has characterized the politics of the Kingdom to this day, yet existing oil state theories do not account for it.

As contingent as early institutional decisions were, the oil-based growth of the state has made the deeply fragmented and heterogeneous constellation of Saudi state organizations increasingly immovable over time. The expanding state apparatus was used to co-opt, and in some cases create, important sections of Saudi society. This allowed the leadership to control these sections politically and to prevent them from organizing independently. But as hundreds of thousands of individuals rapidly entered an underdeveloped bureaucracy, the leadership's capacity to control their daily behavior was often limited. While politically quiescent, many of these newly minted Saudi bureaucrats took to pursuing their own interests on an individual level or at least proved impervious to the imposition of standardized and predictable administrative procedures—two factors abetting the emergence of access brokers around the state. As different sections of the state functioned according to different rules and catered to different social constituencies—ulama, tribes, the urban middle class—the overall coherence of the state apparatus decreased further. Once the leadership had institutionalized stakes in the state's resources in the shape of employment, public services, and subsidies, moreover, such clientelist distribution structures proved difficult to reverse.

As I will argue, the "state autonomy" posited by rentier state theory holds only for a specific historical period. Rentier theory does not explain the smaller-scale relations of power and exchange in the Saudi distributional state that underlie both its declining autonomy and its institutional fragmentation. It is the historically determined relationships of various institutions and their internal power relations, rather than the "average" or aggregate characteristics of the state, that allow us to understand policy outcomes, however.

The post-World War II process of Saudi state formation can be divided into three distinct historical phases, which correspond to the next three chapters. Chapter 2 will describe the strongly personalized and fluctuating institution-building under King Abdulaziz and his first successor, Saud, from the early 1950s on, tracing the early emergence of bureaucratic fiefdoms and clients. Chapter 3 will analyze the rudimentary consolidation and expansion of the state as well as its patterns of rule and authority under King Faisal, who stabilized bureaucracy but was less successful at overcoming its fragmentation. The final chapter of this section, chapter 4, will look at the last phase of rapid state expansion, the oil boom under Khalid and Fahd, in which the segmented Saudi state managed to wrap itself around large parts of society.

My emphasis on relative continuity and cumulative growth is in contrast to Kiren Chaudhry's narrative of radically discontinuous institutional development. Her *Price of Wealth,* the only other work addressing the institutional history of the Saudi state after World War II in detail, posits that the 1970s oil boom dismantled a bureaucracy that had developed quite a high degree of administrative coherence and capacity in the 1950s and 1960s. My account, rather, suggests that problems of bureaucratic fragmentation and low regulatory power were apparent in the modern Saudi state right from its inception—what differed over time was the scale at which the state has operated and enveloped society.

Each chapter broadly follows the same internal structure. They first address main developments on the level of elite politics and grand institutional design, focusing on balances of power between senior princes, ministerial reorganizations, and interagency relations. The chapters then delve into an analysis of the changing features and capacities of the bureaucracy itself in terms of growth, formalization, informal penetration, and regulatory power. On that basis, the chapters then analyze how an expanding state has affected state-society relations in terms of cooptation and brokerage.

What the Chapters Do Not Do

The analysis starts in the early 1950s, as it is then that the royal leadership set up the first nationwide, formal bureaucratic structures with the (at first largely unrealized) ambition of regulating Saudi society and economics—many of which are still with us and feature prominently in the case study chapters. The account touches on preexisting state structures but will not describe them in detail. Although interesting for the broader process of state formation, they are less relevant for the political economy of bureaucracy and economic reform attempts addressed in the case studies. This book looks at "modern bureaucracy," less so regional governors, tribal *shuyukh,* religious elites, and other social authorities, unless they are part of the modern administration.

The narrative is by no means complete. My ambition is not to cover all institutions and important decisions in the making of the modern Saudi state. This would be a book of its own. Rather, I outline several crucial historical junctures and select examples to sketch the general forces of institutional development in the Kingdom, which have resulted in an arrangement that the case studies will explore in further detail.

Chapter 2

Oil Fiefdoms in Flux

The New Saudi State in the 1950s

> *Verily, my children and my possessions are my enemies*
> —Ibn Saud, shortly before his death

The state that King Saud inherited in 1953 from his father, King Abdulaziz, was at once highly centralized and extremely underdeveloped in administrative terms.[1] Abdulaziz ("Ibn Saud" in Western parlance) had by 1934 by and large managed to extend his sovereignty over all of modern-day Saudi Arabia. He did, however, not rule through a modern bureaucratic apparatus but rather through personal links with trusted lieutenants, local intermediaries, and clients.

In the Najd, the central region of Saudi Arabia whence the Al Saud hailed, Abdulaziz ruled through a personalized administrative system, with his son Saud as vice-regent and a few regional governors (*umara*) below him who were directly responsible to the king. In the larger cities there were some judges and financial officers.[2] There was nothing like a central, integrated bureaucracy, and the position and influence of most actors were defined through their personal relationship to the king rather than posts in a formal apparatus.[3] The system completely

---

Epigraph: Philby has reported Ibn Saud uttering this quotation from the Quran; Jones to State, Six months of King Saud, 10 May 1954, U.S. Records on Saudi Affairs, Vol. 2 (Slough: Archive Editions 1997), henceforth USRSA.

1. The main sources for this section are a number of historical doctoral dissertations and standard secondary sources, including Al-Awaji 1971; Solaim 1970; Vasiliev 2000; Lacey 1981; and Philby 1950.

2. Al-Awaji 1971: 43; Al-Ammaj 1993: 62.

3. Abdulaziz ruled "without any written constitution in so far as the rights of the subject and the form of the administration are concerned. In other words the constitutional position is fluid and

revolved around Abdulaziz, who was king, chief legislator, and had the right to review all judicial decisions.[4] Most of the departments at Abdulaziz' court were occupied with the logistics of the court itself and not with broader administration or public services.[5]

In the Western region, taken from the Hashemite dynasty in the mid-1920s, Abdulaziz had inherited a somewhat more elaborate bureaucratic structure, which was to be run by his second oldest surviving son, Faisal. Some local administrative bureaus (*mudiriyat*) existed there with sectoral responsibilities in health or education, but their task was only the execution of local policies, not policy-making.[6]

The Saudi government's regulatory and distributive outreach was strictly limited. Most Saudis had little to do with it, as apart from basic security, the state provided hardly any services to its citizens. There were few formal, national bodies during most of Abdulaziz' reign. A rudimentary Ministry of Foreign Affairs had been set up in 1930 under Faisal; Ministries of Finance and Defense followed in 1932 and 1946.[7] No formal mechanisms of budgeting or governance linked these agencies. A 1947 attempt by Minister of Finance Abdallah Sulaiman to set up a state budget failed completely, as the king vetoed the necessary cuts and called for funds to be made available outside the budget.[8]

Abdulaziz' Early Clientelism

Much of the polity around Abdulaziz was personalized along clientelist lines. The local authorities often were governors from Abdulaziz' following whom he had deployed in the course of his conquests, frequently from collateral branches of the Al Saud and allied Najdi families. Abdulaziz also co-opted preexisting local players into his system of rule, in particular on lower levels of authority, often

---

continuously developing by a process of trial and error. There is no law and institution except the Shar' and the monarchy, which is not subject to modification." Notes on Saudi Arabia for Dr. Hugh Scott, sent 29 August 1944, Philby archive, St. Antony's College, Oxford, Great Britain 165–0229 PHILBY, 1/4/9/3/28, henceforth PHILBY. The few administrative bodies in existence grew without any specific laws governing their functioning; a civil service board in the Ministry of Finance only came into existence with the first Council of Ministers in 1953. *At-taqrir ath-thanawi li'diwan al-muwadhdhafin al'am 1391–93* (Biannual report of the Civil Service Bureau, 1971–73), Institute of Public Administration documentation center, Riyadh, henceforth IPA.

4. Al-Seflan 1980: 81.
5. See the list in Vasiliev 2000: 293–94.
6. Yizraeli 1997: 101; Solaim 1970: 8; Al-Awaji 1971: 45; Al-Ammaj 1993: 63.
7. *Lamha tarikhia mujaza 'an tatawwur al-ajhiza al-hukumiyya (idarat al-buhuth wal-istisharat)* (Concise historical overview of the government bodies' development; Studies and Counseling Bureau, n.p., n.d.), IPA.
8. Young 1983: 23.

following defeat in combat.[9] These would usually be tribal leaders or members of important urban clans, though often not the previously leading ones. This way, Abdulaziz managed to establish the supremacy of his rule without disrupting local structures of governance, creating structures of managed dependence and intermediation.[10]

Many of these early deals would position regional players for decades to come. Several of the early clients of Abdulaziz in outlying regions, for example the merchant families of Al-Gosaibi in the Eastern Province or of Alireza in Jeddah, still are close to the royals to this day. The system was also sufficiently fluid that opposition actors like the Hijazi separatists from the Dabbagh and Sabban families were subsequently co-opted and given local administrative posts. Several Dabbaghs are still serving in high posts today.

Power was more balanced and rule much more indirect than in later decades, with considerably more local autonomy outside the framework of a formal state. In a pattern that was to be reversed in the oil era, some of the big merchant families bankrolled Abdulaziz' government in the 1930s with significant loans, making them allies at least as much as subordinates.[11]

Despite limited resources, the generosity (*karama*) of the king already played a vital role in maintaining his authority. Large dinners and gifts to visiting leaders of local communities helped to cement the roles of the patriarch and his various clients. After World War II, oil income boosted the scale of this system. In the late 1940s, a government-owned DC3 flew to provincial centers loaded with silver coins to hand out to tribes.[12] When U.S. financial advisor Arthur Young met the king in Taif in the early 1950s, some two thousand guards and hangers-on camped around his palace waiting to be fed by him.[13] The king employed specific retainers in charge of doling out grants (*minah*).[14] By then, such patrimonial generosity was already largely based on oil rents originating outside the Kingdom, representing one-way distribution rather than traditional tax-based redistribution.

---

9. Defeated nobility would often become (subordinate) members of the Saudi elite; Vasiliev 2000: 113.

10. In Eisenstadt's (1973: 34) typology it is typical for the center of a patrimonial system to reach out to the periphery largely through existing elites, with no transformative orientation.

11. After the conquest of the Hijaz, the global recession and the war with Yemen meant that the king had to borrow from merchants; Ministry of Petroleum and Minerals 1963: 2.

12. Jidda to State, Survival Prospects for the House of Saud, 6 June 1966, U.S. National Archives and Records Administration, College Park/Maryland, Record Group 59, 250, 5–7, box 2645, folder POL 15–1 SAUD (5/21/66), henceforth NARA.

13. Young 1983: 41. Already in the early 1940s, Ibn Saud traveled with a large entourage, hiring some six hundred cars; cf. Financial Position of Saudi Arabia (1942), UK Public Record Office, Kew, PRO/FO 371/31451, henceforth PRO.

14. Interviews with advisors to Saudi government.

The cooptation of social forces characteristic of rentier states had set in on a small scale, tilting the balance of power as rentier theory posits.

High Politics: Institutionalization under Abdulaziz' Sons

The large-scale development of Saudi oil by U.S. concessionaire Aramco after World War I had led to a rapid increase in oil revenue, from 10.4 million US$ in 1946 to 56.7 million US$ in 1950.[15] While old Abdulaziz' personal rule grew increasingly slack,[16] there arose a need to manage the huge sums of money flowing into the Kingdom, as at least his second oldest surviving son, Faisal, and the American diplomats and oilmen present at the time realized.[17] In the absence of any scheme for budgeting and development, the court, its clients, and their entourages frittered away much of the growing income.[18] There was a further motivation for institution-building: with the unquestioned patriarch leaving the stage, a new generation of princes vied for power and prestige, giving them a stake in the creation of new government posts and ministries.[19]

In 1951 an already weakened Abdulaziz decreed a rudimentary functional differentiation of domestic bureaucratic institutions. The Ministry of Interior (MoI) was split from the fiscal and administrative affairs of the Ministry of Finance, which had hitherto been a small government unto itself responsible for most of the Kingdom's embryonic public services.[20] Differentiation meant spread of the royal family into state posts: the MoI and the health portfolio were given to Faisal's son Abdallah.[21] In December 1953, not long after Abdulaziz' death, the Ministries of Education and Agriculture were created as spinoffs from the Ministries of Interior and Finance, respectively.[22] Aspiring Prince Sultan took charge of agriculture, his full brother Fahd took over education.[23]

There were few functioning mechanisms to coordinate their work. The formal introduction of a Council of Ministers in 1953, before the king's death, changed

---

15. Young 1983: 20. Similar figures are presented in Ministry of Petroleum and Minerals 1963.
16. Letter of 6 August 1949, PRO/FO 371/75507 (E 10334).
17. Embassy to State, Reforms undertaken by the crown prince, 16 November 1952, USRSA, Vol. 2.
18. Letter Philby to Barker, Jidda, 6 August 1953, "misc" file, PHILBY.
19. Yizraeli 1997: 20.
20. Royal decree 5/11/4/8697, cited in *Lamha tarikhia*: 6–7.
21. Hare to State, telegram 714, 4 June 1951, USRSA, Vol. 3.
22. *Lamha tarikhia*: 8–9.
23. At the same time, senior princes in the 1950s set up schools for their own offspring; cf. Vasiliev 2000: 433.

actual practices of governance very little.[24] The new king, Saud, Abdulaziz' oldest surviving son, governed according to the patrimonial style of his father, basing his rule on the royal court and his advisors instead of the cabinet and embarking on grand tours through the country in which he directly disbursed royal largesse.[25] Decision-making on major national questions remained largely informal, and governance was marred by "princely jealousies."[26]

Designing the Bureaucracy: "Form Follows Family"

It was largely intra-family patrimonial politics that determined who would receive what government post—not different from other clan-based, administratively underdeveloped political systems. However, patrimonial politics also determined much of the *institutional design* of the rapidly growing state during Abdulaziz' last years and under his first successor.

Institutions in the 1950s were malleable, often adjusting to the authority and status of the persons or factions leading them; occasionally, they were created from scratch to bolster or weaken specific players. Although the government saw a general trend of functional differentiation, concerns of power balancing often influenced the specifics of bureaucratic expansion and reform. In the summer of 1951, Abdallah bin Faisal was made minister of health and interior for the reported reason that he should be equal in status to Minister of Defense Prince Mish'al.[27] Conversely, the likely motivation for a 1952 plan to set up a Ministry of Air Force, with Mish'al as minister and his full brother Mit'eb as deputy, was to prevent aviation issues from falling under the authority of the new Ministry of Communications under Prince Talal.[28] For a while in 1953, it seemed likely that Saudi Arabia would have two separate government airlines, since Mish'al and Talal could not agree who would control the fleet.[29]

---

24. "In fact the decree [creating the Council] does little more than regularise the present state of autocratic confusion in the hope that by making one autocrat bigger than the rest there will be less confusion." Jeddah to FO, 14 October 1953, PRO/FO 371/104854 (ES 1016/3); cf. *A study of the Council of Ministers,* Local Government Relations Department, Aramco, 9 May 1957, box 9, folder 1, Mulligan Papers, Georgetown University.
25. Citino 2002: 64–65, 78.
26. Jones to State, 13 October 1953, USRSA, Vol. 2; 1954 annual review, PRO/FO 371/113872 (ES 1011/1).
27. Hare to State, telegram 714, 4 June 1951, USRSA, Vol. 3.
28. Embassy to State, telegram 537, 3 April 1952, USRSA, Vol. 3. The order to set up the Ministry of Communications mentioned explicitly that it would be responsible for all transport bar aviation; cf. *Lamha tarikhia:* 8.
29. 1953 annual review, PRO/FO 371/110095 (ES 1011/1).

When Talal resigned as minister of communications in April 1955, the communications portfolio was reassigned to the Ministry of Finance (then under commoner Mohammad Suroor).[30] Talal told U.S. diplomats that this was essentially a solution to avoid picking one of the various princely candidates for the post and offending the others.[31]

Apparently, this sort of accommodation to personal needs and conflicts also happened, on a smaller scale, among senior commoner administrators. For example, Deputy Minister of Finance Mohammad Suroor was given a new post for the supervision of both pilgrimage and broadcasting for the reason that the heads of these two departments were rivals; making Suroor titular head of both meant that neither would be subordinate to the other.[32] When Saudi ambassador to the United States Asad Al-Faqih returned to the Kingdom in 1955, a new post in the Ministry of Foreign Affairs was created according to his own proposal ("Inspector of the Diplomatic and Consular Service"). The establishment as well as the dissolution of an independent Ministry of Economy was similarly tied to the shifting fortunes of its senior administrator, Ahmad Moosli.[33] Although skills often did matter for selecting commoner administrators (such as Oil Minister Abdallah Tariqi or Minister of State Abdallah bin 'Adwan), structures of clientelism and long-term loyalty also played a prominent role, often to the detriment of performance.[34]

Under King Saud, senior appointments and institutional changes increasingly became a function of his rivalry with Crown Prince Faisal. The details of this struggle have been recounted elsewhere, so a few examples of how it affected the institutional setup of the Kingdom will suffice.[35] The defense establishment was a prime battleground in terms of successive appointments and administrative re-engineering. Early on in the struggle, King Saud detached the Royal Guard under Musa'd bin Saud from the Ministry of Defense to weaken Mish'al and strengthen his own sons. The Royal Guard temporarily became the most powerful military

---

30. Royal decree 5/19/1/1416; *Lamha tarikhia*.

31. The return was rated as a possible improvement, as the Ministry of Communications was said to be corrupt and had achieved little; cf. Stein to State, Resignation of prince Talal bin Abdulaziz as Minister of Communications, 30 April 1955, USRSA, Vol. 5. In November 1955, up-and-coming Prince Sultan would become head of the Ministry of Communications, once again recreated as an independent entity; cf. royal decree 5/19/1/720; *Lamha tarikhia*: 8.

32. Embassy to State, Mohammed Suroor to supervise pilgrimage and broadcasting, 22 December 1952, USRSA, Vol. 2.

33. Embassy to State, News summary for September, 6 November 1955, USRSA, Vol. 4; Pelham to FO, 8 August 1953, PRO/FO 371/104854 (ES 1016/1).

34. One rather embarrassing example was Saudi UN Ambassador Baroudi, who was useless by the admission of his boss Omar Saqqaf but to whom Faisal was obliged; cf. Saudi Arabia: Government Officials, FCO 8/1168. Similarly, Ibrahim Suwayyel had no visible qualification for being minister of agriculture apart from being loyal and trusted; biographical sketch (n.d.), Mulligan Papers, box 2, folder 4.

35. Yizraeli 1997; Samore 1983.

unit.³⁶ In May 1955, Saud decreed the formation of a modern National Guard to be headed by another of his sons, Khalid bin Saud. This again was interpreted as a step against Mish'al, who had been able to prey on National Guard resources before, as it had been a relatively weak institution headed by a commoner.³⁷ In 1964, after Saud's final defeat by Faisal, the Royal Guard was attached to the Ministry of Defense, implying its disbandment as an independent institution.³⁸ Institutional design followed power politics.

Institutions as Fiefdoms

Institutions, being tokens in political games, were also used for building personal fortunes and expanding one's following. Proprietary patterns emerged early on: In his later years, Abdulaziz had "granted" several of the new agencies to specific factions among his sons, a process in which maternal lines played an important role.³⁹ Examples include the already mentioned National Guard, the governorships of Riyadh and other major regions, and the Ministry of Defense, controlled successively by full brothers.⁴⁰ Whereas nonroyal notable families maintain fewer hereditary claims to senior posts today than in the 1950s, different factions of the Al Saud family have defended their patrimonial control over specific organizations ever since.⁴¹

Officials gradually acquired personal followings through what they could do for their followers.⁴² Princes in bureaucratic positions quickly surrounded themselves with hangers-on, advisors, and business partners.⁴³ As budgets grew rapidly,⁴⁴ most of the senior figures used the institutions they controlled to distribute

---

36. Faisal's Character (n.d.), Mulligan Papers, box 1, folder 70.
37. Yizraeli 1997: 153.
38. Jidda to State, The Power Structure in Saudi Arabia, 23 March 1965, NARA RG 59, 250, 5–7, box 2642, Folder Pol 1 General Policy Background SAUD.
39. Yizraeli 1997: 37.
40. The governorship of Riyadh went from Sultan to Naif in 1953, then to Salman in 1955. The Ministry of Defense was controlled by the full brothers Mansur (who died in 1951), Mish'al (his successor), and Mit'eb (the latter's deputy).
41. Lipsky 1959: 96.
42. Ibid.
43. Examples of hangers-on include Saud's confidant 'Id bin Salem (Harry W. Alter, The Royal Cabinet, 24 August 1960, Mulligan Papers, box 3, folder 61); Adnan Khashoggi, who reportedly was given his first break by Talal and Nawaf before becoming a client of Fahd and Sultan (Adnan Muhammad Khashoggi, October 1971, Mulligan Papers, box 1, folder 70); Ibrahim Shakir, who was a client of Minister of Finance Abdallah Sulaiman (Consulate Basra to USSD, Enclosure No. 1 to Despatch No. 111, 4 May 1950, USRSA, Vol. 2); Faisal's brother-in-law Kamal Adham; and the Alireza merchant family, all supported by Faisal; cf. Field 1984: 110.
44. In 1953, when boys' education was elevated to ministerial status under Fahd, the budget was 20 million Saudi riyals. Within four years, expenditure increased to 88 million SR; cf. Supporting study 3:

personal favors and to conclude business deals.⁴⁵ Minister of Defense Mish'al, for example, was said to have used his privileged post to keep millions of freshly minted riyals for himself;⁴⁶ he has been known as one of the richest princes ever since, reportedly due to extensive deals in state land.⁴⁷ His full brother and deputy Mit'eb held the national fishing concession, whose rights he partially resold, had several companies registered under his name, and has also become known for large land holdings.⁴⁸ Rent-seeking was rampant, enabled by an unaccountable state based on external income, just as rentier theories would predict.

It was often social structures not captured by macro-oriented rentier theory that would determine *how* rents were sought, however. "Gatekeepers" emerged early on to control access to princes and their institutions. One aide of Talal, for example, was responsible for all procurement under Talal's tenure as ambassador in France (Talal, whose Ministry of Communications was reportedly corrupt, also was one of the richest princes despite his tender age).⁴⁹ Eight or nine different groups were reported to exist in the entourage of King Saud, all struggling for access and the best positions of brokerage.⁵⁰ Different royals tended to recruit their gatekeepers from different circles: Saud tended to cling to the old foreign advisors of his father, whereas Faisal had more Hijazi among his clients.⁵¹ Fahd would surround himself with young, educated Najdis.⁵²

Most of the princely gatekeepers had an urban background in common, while tribal elites gradually lost out. The greater the funds circulating in the underdeveloped but rich state, the more "brokers" of state resources would emerge, also

---

education case, Ford Foundation archive, box 1, Institute of Public Administration, Riyadh, henceforth FF/IPA. The Ministry of Agriculture under Sultan quickly acquired large amounts of machinery after its establishment in 1953, even though it lacked the skills to deliver services; cf. Supporting study 2: the agriculture case, FF/IPA, box 1.

45. The use of public positions for private gain was condoned and widespread at the time; cf. Lipsky 1959: 103. Philby vividly denounced rapacity and corruption of officials in manipulating government procurement and contracting; cf. The new reign in Sau'di Arabia (21 January 1954), GB 165–0229 PHILBY, 1/4/9/3/39. Several princely officials were engaged in commerce; cf. The transformation of Arabia (talk at the Royal Geographical Society), GB 165–0229 PHILBY, 1/4/9/3/38. When Minister of Commerce Alireza resigned, he complained bitterly about princely greed; cf. Morris (UK embassy Washington) to FO, 25 November 1958, PRO/FO 371/132656 (BS1015/16).

46. Several regional governors reportedly did the same; cf. Young 1983: 78.

47. Field 1984: 101–2. His brother Mit'eb is comparably rich; cf. Business International 1985: 7.

48. Economic Report for November 53 to January 1954, PRO/FO 371/110103 (ES1101/1).

49. Copy of a letter from Mr. P. Smith, former tutor to a member of the Saudi royal family (28 December 1958, Rome), PRO/FO 371/114875 (BS 2181/1).

50. Pelham to FO, 28 March 1954, PRO/FO 371/110098 (ES 1017/2).

51. The Royal Cabinet, 20 August 1960, Mulligan Papers, box 3, folder 61; Field 1984: 110; Muhammad Omar Tawfiq (n.d.), Mulligan Papers, box 2, folder 4; Faisal's Character (n.d.), Mulligan Papers, box 1, folder 70; Huyette 1985: 69.

52. Interviews with former senior bureaucrats.

below the senior level. As we will see, brokerage became a veritable national vocation in the 1970s and 1980s.

The State of the State in the 1950s: Formal and Informal Authority

As family politics combined with a lack of administrative tradition, the unprecedented growth of the bureaucracy in the 1950s was not accompanied by much rationalization. Although some new agencies helped to increase the provision of public services,[53] the sudden availability of resources also led to uncontrolled, patronage-based bureaucratic sprawl—rather than the legal-rational state building presented in Kiren Chaudhry's account. As rentier theory would lead us to expect, availability of outside income reduced the need to develop powerful and cohesive regulatory institutions.

Many of the young Saudi administrators were completely inexperienced, especially the royal ones, who without exception had received only a court education.[54] Mish'al was reported to be quiet, dignified, and serious about his job, but when it came to institution-building, "through lack of education, knowledge and experience, performance results have been practically nil as the Ministry remains unorganized and its procedures are on an entirely personal basis. He leans heavily for advice and recommendations on foreign counsel."[55] Two Egyptian advisors played a particularly prominent role.[56] With illiteracy in Saudi society estimated at about 95 percent in 1956, qualified Saudi administrators were almost impossible to come by.[57]

Informal structures of authority determined the actual importance of institutions. The bits of the bureaucracy that mattered were run by important princes or by commoners close to the court. A formally important institution like the Ministry of Commerce, although headed by influential merchant Mohammad Alireza, proved to be largely powerless.[58] Undercut by successive ministers of

---

53. Biographical sketches of Fahd and Sultan, Mulligan Papers, box 1, folder 70. In 1953, when boys' education was elevated to ministry status under Fahd, 55,000 pupils were enrolled. Within four years, the number of students increased to 224,000; cf. Supporting study 3: the education case (12/29/72), FF/IPA, box 1.
54. Various biographical sketches, Mulligan Papers.
55. Department of the Air Force, Visit of HRH Prince Misha'al Abdulaziz al Saud, Minister of Defense, Saudi Arabi [sic], including 2 biographies, 6 November 1951, USRSA, Vol. 2.
56. One of the advisors apparently received four times his previous official salary, demonstrating the scope for arbitrary patronage. Ibid.
57. Lipsky 1959: 277.
58. Toomey to State, Saudi Press Criticizes the Minister of Commerce, 2 October 1958, USRSA, Vol. 5.

finance with better royal access and with no commercial regulations emerging from the cabinet, he complained that he had no de facto role at all.[59] Conversely, a previously inconsequential post such as that of comptroller general suddenly became important when it was filled by senior prince Musa'd bin 'Abdulrahman.[60]

Institutions were personalized also on lower levels, where formal procedures remained unclear. Large parts of economic life remained unregulated, and dealings with the bureaucracy happened on a personalized basis.[61] Although layers of idiosyncratic red tape often expanded around individual offices, there were no encompassing systems of civil service training, seniority, or even document filing in the 1950s.[62] Recruitment was personalized, with kinship and tribal relations frequently used as employment criteria.[63] The pursuit of personal gains was widespread, and many offices were characterized by inactivity.[64]

The U.S. embassy at the time was trying to keep track of administrative development and drew charts of agencies, but "the diagrams themselves have proved to be most confused, since this Government does not lend itself at all well to such schematic presentation,... the organization is so confused, particularly among the lower echelons, that an unwarranted level of research would be required in order to fill out charts at all these levels... since organization means little to this Government, it is very flexible and changes with such rapidity that charts are outmoded almost before they can be prepared."[65] Aramco during this period kept a roster of Saudi public personnel on which the government itself depended due to lack of oversight over its fluctuating institutions.[66] Several agencies, such as the

---

59. Ministry of Commerce, 24 August 1960, box 3, folder 61; biographical sketch Alireza family (n.d.), box 1, folder 170, Mulligan Papers.

60. Embassy to USSD, biweekly report for 1st half of May, 11 June 1957, USRSA, Vol. 4.

61. Only one firm, for example, had enough clout to register trademarks at the responsible office in Makkah in the mid-1950s (the office at the same time fought a battle of obstruction against the new Ministry of Commerce that was supposed to be formally responsible). The "absence of commercial legislation" was a fundamental problem for foreign companies; cf. Heath to FO, 7 February 1955, 29 June 1955, 26 October 1955, PRO/FO 371/113896 (ES1471/1; ES1471/3; ES1471/4). While there was a department in the Ministry of Commerce responsible for the implementation of commercial law, such a law seemed not to exist; cf. Ministry of Commerce, 11 March 1957, box 9, folder 1, Mulligan Papers.

62. Vitalis 1999: 661; Supporting study 1: personnel management (17/12/72, draft), FF/IPA, box 1. *At-taqrir ath-thanawi.*

63. Early reports providing basis of administrative reform project, Civil service in Saudi Arabia (August 1963), FF/IPA, box 18.

64. Lipsky 1959: 103

65. Embassy to State, transmittal of charts of Saudi Government Organisation, 14 November 1951, USRSA, Vol. 2; cf. Vitalis 2007: 203.

66. Lippman 2004: 50.

Ministry of Economy, created in July 1953, existed only on paper,[67] as did senior ministerial positions.[68]

The bureaucratic capacity of many oil-based institutions was low, which comes as no surprise to rentier theorists. What rentier theories do not address, however, is the meso-level state fragmentation brought about by rapid, oil-enabled bureaucratic expansion. As far as institutions mattered, their day-to-day operations were often carried out more or less autonomously.[69] Administrative sprawl and the personalized nature of authority meant that coordination between agencies was lacking, with different institutions often producing directly contradictory decisions and jurisdictions remaining unclear.[70] If anything, institutions communicated with the king and not with each other. As early as 1952, six different entities were supposed to be in charge of economic planning.[71] Three different agencies claimed responsibility for cereal imports.[72] Government contracting was in a state of chaos, as different channels of procurement existed in parallel.[73] Agencies ignored the rules and policies of other agencies[74] or blamed one another for policy failures.[75] A pattern was set, one that Saudi policy-makers would encounter time and again in future decades. Subsequent efforts of administrative rationalization would often be offset by the simultaneously increasing complexity of the state.

The formal operation of institutions was further complicated by the existence of additional informal fora of decision-making. A U.S. embassy dispatch, describing the "*majlis*" system existing around all ministers, grumbled that "theoretical channels of authority are ignored and matters presented to the Government for its consideration tend to disappear for indeterminate periods of time into the

---

67. "Adding one more plan to the plethora of schemes for strengthening the governmental organisation and administration for economic affairs, the formation of a Ministry of Economy under the incumbent Minister of Finance appears to be no more than a paper reorganisation." Jones to State, establishment of Saudi Ministry of Economy, 19 July 1953, USRSA, Vol. 3. All decrees are listed in *Lamha tarikhia*.
68. Discussion with Hafiz Wahba, 23 December 1961, Mulligan Papers, box 3, folder 10.
69. "Hitherto each Minister has done whatever he found within his pleasure and his power," commented a 1954 dispatch. While "the old inter-departmental jealousy and confusion still exist[ed]," 1953 had seen some involuntary decentralization and increasingly slack reactions of agencies to central orders due to Abdulaziz' weakening and death; cf. 1953 annual review, PRO/FO 371/110095 (ES 1011/1).
70. El-Farra 1973: 73; Jones to State, Possible conflict of authority between the crown prince and the minister of finance, 1 November 1953, USRSA, Vol. 3.
71. Embassy to State, now six different entities supposed to be in charge of economic planning, 20 December 1952, USRSA, Vol. 2.
72. Jedda Economic Report November 1952 to January 1953, 2 February 1953, PRO/FO 371/104859 (ES1101/2).
73. PRO/FO 371/104867 (ES 1122/1).
74. This included even formally subordinate offices such as the patent office under the MoF; Heath to FO, 10 May 1954, PRO/FO 371/110210 (ES 1471/1).
75. Appendix B to letter of 20 July 1957, Operational Considerations Arabian Research Division, Mulligan Papers, box 2, folder 57.

nebulous depths of this or that 'diwan.'"[76] The persistence of such informal structures led to great de facto centralization of authority, further contributing to the inertia of the bureaucracy on lower levels.[77]

As far as formal posts went, many of them were just depositories for clients of ministers or princes. The government was extremely reluctant to fire employees and was "full of officials and advisors who are known to be excess baggage, but they remain on the payrolls and just hang around, no one ever considers discharging them."[78] In these early years of the Saudi state, oil-funded bureaucratic positions came to be seen as entitlements.

Despite national budgets that were still tiny by international comparison, senior regime figures frequently used institutions for "parking" clients and in some cases for political cooptation. The latter pattern was demonstrated vividly by the Majlis Ash-Shura, a consultative-legislative council Abdulaziz had created in the Hijaz in the 1920s to represent merchant and notable interests. Although the court progressively cut it off from substantial policy deliberations, it continued to debate issues of national policy year after year without exerting any discernible influence or drawing any public interest.[79] By the early 1950s, it was described as a "catch-all for government officials nearing retirement or for whom there are no other jobs... with little work and almost no responsibility."[80] Membership in the body, "commonly known as the Council of Do-Nothings," was for two years by statute but effectively lasted for life.[81]

State-Society Relations: Local Mobility and Clientelism

While the state would not reach out to Saudi society at large, it offered great chances for the select set of personal clients of the Al Saud who happened to be in the right place at the right time. Access to the expanding court, with its growing needs for material supplies, offered great opportunities of social and economic mobility and for new brokerage roles. Many of the big Saudi merchant families established their privileged positions as suppliers to King Abdulaziz. Some of

---

76. Jones to State, 23 September 1953, USRSA, Vol. 3.
77. Lipsky 1959: 109–10.
78. Embassy to State, transmittal of charts, 14 November 1951, USRSA, Vol. 2.
79. Its resolutions said nothing about issues of implementation, and it did not receive any budget drafts from the government. Minister of Finance Abdallah Sulaiman probably played an instrumental role in blocking Hijazi government organs; cf. Hare to State, transmittal report of the Saudi Consultative Council for the Year 1369, 26 October 1950, USRSA, Vol. 2 (see also the same dispatch for the next year, 8 December 1951); Consultative Council, 29 December 1956, Mulligan Papers, box 9, folder 1.
80. Embassy to State, transmittal of charts, 14 November 1951, USRSA, Vol. 2.
81. Embassy to State, enlargement of consultative council, 28 January 1953, USRSA, Vol. 2.

them received exclusive trade agencies by the king, agencies they hold to this day. Big Saudi business players like Jomaih, Rajhi, and Juffali first appeared at this time, emerging from a humble Najdi background. Jomaih, for example, was the grocer of the king from the village of Shaqra, and Rajhi was a money-changer in the Riyadh *suq*. Positions of privileged access were often transferred across generations, as in the case of the king's physicians Rashad Pharaon and Mohammad Khashoggi, whose offspring would later use their royal links to broker state contracts. Saud, who liked big-ticket construction developments, had his own favored clients among the emerging Saudi contractors, and his sons partnered with certain merchants.[82]

In a related pattern, auxiliary positions at the court could be the starting point of senior public service careers. Without any formal education, Abdallah bin 'Adwan, for example, worked his way up to leading positions in the all-important Ministry of Finance through the intervention of Crown Prince Saud, whose bodyguard he had been.[83] Illiterate 'Id bin Salem progressed from a simple mechanic to head of the royal garage, keeper of the privy purse, and one of Saud's leading advisors.[84] Interpreters Abdallah Tariqi and Mohammad Ibrahim Mas'ud (the latter with the U.S. embassy) would become minister of oil and minister of state, respectively.[85]

Abdallah Sulaiman's Bureaucratic Empire

It is the Sulaiman Al-Hamdan family that probably has combined both patterns—merchant and bureaucratic mobility—most masterfully. Their story illustrates both the contingent and personalized nature of early institution-building and the great opportunities it brought for regime clients.

Abdallah Sulaiman was a Najdi of humble background who worked his way up from clerk to being the first Saudi minister of finance and a close confidant of Abdulaziz. Through the early 1950s, Sulaiman's ministry combined numerous administrative duties; its subordinate departments included health, public works, education and agriculture. Their workings were usually opaque and left great scope for patronage. When Crown Prince Saud issued a set of financial regulations in 1952, these seemed to express little more than "wishful thinking,"

---

82. Saud's son Saab, for example, was involved in plans for the first domestic oil refinery; cf. *Middle East Business Intelligence (MEED)*, 3 June 1960: 249; on contracting, see Yizraeli 1997: 138.

83. Hare to State, promotions within ministry of finance, 2 March 1953, USRSA, Vol. 3.

84. Lacey 1981: 308, 337; Holden and Johns 1982: 182; The Royal Cabinet, 24 August 1960, Mulligan Papers, box 3, folder 61.

85. Young 1983: 39, 61; Jidda to State, 16 October 1964, NARA RG 59, 250, 5–7, box 2644, folder POL 7 Visits. Meetings. SAUD (1/1/64).

according to a U.S. assessment, as they would "do little to alter the present widespread system of letting contracts on a basis of personal favoritism rather than expected performance and cost."[86]

At the MoF-affiliated customs, Ottoman-inspired clearing procedures involved up to thirty different signatures; kickbacks were common. The situation in the postal services was similar, and a British diplomat commented sardonically that "the Director of Posts proudly proclaims that he had British training. There are naturally Saudi improvements on this; the postage stamp clerk, for instance, will take the money for letters and stick the stamps on later; after eighteen months in the job he can build himself a large house."[87]

The king gave large administrative discretion to Sulaiman, who was based in Jeddah.[88] His brother Hamad and his son Abdulaziz were vice and deputy minister, respectively; the higher rank for his brother, which still exists today, was created specifically for him. Apart from profiteering activities, Hamad was reported to be largely inactive.[89] His son Sulaiman was in turn installed as assistant deputy minister, and to complete family control, Abdallah Sulaiman also brought in his reportedly inept nephew as vice-regent during his absences.[90]

An inventive broker-cum-administrator, Sulaiman had started to procure government supplies through figureheads in the 1920s. His friends among the merchants lent him money without interest, which the MoF in turn used to purchase goods for the court from them for inflated prices.[91] Sulaiman received the lucrative cement franchise for the Hijaz in the late 1940s.[92] He owned palaces and large tracts of land and had his own entourage of some four hundred people. After retiring in 1954, he ran hotels and several trading companies. His descendants still are among the most prominent merchants of Jeddah.

Several big businessmen were "made" or at least boosted by the MoF during Sulaiman's time, including some who served under him.[93] The case of Hasan

---

86. Embassy to State, financial regulations issued by crown prince, 20 December 1952, USRSA, Vol. 2.

87. Philips to Eden, 16 April 1953, PRO/FO 371/104860 (ES1105/1); Young 1983: 50ff.

88. Notes on Saudi Arabia for Dr. Hugh Scott (sent 29 August 1944): 6; GB 165-0229 PHILBY, 1/4/9/3/28.

89. Hare to State, telegram 31x[illegible], 17 July 1952; USRSA, Vol. 3; Pelham to FO, 8 August 1953, PRO/FO 371/104854 (ES 1016/1). The abuse of his official position by Sulaiman's brother Hamad and his sons was described as "gross and scandalous" as early as 1946; List of Personalities in Saudi Arabia, 1946, PRO/FO 371/52832.

90. Pelham to FO, 8 August 1953, PRO/FO 371/104854 (ES 1016/1); Jedda Economic Report November 1952 to January 1953, PRO/FO 371/104859 (ES1101/2).

91. Vasiliev 2000: 298-99.

92. Field 1984: 108.

93. In the early 1950s, for example, Siraj Zahran and Mohammad Zaidan, both subsequently prominent business names, were assistants to Mohammad Suroor, advisor to Sulaiman and head of the purchasing

Sharbatli is a colorful example of how upstart clients could gain status and be co-opted into nominal state functions. The diplomatic note on him from the U.S. embassy in Jeddah bears a lengthy citation:

> "Climaxing a rags-to-riches career illustrative of the fluidity of Hejazi mercantile society, Hasan Sharbatli, Jidda merchant and public benefactor, recently received from the King the title of Honorary Minister of State," making him the ninth of this kind. The title "implies no necessary assumption of government powers.... Sharbatli, who only ten years ago was a fruit-vendor and small-time auctioneer in Jidda, reportedly gained entry into the favored circles of the Ministry of Finance by sending to Hamad al SULEIMAN, brother of the minister, a gift of fruit. So pleased was the recipient that he recommended the appointment of Sharbatli as Government Purchasing Agent.
>
> In this capacity, Sharbatli evolved an eminently satisfactory relationship with the government. Selling a hypothetical 5,000 riyals worth of fruit, he would then bill the Ministry for 50,000 riyals and subsequently share the profits with his sponsors in the Ministry."[94]

Sharbatli accumulated a total reported credit of 43 million Saudi riyals with the government. Unable to recover the debt for himself, Sharbatli became one of the biggest charitable benefactors in the Kingdom, being allowed to use some of the sums for this purpose. He had become one of the richest men in the Kingdom, had a virtual monopoly on pilgrim travel, and reportedly gave 500,000 riyals for the establishment of the Saudi Air Force.[95] He also set up his own bank.[96] Sharbatli's case illustrates that, although the royal family was squarely at the center of the growing Saudi system, more or less random networks of commoners within and around the administration enjoyed great distributional leeway and mobility at an early stage—and the chance to position themselves for the decades to come.

This is in marked contrast to the later mobility closure once the state had fully formed in the 1980s. Rentier theories do not capture this kind of historical path dependency. Oil at an early stage of state development can be a great enabler for

---

department in Riyadh; cf. Embassy to State, transmittal of charts of Saudi Government Organisation, 14 November 1951, USRSA, Vol. 2. Other high-ranking MoF officials during Sulaiman's time, like Lebanese Najib Salha, had enough maneuverability in their administrative units to establish their own clients. Salha reportedly "made" the Kaaki family (who were to found the large National Commercial Bank) during World War II; cf. Revision of Leading Personalities, PRO/FO 371/52832.

94. Hare to State, appointment of Hasan Sharbatli as Minister of State, 28 July 1952, USRSA, Vol. 2.

95. He is also described as "completely illiterate" and "possessed of a dour personality," indicating American disapproval of this particular variant of social mobility. The note ascribes an unknown Najd origin to him, which however is inaccurate (the Sharbatlis are from Egypt; interviews in Riyadh).

96. Foreign Office minute, 24 January 1958, PRO/FO 371/133156 (ES 1111).

shaping institutions and staking out individual positions. But as we will see, later on, when claims on resources have been meted out, the system can become highly immobile.

The story of Sulaiman also demonstrates how the local power of commoners, in the final analysis, depends on royal patronage. The minister was a favorite of King Abdulaziz but not of his son Saud. When the Ministry of Commerce was created in 1954 under Mohammad Alireza, a Hijazi stalwart of the Al Saud, this was taken as a sign that King Saud wanted to weaken the Ministry of Finance under Sulaiman.[97] Although his family remained in business, Sulaiman resigned the same year and was replaced by a follower of Saud. Once out of favor, even the most senior and trusted commoner with a large personal clientele and many merchant friends could count on little independent support.

An Early Island of Efficiency: SAMA

Before his exit, however, Sulaiman had surprisingly gotten involved in probably the most significant effort at administrative rationalization in the 1950s: the creation of the Saudi Arabian Monetary Agency (SAMA). In the early 1950s, monetary chaos reigned in Saudi Arabia, with both silver and gold coins circulating, each following different price shifts, fragmenting markets, and preventing coherent national budgeting.[98] After French and American plans for a revenue commission to help with Saudi budgetary matters had come to nothing, the U.S. embassy and Aramco successfully lobbied with the king to accept U.S. consultants into the Kingdom. In 1950, Abdulaziz approved the assistance of five U.S. experts, covering monetary, customs, tariff, budget, and accounting matters.

The experts helped with the December 1952 tariff reform, in which ten different surcharges were unified, basic goods exempted, and luxury goods given higher tariffs. Exceptions for officials, which had been predictably abused, were abolished, and export duties were removed.[99]

U.S. advisor Arthur Young managed to convince Sulaiman and the king of the need for basic monetary and banking regulation—none of which would immediately infringe on their powers of patronage. The king gave Young a few hours to draw up a plan for what would become SAMA, the first Saudi agency to have a research department. Reflecting the Saudi state's lack of data-gathering capacity,

---

97. Wadsworth to State, 16 March 1954, USRSA, Vol. 3.
98. Young 1983: 14.
99. Young 1983: 50–51; Jedda Economic Report November 1952 to January 1953, PRO/FO 371/104859 (ES1101/2).

Young earlier had had to ask his Lebanese assistant to go to the local suq and write down basic prices to establish a crude price index.[100]

Due to the inexperience of the private sector, the government was to provide the entire capital for SAMA, which would derive its income by charging the government for its services. The first head of SAMA was also an American, and accountants were hired from Lebanon.[101] No Saudis, it appeared, would have had the required training.

Young had managed to convince King Abdulaziz and Sulaiman of the pressing need for institution-building in a strategic sector. The SAMA charter and the use of foreign technocrats bolstered the relative autonomy of the body. The institution performed remarkably well, stabilizing exchange rates, creating a workable Saudi currency, and a system of balance of payments accounting.[102]

After the MoF had temporarily sidelined SAMA, another outsider took over the governorship in 1957: Anwar Ali, a Pakistani who had come to the country with an IMF mission, and who held the post as a confidant of King Faisal until his death in 1974.[103] Unlike most other agencies, SAMA was kept outside of the regular civil service structures, with its own payment and recruitment schemes.[104] Drawing on royal protection and growing resources, it could build up its own research and training capacities. SAMA today is regarded as possibly the best central bank in the Middle East.[105]

The sudden oil riches gave the Saudi elite great leeway in their design of state institutions. Although temptations of rent-seeking were strong, personal agency intervened to determine specific outcomes. In the important case of SAMA, growing state resources were used to build an efficient administration that has lasted to this day. Contrary to the forecasts of orthodox rentier theory, oil does not automatically equal institutional decay, but it can enable the opposite, depending on what elites decide. If anything, oil tends to widen the menu of institutional choice. As the history of the modern Saudi state demonstrates, oil-based bureaucracies can be heterogeneous; it is problematic to treat them as aggregates, as the literature has done thus far. We will understand more about their capacities and

---

100. Young 1983: 47.
101. Embassy to State, financial regulations issued by crown prince, 20 December 1952, USRSA, Vol. 2.
102. While fiscal policy in the early 1950s was generally incoherent, "the Government [was] saved from some of its own follies by the Monetary Agency which under the able guidance of an experienced American banker acts as a national bank"; 1953 annual review, PRO/FO 371/110095 (ES 1011/1).
103. Beeley to FO, 15 June 1955, PRO/FO 371/114888 (ES1111/2).
104. Supporting study 1: personnel management, FF/IPA, box 1.
105. Lippman 2004: 110; Standard & Poor's 2007: 13.

limitations if we look at them on the meso-level, that is, on the level of individual institutions.

Conclusion: A State Created above Society

The Saudi state in the 1950s was divided into different fiefdoms from the very start, each acquiring its own clients, be it official employees, hangers-on and brokers, or business partners in society. Although fragmented, the setting remained highly fluid and under-institutionalized. Institutional change was often a result of personal shifts of fortune within the elite. Elite decisions—both good and bad ones—greatly affected the setup of the bureaucracy as well as chances of access and social mobility. The royal family remained the central hub of the strictly hierarchical polity, as it ultimately controlled all resources of patronage. Yet the various spokes attached to and fed by this hub could change quickly and end up looking very different from one another: slush funds for personal enrichment in some cases, repositories of passive bureaucratic clients in others, kernels of administrative efficiency in yet others. Oil surpluses were used to create very different types of organizations with different purposes.

How could the diverse sprawl of Saudi institutions in the 1950s happen in such an unencumbered way, governed by personal interests at least as much as administrative or developmental needs? In the 1950s, the national income was still extremely small by the standards of modern statehood, but the aggregate societal demands on it were even smaller.[106] Although objective development needs might have been great, society was unprepared for the riches. It is difficult to prove a negative—the absence of social constraints—but the documentary record seems to permit only one conclusion: The Kingdom knew no public space to negotiate over national budgets, which were fed externally and grew suddenly. Few Saudis would have been knowledgeable enough to have coherent ideas of administrative development, and few were in direct contact with the state.[107] Levels of political mobilization in society were low.[108]

The different elite groupings in the policy debates of the 1950s were all led by princes.[109] To the extent that some royals attempted to mobilize wider social

---

106. Huyette 1985: 106.
107. Lipsky 1959: 97.
108. Internal Cohesion in Saudi Arabia, Joint Research Department Memorandum, 14 July 1966, NARA RG 59, 250, 5–7, Box 2642 [folder label missing].
109. Yizraeli 1997: 22.

support against their rivals, they failed.[110] Saudi Arabia had no coherent classes to demand services of the state. Hijazi merchants were probably the only relatively well-organized and educated lobby group on economic issues. But many merchants were quickly tied up in clientele structures,[111] as shown by the examples of the MoF and the Majlis Ash-Shura.[112] There are no records of lobbying for coherent developmental expenditure or administrative advancement,[113] and chambers of commerce were state-dependent and "rarely unite[d] to exert pressure for amendments in government policy."[114] Having lost their traditional economic base, the tribes also had become state-dependent at this stage. They were co-opted through royal subsidies, which were doled out in patrimonial style by Abdulaziz and, even more so, his successor Saud, without much modern state infrastructure interfering—not that it occurred to anyone to desire this.[115] A small coterie of progressive technocrats enjoyed a brief moment of political prominence as advisors of King Saud in 1961, but with little independent social support to draw on, they were quickly dismissed when their royal patrons tired of them.[116] U.S. observers in the early 1960s detected no coherent public opinion or civil society to speak of.[117]

In fact, no group outside of a few princes and senior ulama possessed the right to be consulted on policy.[118] The ulama around the Saudi court were almost certainly reluctant to have a modern administration established. But once it was clear

---

110. Saud's attempts to rouse tribes against Faisal failed; cf. Jidda to State, King-Faysal Confrontation, 3 January 1964, NARA RG 59, 250, 5–7, box 2645, folder POL 15 Government SAUD (1/1/64). When reformist prince Talal tried to build up a popular following during a tour in the Eastern Province, merchants proved uninterested, and young professionals refused to engage with him; cf. Saudi National Legislative Council, 2 August 1961, Mulligan Papers, box 3, folder 8.

111. Many merchants were co-opted into administrative structures through posts: "The local merchants of the country [have] been allowed to take an active part in the direction of its affairs as officials." Cf. The new reign in Sau'di Arabia (21 January 1954), GB 165–0229 PHILBY, 1/4/9/3/28/39.

112. "The merchants all, to a greater or less [sic] degree, depend on the spending power of the court." Cf. Brief for Shuckburgh for talks with US beginning 12 January 1956, PRO/FO 371/120754 (ES 1015/3).

113. Saudi merchants were rated as individualistic, with limited interest in collective organization; cf. Ministry of Commerce, 11 March 1957, Mulligan Papers, box 9, folder 1.

114. Chambers of Commerce in Saudi Arabia, 14 April 1956, Mulligan Papers, box 2, folder 51; Jidda to State, The Power Structure in Saudi Arabia, 23 March 1965, NARA RG 59, 250, 5–7, box 2642, folder Pol 1.

115. Niblock 1982: 86–96; El-Farra 1973: 28, 135; Al-Rasheed and Al-Rasheed 1996. Tribal leaders were usually controlled through regional governors; cf. Lipsky 1959: 116.

116. Vitalis 2007.

117. Ibid.: 141, 143.

118. Jidda to State, The Power Structure in Saudi Arabia, 23 March 1965, NARA RG 59, 250, 5–7, box 2642, folder Pol 1. In the mid-1940s, significant parts of the mercantile community did not pick up on the opportunities created through the new oil earnings; cf. memorandum of 11 September 1946, PRO/FO 371/52834 (E 9536).

that a state would be built anyway, they would have no particular stakes in the royally driven splits, mergers, and reshuffles of the secular bureaucracy. Probably the royal family's most important policy constituency, the ulama would be successively co-opted and bureaucratized through the creation of new educational and judicial institutions. This would give a bureaucratic underpinning to the official Wahhabi doctrine that had all along demanded submission to the imam, the political leader of the community of believers, that is, the king.[119]

In line with what some rentier theorists have argued, a dominant state could keep society fragmented through dependence on its resources. All told, therefore, the new external income left much space for regime elites to pursue their individual patronage interests and to negotiate distribution and institutional design *within* the state. There were of course demands of immediate clients and partners outside of the formal state but, as far as we can tell, not really of larger groups in society. Distribution through the state, although creating new expectations and demands, was unequal, local, and granted by the elites. Much of the modern state was originally created above society, following its own peculiar logic of fiefdoms and their lateral sprawl.[120]

Insofar as there was lobbying for formalization and modernization, this emerged to a significant degree outside the domestic setting, strictly speaking. Finding it difficult to deal with the patrimonial Saudi state and its unordered fisc, Aramco and the U.S. embassy were probably the most significant lobbyists for bureaucratic rationalization in the Kingdom at the time, ahead of all domestic groups.[121]

This history of a state created above society confirms rentier theory expectations of high state autonomy. Where the pre-oil political framework is underdeveloped, elites are pretty free to use the oil income as they want. As the following chapters will show, however, in the Saudi case this autonomy declined strongly as the state grew and increased its distributional obligations. Once again, the lack of the time dimension in rentier theories becomes obvious. Despite its continued reliance on external income, today's Saudi regime is much less autonomous in changing bureaucratic structures or distributional policies than the early post–World War II regime. Early decisions on institutional design, often taken for contingent reasons, can greatly impact the shape of the state apparatus much further down the line, when structures congeal. Existing theories of oil states do not account for this path dependency.

---

119. On the ulama's effective political abstinence, see Lacroix 2007; Al-Fahad 2004.
120. As Eisenstadt (1973: 56) points out, in patrimonial systems (such as the early Saudi state), elite conflicts tend not to lead to broader crises as long as broader social groups and demands are weak.
121. Aramco generally urged the princes to streamline their government (Yizraeli 1997: 102), and so did the embassy; cf. Embassy to State, Reforms undertaken, 16 November 1952, USRSA, Vol. 2.

Chapter 3

The Emerging Bureaucratic Order under Faisal

Trying to build up Saudi Arabia as an anticommunist bulwark in the Middle East, the U.S. government and embassy were highly concerned over the administrative chaos of the 1950s and voiced their worries repeatedly. However, American pressure was largely ineffective under Saud, who allowed his state to sprawl incoherently. It was Faisal's reign that brought some degree of order into the government apparatus. But again, this was at least as much driven by royal family politics as by perceived development needs. Faisal's struggle with Saud was at its apex between 1958 and 1962, exactly the period during which institutions were reshaped the most dynamically, by and large following the shifting balance of power between the Faisal and Saud factions.

After 1962, a stable coalition of elite forces combined with bureaucratic growth and institutionalization to "lock in" a permanent constellation of actors and organizations at the core of the state. But although the fluidity of the state apparatus decreased, personalization of power at the top and persistent overcentralization continued to prevent meaningful coordination between different agencies; while gross nepotism was curtailed, informal structures of favoritism and brokerage persisted. In many ways, the more formal state under Faisal only served to institutionalize the clientelist nature of the bureaucracy.

Some of the institutional balancing games between Faisal and Saud in the field of security have already been adumbrated. After 1958, the struggle between the two brothers went deeper than that, however. The very nature of the Saudi was at stake in the conflict, with fundamentally different institutional concepts being deployed by both players, reflecting the fluidity of Saudi governance structures at the time. In this unstable period, Saudi Arabia experienced not only cabinet reshuffles every few months but also repeated reorganizations of core state institutions.

High Politics: Faisal's Order by Cabinet

The Saudi drive for political order began with a May 1958 decree that enhanced the status of the Council of Ministers, with Faisal as prime minister.[1] The reason for the decree was that a coalition within the Al Saud wanted to enhance Faisal's standing against Saud, who was, among other things, perceived to be favoring his sons over his brothers and incapable of handling the state budget, which was in severe deficit. Royal expenses in 1957 were estimated at 40 to 60 percent of the budget, large parts of which Saud distributed to his followers and the public.[2]

Although a step toward formal government, the May 1958 institutional redesign most of all reflected the personal balance of power between Faisal and Saud. Until Saud temporarily regained power in 1960, Faisal achieved a certain separation of state and royal family affairs[3] and, with IMF assistance, a degree of budgetary control and monetary reform.[4] This he accomplished through a centralization of authority in his own person, however. In addition to being minister of foreign affairs and minister of finance, in April 1959 he also assumed the post of minister of interior, previously held by his reportedly ineffective son Abdallah. Although this enabled Faisal to exert more control over regional fiefdoms such as the Eastern Province under the Jiluwi clan and the Riyadh governorate under Salman, it also created a bottleneck of decision-making.[5] Rated a thorough individualist by Aramco researchers, Faisal often sat on decisions, procrastinating for months or not making them at all. Several liberal Saudis, while conceding that Faisal was more likely to stick to formal rules than Saud, deplored such overcentralization.[6]

Institutions at the very core of the state were pliable weapons in the struggle. As king with a court, Saud had larger quasi-traditional, patrimonial resources at his disposal and used them to increase his following through royal handouts and

---

1. Embassy to State, Biweekly Report, 17 May 1958, USRSA Vol. 4.
2. Yizraeli 1997: 122; Lippman 2004: 105.
3. An achievement with which many Saudi technocrats credit him; cf. Al Saud 1982: 94–95.
4. SAMA's first annual report in 1960 stressed the achievement of budgetary order, contrasting it with the preceding chaos; cf. *MEED,* 23 January 1959: 45. Clearer and more meaningful budgeting was also lauded by the Americans; cf. Embassy to State, biweekly reports, 5 January 1959, 4–11–1958, USRSA Vol. 4.
5. Sweeney to State, telegram, 14 April 1959, USRSA Vol. 5; Joint Weeka 33, 1 September 1964, NARA RG 59, 250, 5–7, box 2643, folder POL 2–1 Joint Weekas SAUD (7/1/64).
6. Heath to State, Over-centralization of authority in Saudi Government, 22 July 1959, USRSA Vol. 5.

tours through the country. Faisal, conversely, promoted more modern institutions of rule to boost his otherwise inferior standing, trying to curtail the royal budget and to enforce cabinet rule. There was no superior constitutional (or societal) framework allowing the struggle to be conducted *within* given institutions. On several occasions, the court and the Council of Ministers created parallel bureaucratic bodies to achieve practically the same ends.[7]

The brothers' attitudes toward government reform could be quite calculating, as their institutional track record shows: "traditional" King Saud would promote modern bodies—like a new Majlis Ash-Shura with legislative powers—and "modernizing" Prince Faisal would cling to "traditional" institutions—like his Hijazi vice-regency—when it suited their power interests.[8] That the vice-regency hampered national administrative integration did not seem to faze Faisal as long as he could use the post as a bargaining chip to attain the premiership.[9]

Similarly, Saud's temporary adoption of progressive, constitutionalist technocrats into senior positions in late 1960, vividly described by Robert Vitalis, was a mere "power play" that had little to do with ideology and that lasted a bare two months while his other advisors were sick or distracted.[10] Once the senior princely partners of Saud decided to remove them, the constitutionalists had no popular constituency to fall back on.[11]

If he found it politically expedient, Faisal was quite capable of hampering bureaucratic rationalization. The Council of Ministers, then, was most of all a useful instrument to help him gain power.[12] It is nonetheless true that after Saud's fall from grace in 1962, the council improved policy coordination somewhat—not least because the royal diwan as parallel political actor had disappeared. Faisal was generally willing to devote more time to the formal workings of government, seeing a degree of budgetary control and more clearly defined administrative structures as instruments of stability. Most of his reformist ambitions beyond the establishment of basic order petered out soon after Saud had been sidelined in 1962, however.[13]

---

7. In 1960, a special advisory commission was created as a royal counterpart to the Economic Development Board, which reported to the Council of Ministers. In 1961, plans were bandied about for two separate national broadcasting authorities; cf. Resignation of Amir Nawwaf, 1 August 1961, Mulligan Papers, folder 8, box 3.
8. Heath to State, telegram, 27 November 1959, USRSA Vol. 5; Yizraeli 1997: 53.
9. Ibid.
10. Vitalis 2007: 221.
11. Interviews in Riyadh with contemporary witnesses.
12. Predictably, the austerity starting with the 1959 budget did not imply rationalization all across the board but hit mostly the budgets of Saud and his minister sons, whereas Faisal's allies saw constant or increasing allocations; cf. Yizraeli 1997: 129.
13. Feisal's policy statement: an eight-month review, supplement for the period 31 July–21 Oct 63, Mulligan Papers, box 4, folder 11.

Consolidation of Fiefdoms: The Cabinet Deal

Throughout the 1960s, the Saudi budgets once again continued to expand—quadrupling within less than a decade—and the bureaucratic apparatus grew with them, offering new mobility opportunities for commoners.[14] Saudi civil employees increased from 40,000 in 1960 to 97,000 in 1970.[15] The Ministries of Petroleum, Hajj, Labor, and Social Affairs and Information were created between 1960 and 1963.[16] After 1963 and before the post-1973 oil boom, however, the growth of the budget implied the growth of existing institutions rather than the creation of new ones. The expansion of fiefdoms would continue, but in a more orderly manner.

The cabinet that first assembled in October 1962 became the gravitational center of Saudi politics, representing a post-Saud distribution of power that grew increasingly immovable. It was based on a dominant coalition within the Al Saud and was the end result of the Saud-Faisal struggle, in which Faisal's senior allies were rewarded with ministerial posts: Fahd and Sultan, young and "modernizing" allies of Faisal in the struggle against Saud, were given the portfolios of Interior and Defense, respectively. Faisal kept the post of prime minister, which he would fuse with the kingship from 1964 on, a structure his successors have maintained.[17]

Fahd and Sultan brought full brothers as deputies into their institutional patrimonies in subsequent years. Sultan has kept his ministry ever since, while Fahd handed his portfolio over to his younger brother Naif on becoming crown prince in 1975, who in turn made his younger full brother Ahmad deputy. Prince Abdallah, another ally of Faisal, was given cabinet status and control of the National Guard in 1963, which he also has kept until today, installing several of his sons in leading positions. In 1967, the special post of second deputy prime minister was created for Fahd, underlining his ambitions to be next in line after his older half-brother Khalid, a moderate figure whom Faisal had made crown prince for reasons of seniority and intra-family balance.

---

14. From 1.4 billion SR in 1960 to 5.8 billion in 1968/69; cf. *MEED*, 8 January 1960: 8; 4 October 1968: 976.

15. Total state employees increased from 52,000 to 124,000, indicating the growing role of expatriates; cf. Krimly 1993: 231. Other sources cite higher figures, up to 140,000. Counts seem to differ according to whether unclassified (menial) jobs and expatriates are included and whether one looks at budgeted or actual employees; see Vasiliev 2000: 462; Huyette 1985: 106; Al-Ammaj 1993: 145; Al-Hamoud 1991: 382ff. The relative shifts and growth trends in all sources are comparable, however.

16. *Lamha tarikhiyya mujaza 'an tatawwur al-ajhiza al-hukumiyya (idarat al-buhuth wal-istisharat)* (Concise historical overview of the government bodies' development; Studies and Counseling Bureau, n.p., n.d.), IPA.

17. Huyette 1985: 74.

All of the major princes jealously guarded their administrative territories, expanded their array of clients, and defended their wide-reaching autonomy. Their rapidly growing ministries became irreducibly identified with them as individuals. The 1963/64 budget included disproportionate increases in the items for Defense, the National Guard, and Interior commensurate with the status of their new patrons.[18] Growing budgets allowed for ambitious programs, most notable among them Sultan's expansion of the Saudi air force.[19]

Military procurement was usually conducted past the Council of Ministers,[20] and the senior princes in the security apparatus could pursue their pet projects independent of the budgetary process.[21] Staff numbers, business opportunities, and networks of brokers expanded concurrently, even if the technical and administrative capacity to absorb new resources was often lacking. Corruption at Defense reportedly was rampant,[22] involving key ministry officials and external brokers such as Adnan Khashoggi, son of a physician to King Abdulaziz.[23]

The defensive reflexes of the growing security fiefdoms were strong. Attempts to set up a public works department that had existed on paper since 1963 went nowhere because Defense and Interior were expected to resist, defending their large stake in infrastructure procurement and housing.[24] Generally, however, tradeoffs and budget fights between military and civilian agencies were restrained, as growing budgets provided all players with continuously bigger shares of an expanding cake.[25] The state could conveniently expand in different directions, bankrolling parallel structures without facing tough choices of prioritization.

---

18. Saudi Arabian Budget for 1383/84, 21 January 1964, NARA RG 59, 250, 5–7, box 906, folder FN 14 Public Debt SAUD (1/1/64).

19. A volume of 100 million pounds was mentioned in 1965; *MEED*, 19 November 1965: 526. In 1966, the number had climbed to over 400 million US$ (more than 140 million pounds at the 1966 exchange rate); cf. Jidda to State, 14 February 1966, NARA RG 59, 250, 5–7, box 2645, folder POL 15 Government SAUD (1/1/64). Military salaries were high compared to other Middle Eastern states, reflecting Sultan's patronage powers; cf. Jidda to State, The Power Structure in Saudi Arabia, 23 March 1965, NARA RG 59, 250, 5–7, box 2642, folder Pol 1 General Policy Background SAUD.

20. Memo of conversation: Saqqaf and Talcott W. Seeley, 28 November 1965, NARA RG 59, 250, 5–7, box 2643, folder Pol 2 Saud (10/1/65).

21. The actual expenditure for civil projects in the 1965 budget, for example, was 33 percent lower than planned, while actual security spending was 60 percent higher; cf. Supplementary Analysis of the Saudi National Budget, 21 February 1968, NARA RG 59, 150, 64–65, box 816, folder FN 15 SAUD (1/1/67).

22. Jidda to State, The Power Structure.

23. Well-connected Saudi Expressed Views Regarding Recent Defense Contracts, Reform, and Royal Family Succession, 9 June 1968, NARA RG 59, 150, 64–65, box 2470, folder POL 2 SAUD (1/1/68).

24. Former Ambassador al-Khayyal Heads Public Works Department, 7 April 1964, NARA RG 59, 150, 64–65, box 2645, folder POL 15 Government SAUD (1/1/64).

25. The Saudi Army Officer's Role in National Affairs, 27 November 1965, NARA RG 59, 150, 64–65, box 1675, folder DEF 6 SAUD.

Rivalries seem to have been more pronounced, and more personal, between security agencies, in particular the National Guard and Defense.[26] Prince Abdallah insisted on negotiating directly with the U.S. government over weapons procurement, demanding that neither Defense nor the Saudi embassy in Washington be informed. According to the U.S. embassy, Faisal wished that the National Guard and the military remain separate in order to balance their interests.[27]

Thanks to their inextricable embeddedness in ever larger and essential parts of the state, after the 1960s senior princes would become increasingly difficult to dislodge—even after kings' deaths heralded a relative shift of forces and a succession of political crises put the regime under heavy pressure.

Policy-Making Structures: Faisal's Centralization

Beyond the "sovereignty ministries," most of the other posts in the 1962 cabinet were controlled by commoners, as Faisal strove to limit the number of princes in the cabinet to make it manageable. This is another pattern that would become "locked in" in the long run, as most of the cabinet posts reserved for nonroyals in 1962 have been held by commoners to the present time.[28]

Commoners under Faisal were allowed to run their own institutions but had little say on larger policy issues. The post-1962 consolidation of authority did not automatically lead to smoother decision-making, as authority remained centered around the Al Saud and their king. During the 1960s, Faisal sat atop the bureaucratic apparatus and controlled all major national decisions, allowing for little autonomous intra-agency coordination or initiative. A 1969 U.K. diplomatic report commented: "Though a few sectors of the bureaucratic structure are effective, the decision-making process is still concentrated in a traditional system: the personal rule of a King who is his own Prime Minister and makes decisions after consulting his advisers. The strain on Feisal is heavy, and as he grows older, he is becoming more rigid."[29] By 1971, Faisal had left the Council of Ministers virtually unchanged for eight years.[30]

The cabinet under Faisal's control tended to get bogged down by excessive centralization, even deciding bureaucratic minutiae such as individual promotions.[31] At the same time, meetings between the king and his advisors were still far more

---

26. Interviews with Western diplomat present in Saudi Arabia at various times since the late 1960s.
27. Joint Weeka 18, 4 May 1965, NARA RG 59, 250, 5–7, box 2644, folder POL 2–1 Joint Weekas SAUD (1/1/65); National Guard Suspicion of Regular Saudi Armed Forces Loyalty, 14 September 1965, NARA RG 59, 250, 5–7, box 1675, folder (12–5 SAUD).
28. Yizraeli 1997: 96, 110.
29. Morris to Stewart, The Saudi Arabian Internal Scene, January 1969, PRO/FCO 8/1165.
30. The much-hailed first five-year plan of 1970 was practically silent on administrative reform issues, listing only a few steps already taken; cf. Central Planning Organization, *Development Plan* (1970): 45ff.
31. Al-Awaji 1971: 206.

important than formal sessions of the council, undermining its collegiate function.[32] Even Faisal's closest counselors were often afraid of openly confronting him with policy problems and were very aware of their subordinate status.[33] Posts and institutions were granted by the Al Saud, from whom all authority and resources originated.[34]

The State of the State in the 1960s: Overlapping and Disjointed

Overcentralization plagued the administration also below the level of the Council of Ministers.[35] Omar Saqqaf, minister of state for foreign affairs, opened all mail addressed to his ministry personally.[36] Saudis had to travel to the capital to resolve routine administration issues (which incidentally favored the Najdis of the central region, who were strongly overrepresented in the state apparatus anyway).[37]

Although the cabinet allowed for a basic division of labor,[38] excessive centralization often went together with a lack of coordination between different institutions. Parallel administrative structures and duplication of jurisdictions persisted. The archives are replete with references to the state's fragmentation, lamenting poor inter-Ministry coordination and conflicting personal ambitions.[39] The distribution of responsibilities among the Ministry of Finance, SAMA, and the Central Planning Organization, for example, was unclear, and serious differences repeatedly arose among them.[40] Different ministries operated on the basis of different

---

32. Morris to Stewart, Internal Scene, January 1969, PRO/FCO 8/1165.
33. J.B. Armitage to R. McGregor Esq., AD, FCO, 31 July 1971, PRO/FCO 8/1733; Morris to Acland, 12 July 1971, PRO/FCO 8/1733. The king reportedly only heard praise, as no one dared to speak out in his presence; Secret Memorandum of Conversation (Abdul Hamid Derhali, Ministry of Petroleum), 5 June 1955 [probably a typo—should read 1966], NARA RG 59, 250, 5–7, box 2645, folder POL 15–1 SAUD (5/21/66).
34. Faisal's concentration of authority may have been a necessary condition for limiting the greed of other family members. Whatever the reason, it did impede collegial decision-making and burden-sharing in an increasingly complex government.
35. References to overcentralized decision-making in Saudi ministries are scattered liberally throughout almost every Ford Foundation report cited in this chapter (labeled FF/IPA). All authority tends to be vested in two or three persons, with too many units reporting to one minister or director (see subsequent footnotes).
36. Holden and Johns 1982: 266–67.
37. Sixty-one percent of Al-Awaji's late-1960s sample of 271 high-level bureaucrats came from Najd; cf. Al-Awaji 1971: 169–76.
38. Since the late 1950s, there had been a gradual shift from absolute monarchy toward cabinet rule; Internal Cohesion in Saudi Arabia, Joint Research Department Memorandum, 14 July 1966, NARA RG 59, 250, 5–7, Box 2642 [folder label missing].
39. Weekly Dhahran summary to 1 February 1967, NARA RG 59, 150, 64–65, box 2470, folder POL 2 SAUD (1/1/67); cf. also Economic development planning in Saudi Arabia: an assessment, 4 August 1964, NARA RG 59, 250, 5–7, box 762, folder E5 Ec Dev SAUD (1/1/64).
40. Vasiliev 2000: 404; Report on a seminar on program budgeting, 30 May 1968, memorandum prepared by participants, FF/IPA, Box 2.

territorial units, making their policies geographically incompatible.[41] Local units often reported to more than one agency, resulting in jurisdictional conflicts, and units with similar functions reported to different agencies.[42] The functional separation between various units within an agency was often left similarly undefined.[43]

Due to lack of follow-up on the all-responsible highest level, several agencies in the 1960s still existed only on paper, including the Public Works Department, the Ministry of Justice, and the Industrial Bank.[44] The system was so underdetermined that at one point, confusion broke out over whether Faisal himself actually still was minister of foreign affairs or not.[45]

Government agencies, as far as operational, frequently were islands of their own, and their communications, as far as extant, were upwards to their royal superiors only.[46] There was no master plan or central coordinating institution for administrative and economic development. While the government kept expanding, there was no systematic review of the sprawling array of government agencies.[47] "Instead, there are several organizations, committees and study-groups, which are simultaneously involved in intertwining activities," suffering from "overlapping and fragmentation" and producing policy as "sporadic reaction or response."[48] Policy initiatives, if any, were undertaken by individual agencies and not coordinated as part of a national plan.[49] Saudis, Western consultants complained, did not appreciate the need for interagency analysis.[50]

By 1970, the government had never been able to close its books on time because agencies submitted their financial statements to the Ministry of Finance either too late or not at all.[51] Budgets were de facto put together by individual ministers

---

41. Ernest T. Spiekerman to Minister of State, Ministry of Finance and National Economy, final report (29/10/72), FF/IPA, Box 1.

42. Awaji 1971: 208; "duplication and conflicts in authority and responsibility" existed in particular before the introduction of the 1971 civil service code; Binsaleh 1982: 65. See also Ministry of Communications: general report and recommendations (August 64), FF/IPA, Box 11.

43. Supporting study 2: agriculture case, FF/IPA, Box 1; Supporting study 3: education case, Box 1; Report on health administration, in: A progress report: The administrative reform program, FF/IPA, Box 1.

44. Man to Gordon-Walker, 24 December 1964, regarding the 64/65 budget; Man to Stewart, 13 December 1965 regarding the 65/66 budget; PRO/FO 371/179886.

45. Morris to Acland, 12 July 1971, PRO/FCO 8/1733.

46. Typically, Omar Saqqaf complained to Western diplomats that other ministries would contact foreign governments directly instead through his ministry; Morris to McCarthy, 12 February 1969, PRO/FCO 8/1165.

47. A summary report of the Ford Foundation to the MoF decried the uncoordinated sprawl of agencies and the fragmentation of responsibilities; Spiekerman, final report, FF/IPA.

48. Al-Awaji 1971: 162; *At-taqrir ath-thanawi li'diwan al-muwadhdhafin al'am 1391–93* (Biannual report of the Civil Service Bureau, 1971–73), IPA.

49. Economic development planning, NARA.

50. Lippman 2004: 145.

51. Some ministries never submitted their financial reports, while others submitted reports that were "late and unreliable"; Financial management program and system, September 1970: 5, FF/IPA, Box 1.

rather than by the Council of Ministers, as they could transfer funds from one line item to another and there usually were no closing accounts to check. Abuse was reported.[52] Efforts to consolidate the organization of government by a large Ford Foundation consultancy mission[53] yielded scant results; local representatives counterparts of the mission reported meeting stiff bureaucratic resistance.[54] Bolstered by expanding budgets, the regime was never forced to rationalize its mushrooming bureaucracy but rather could allow its various components to grow in parallel.

Once elite politics was consolidated, Faisal felt little pressure to pursue a coherent development strategy. He was not a forceful modernizer and had in fact postponed action on all but one point of his ten-point reform program of 1962.[55] Although he had spoken of the need for economic planning during his struggle with Saud, there was little follow-up.[56] The Supreme Planning Board created in 1961 never proceeded beyond discussing general ideas and never managed to overrule other ministries,[57] which were reluctant to report to it.[58] Development expenditure in the 1960s often stayed well below the budgeted amounts.[59] SAMA's 1963 annual report pointed to conflicting opinions among different ministries as one of the main reasons that a sizeable 107 million SR budget for studies had gone largely unused.[60]

Harold Folk, an advisor sent by the World Bank to consult on economic development, was dismissed by Faisal in 1963. Few ministers had accepted his idea of integrated development.[61] With much more money on hand than local expertise, numerous international agencies and consultants were active in the Kingdom, advising or setting up various agencies.[62] No one coordinated their activities, however, and results were often mediocre.[63]

---

52. Reorganization of the Budget (Taylor, July 1963): 11–12, in: Early reports providing basis of administrative reform project, FF/IPA, Box 16.
53. *MEED*, 20 December 1963: 574.
54. Possible Administrative Improvements in the Passport Office and the Customs Department, 2 October 1967, NARA RG 59, 150, 64–65, box 2472, folder POL 15–1 SAUD (1/1/67).
55. Yizraeli 1997: 95.
56. *MEED*, 1 July 1960: 297.
57. Based on U.S. recommendations, the Planning Board was never accepted by the ministries; cf. Morris to Douglas-Home, 3 February 1971, PRO/FCO 8/1742.
58. Economic development planning, NARA.
59. Faysal's policy statement: an eight-month review, supplement for the period 31 July–21 October 1963, Mulligan Papers, box 5, folder 11.
60. *MEED*, 20 September 1963: 416.
61. Yizraeli 1997: 141.
62. These included World Bank, UNESCO, ILO, A.D. Little, Dioxadis, Italconsult, Stanford Research International in addition to various official U.S. institutions; *MEED*, 23 September 1966: 457; 5 October 1967: VII; 21 March 1969: 389.
63. The Ford Foundation mission on administrative reform, for example, was generally rated as a failure when it was shut down in the early 1970s; cf. Morris to McGregor, Saudi Economic Situation, 1970, PRO/FCO 8/1488.

Some parts of the state apparatus—the College of Petroleum and Minerals in the Eastern Province,[64] the Institute of Public Administration[65] in Riyadh, or the Central Bank SAMA in Jeddah—functioned well, but they did so largely autonomously. The Petroleum College, under the patronage of powerful oil concessionaire Aramco, and SAMA, under the patronage of King Faisal, profited from autonomous legal status and recruitment structures.[66] Again, the persons heading them often played a large role in determining their success.[67] Contingent as these results were, they did not form part of larger national development strategies.

Faisal's Civil Service: Balancing Regularization and Nepotism

Although facing an acute shortage of qualified administrators, Faisal managed to assemble some good commoner ministers around him, which might explain that at least some decisions did get carried out.[68] However, as one U.K. diplomatic source explains, "Elsewhere, as so often in developing countries, it is the bureaucratic bindweed that has taken root and proliferated most quickly, its function being to choke decision and action."[69] Aramco analysts considered the increasing bureaucratic inertia a larger administrative problem than the established *bakshish* culture.[70] Although the bureaucracy would react quickly to orders coming directly from the king, its quotidian operations could be painfully slow and convoluted.[71] In 1964, the Ford Foundation discovered that 331 separate steps were needed to put an individual on the government payroll.[72]

The "bureaucratic bindweed" did go along with some regularization of the civil service. Faisal gradually formalized intra-agency structures, although he

---

64. A 1964 diplomatic dispatch mentions the "somewhat incredible accomplishment by Dean Salih Amba and his staff"; Weekly Summary (Dhahran), 7 October 1964, NARA RG 59, 250, 5–7, box 2643, folder POL 2 Saud (8/26/64).
65. Teaching in the Institute of Public Administration, April 1969 report, FF/IPA, Box 9.
66. Memo on the College of Petroleum and Minerals, 3 April 1965, Mulligan Papers, box 3, folder 23.
67. The dean of the Petroleum College Salih Amba, for example, played an important role in recruiting promising young academics and keeping the institution free of religious forces, who had encroached on other Saudi universities; cf. The Rehabilitation of Salah Ambah, 18 April 1973, NARA RG 59, 150, 66–67, box 2586, folder POL 23 SAUD (1/1/70).
68. In the face of an absurd degree of centralization, the Council of Ministers did manage to take a good number of decisions; Rothnie to Wright, 22 August 1974, PRO/FCO 8/2332.
69. Mr. W. Morris to Mr. Stewart, The Saudi Arabian Internal Scene, January 1969, PRO/FCO 8/1165.
70. Malcolm Quint, Administration of justice, 14 August 1960, Mulligan Papers, box 3, folder 61.
71. Jidda to State, Survival Prospects for the House of Saud, 6 June 1966, NARA RG 59, 250, 5–7, box 2645, folder POL 15–1 SAUD (5/21/66).
72. Weekly summary from Dhahran consulate, 6 May 1964, NARA RG 59, 250, 5–7, box 2642, folder Pol 1 Gen. Policy Background SAUD.

moved slowly on that account once Saud was gone. As early as June 1958, a royal decree was issued reforming the civil service. It introduced a classification of employees and salary schedules, criteria for hiring and terminating employment, and prohibitions on the use of public offices for private gain. It also stipulated severe penalties for violations, including prison sentences. Bureaucrats often undertook private business activities at the time, and American observers considered the decree as a serious attempt by Faisal to decrease nepotism.[73] In 1963, the Civil Service Bureau was upgraded to the ministerial level and accorded disciplinary functions.[74] Faisal also limited princely interventions in bureaucratic life and forced royals to pay their bills.[75]

Saudi society, of course, set limits on across-the-board bureaucratic rationalization. To start with, overcoming the shortage of qualified personnel was an uphill struggle in a grossly undereducated society.[76] In the early 1960s, the Institute of Public Administration in Riyadh was set up with UN assistance and started imparting basic administrative skills to more and more Saudis.[77] Although the quality of IPA courses was quite high,[78] the preexisting vacuum in this area meant that its impact on government could only be gradual.[79] At the end of the 1960s, the pool of skilled administrators was still small.[80] Many of the new employees recruited during the decade were unqualified, offices were overstaffed,[81] and many bureaucrats remained inactive in their jobs.[82] The state was expanding too quickly for an undereducated society, giving bureaucratic recruitment the character of cooptation almost by default.

According to official statistical yearbooks, only 8,309 students graduated from secondary schools between the 1960/61 and 1970/71 academic years. This means that the majority of the more than new sixty thousand civil employees hired

---

73. Sweeney to State, telegram, 26 June 1958, USRSA, Vol. 5
74. *Taqrir thanawi*. Saudi Arabia saw an "increasing substitution of institutions for personal rule" once Saud had left; cf. Jidda to State, King-Faysal Confrontation, 3 January 1964, NARA RG 59, 250, 5–7, box 2645, folder POL 15 Government SAUD (1/1/64).
75. Survival Prospects, NARA.
76. The shortage of qualified personnel was repeatedly deplored by diplomats; cf. Heath to State, telegram, 30 July 1958, USRSA, Vol. 5.
77. *Lamha tarikhiyya*.
78. In contrast to their assessment of most other agencies, the Ford Foundation found that "teaching in the Institute is very, very good indeed"; Teaching in the Institute of Public Administration, April 1969 report, FF/IPA, Box 9.
79. Some five thousand civil servants had received training by 1973; Supporting Study 1 (continued): The Institute of Public Administration (1/6/1973): 9, FF/IPA, Box 1.
80. Craig to Stewart, Saudi Arabia: 1968 annual review, PRO/FCO 8/1166.
81. Al-Awaji (1971: 220) mentions "over-staffing of government offices with many half-educated and drop-outs."
82. Interview with James Craig, former UK ambassador to Saudi Arabia, 1979–84, who had also been posted in the Kingdom in the late 1960s, Oxford, June 2005.

during the decade had at best elementary or intermediate schooling, making their utility as would-be administrators questionable.[83] The total of graduates from nonreligious Saudi colleges amounted to 1,054.[84]

Once in place, no matter how (un)qualified, the ranks of new bureaucrats displayed remarkable inertia. The annual turnover of civil service in the mid-1960s was "phenomenally low," at 0.1 percent, compared to 5 to 10 percent in other countries. Bureaucratic sinecures, once attained, were practically never given up.[85] Impassive bureaucrats required orders from above for even routine activities while hesitating to ask their superiors for clarification.[86] Absenteeism was rampant,[87] but "discharge of a civil servant for anything short of a major felony [was] virtually impossible."[88] Superiors preferred to transfer useless or delinquent employees over letting them go.[89] American observers rated the Saudi bureaucracy as more bloated than that of neighboring countries, but trimming the growing ranks of clientelist employment appeared impossible.[90] This situation would remain fundamentally unchanged until today.

Beyond Faisal's establishment of basic order, progress in consolidating the bureaucracy proceeded only gradually. Ingrained nepotism was played out within more clearly defined structures. Attempts by Ford Foundation consultants to create a merit-based public service did not yield significant results. The Civil Service Bureau was trained and extensively consulted by foreigners but had little authority despite its augmented formal status. Its head, Omar Zaini, lacked clout over other agencies, which continued to hire separately.[91] Despite general overstaffing, employees were so unequally distributed that some agencies lacked workers,

---

83. Calculated acc. to Knauerhase 1977: 219.

84. Religious colleges and institutions included, the total was 4292; calculated acc. to Knauerhase 1977: 221. Some of those graduates were, moreover, absorbed by the private sector. A 1968 establishment survey showed that 321 out of a total of 52,588 private Saudi employees had college education (190 out of 16,206 in Jeddah and 68 out of 9,370 in Riyadh); cf. *Nushra ihsa'iyya 'an al-mushtaghilin fi mu'assasat al-qita' al-khas bi-khams wa 'ashrin madina min mudun al-mamlaka* (1388/1968), folder 193 (331,2), IPA (Statistical report on the workers in private sector establishments in 25 cities of the kingdom).

85. Saudi Civil Service: Myers Report on Government Pension Plan, 29 November 1965, NARA RG 59, 250, 5–7, box 2645, folder Pol 15–4 Saud.

86. Monthly commentary February 1969, NARA RG 59, 150, 64–65, box 2470, folder POL 2 SAUD (1/1/69).

87. Craig Interview; Jidda to State, 14 February 1966, RG 59, 250, 5–7, box 2645, folder POL 15 Government SAUD (1/1/64).

88. British National Guard report (enclosed with letter from UK embassy in Washington to NESA Affairs Bureau in State Department, 6 December 1965), NARA 59, 250, 5–7, box 1675, folder DEF 6 SAUD.

89. *Taqrir thanawi*: 6.

90. Corruption in the Saudi bureaucracy: the young college graduate, 7 July 1964, NARA RG 59, 250, 5–7, box 2643, folder POL 2–3 Politico-Economic Reports SAUD (1/1/64); Political Economic Summary of Saudi Arabia, 20 October 1964, NARA RG 59, 250, 5–7, box 2644, folder POL 2–3 Politico-Economic Reports SAUD (1/1/64).

91. Lippman 2004: 146; Supporting study 1: personnel management, FF/IPA, Box 1.

while bureaucrats elsewhere had nothing to do.[92] The Civil Service Bureau lacked links to educational institutions, information on available human resources, and a proper employee classification system.[93]

Although the IPA managed to improve somewhat the skills of administrators with its basic training courses, skills as such did not necessarily result in a bureaucratic "esprit de corps."[94] The widespread presence of Arab expatriates in the bureaucracy, especially Egyptians, contributed little to its coherence, as they usually lacked authority and were reluctant or often not allowed to engage with other administrative units. Egyptian-inspired administrative procedures tended to create great amounts of paper while simultaneously slowing down its flow.[95]

Within basic rules, the administration continued to suffer not only from unclear job descriptions and responsibilities[96] but also from "advancement procedures that were usually based on loyalty to the boss or friendship, kinship, partiality, or personal relations."[97] Qualifying exams, insofar as they were administered, were not standardized across institutions and indeed were often tailor-made for specific individuals. The same was true with formal job requirements. Vacant positions were often only open to the employees of a given institution, consolidating existing fiefdom structures.[98] Though more clearly delimited by a formal framework, structures of bureaucratic clientage persisted. Posts were "granted" to bureaucrats who in many cases were not doing much.

Ford Foundation consultants saw the value system of Saudi administrators as highly deficient, decrying the exaggerated sense of self-importance in combination with a lack of public spirit. Public property supposedly was not seen as a meaningful concept.[99] According to a Saudi account from the late 1960s, red tape created opportunities for rent-seeking and corruption, and the power to (dis)approve was extensively used as "bargaining mechanism."[100] Bureaucratic secrecy was the norm, and little-known regulations, usually dormant, would be applied randomly to put pressure on applicants.[101]

---

92. *Taqrir thanawi*: 4.
93. Ibid.
94. IPA efforts were often hampered by more traditional-minded ministers; cf. Huyette 1985: 109.
95. One hundred and forty-two steps were needed for an appropriation in the Ministry of Agriculture; cf. Supporting study 2: agriculture case, FF/IPA, Box 1, p. 7.
96. Awaji 1971: 215–16.
97. Binsaleh 1982: 65. Kinship seems to have played a particularly important role; Al-Ammaj 1993: 61. The Civil Service Bureau itself mentions favoritism in hiring; cf. *Taqrir thanawi*: 6.
98. Awaji 1971: 139–40.
99. Lippman 2004: 144.
100. Awaji 1971: 211.
101. Even decisions by the Council of Ministers were sometimes considered secret; cf. Regulations for SAMA, 2 February 1958, Mulligan Papers, box 2, folder 58.

Bureaucrats often diverted public services according to their own regional background and, on occasion, their own business interests.[102] Brokerage of state resources through kinship networks seemed to expand along with the state. The 1969 annual review by the British embassy in Jeddah decried the "prevalence of corruption at all levels,"[103] while U.S. diplomats described the "money-grabbing habits of government officials at many levels which encourage agents."[104] While bureaucratic clientelism helped to make larger independent groups in society dependent on the state, it also allowed social forces and interests to penetrate the administration on the micro-level.

Faisal's own commitment to orderly and rule-bound administration should not be overstated. Compared to Saud, he was austere, but that did not mean that he did not have his own trustees, brokers, and business clients whom he would pay off through various channels, as was indeed expected of him. It was an insider deal of Faisal's brother-in-law Kamal Adham that reportedly prompted the clash between Faisal and Minister of Petroleum Abdallah Tariqi which led to Tariqi's dismissal in 1962.[105] Faisal's own sons, Abdallah in particular, were active in numerous business ventures, and several of them partnered with well-known contract broker Ghaith Pharaon.[106] Princes were strongly present in trade and the important cement sector.[107] The Alireza merchant family, close to Faisal since his time as Hijazi vice-regent, was amply supplied with defense-related contracts.[108] Faisal also helped out the Bin Laden family through road contracts after the head of the family, Mohammad Bin Laden, died in a plane crash in 1968.[109]

Throughout the 1960s, Saudi business continued to grow. Many of the early starters from the 1940s and 1950s proved nimble or well-established enough to survive leadership changes and greatly expanded their territory. The oil sector tended to dominate economic development, however, with the private sector consistently providing less than a third of GDP, mostly through activities driven

---

102. Awaji 1971: 23.
103. 1968 Annual Review, PRO/FCO 8/1166.
104. Interior Minister Prince Fahd inveighs against influence peddlers, telegram 2 October 1970 (210700Z), NARA RG 59, 150, 66–67, box 2585, folder POL 15 SAUD (1/1/70).
105. Lacey 1981: 339–40. The royals reportedly needed Tariqi as he was qualified and not venal, a very rare combination at the time; cf. Morris to FO, 15 September 1958, PRO/FO 371/132656 (BS 1015/55).
106. The Sons of King Faisal, 15 November 1966, Mulligan Papers, box 1, folder 70.
107. On Faisal and royal concessions, see Vitalis 2007: 215. In the mid-1960s, princes reportedly just presented their private bills to the Ministry of Finance for reimbursement; cf. Survival Prospects, NARA.
108. Field 1984: 110; *MEED,* 26 October 1967: ii, 717; Survival Prospects, NARA.
109. Business International 1985: 183.

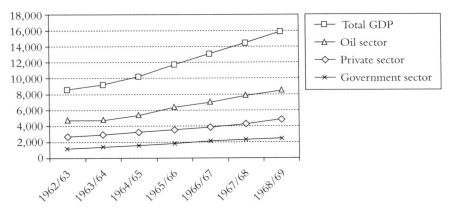

Fig. 3.1 Composition of Saudi GDP (SR million)
Source: Calculated from table 4.12, Knauerhase, pp. 80f.

by public spending such as commerce, transport, and real estate (see figure 3.1).[110] Only eighty-eight enterprises had one hundred or more employees in 1971.[111]

Personal relations continued to be crucial. The distribution of oil income to clients and hangers-on through personalized deals has been widespread in all Gulf Cooperation Council states.[112] Although distribution was relatively well-controlled under Faisal, his rule was no exception to the basic pattern. Contacts in the secretive administration were needed to obtain contracts, and businessmen took to paying retainers of up to 1000 Saudi riyals a month to bureaucrats who might supply them with project information, get their invoices paid, and generally protect their interests.[113] As a complementary mechanism, merchant families would often deploy their offspring in the growing administration. The combination of the state's increasing riches with persistent bureaucratic opacity spawned widespread brokerage.

Faisal knew of the systems of middlemen but turned a blind eye, presumably as he understood the utility of clientelism in maintaining allegiance to and the basic cohesion of the system.[114] In the course of the 1960s, Faisal had contained the worst excesses of waste and personal enrichment, but there was considerable slack left in

---

110. Knauerhase 1977: 80–81. Agriculture, though still a major sector at the time, grew much below average; cf. Knauerhase 1977: 60–61; Industrial Studies and Development Center, *Survey of industrial establishments in Saudi Arabia, Vol. 1 Riyadh (1389/1969)*, IPA.
111. Knauerhase 1977: 139.
112. Field 1984: 96ff.
113. Corruption in the Saudi bureaucracy, NARA.
114. Rothnie to Wright, 22 August 1974, PRO/FCO 8/2332.

the system.[115] Like his predecessors, Faisal would also maintain certain persons to whom he felt personally obliged in their public positions regardless of their qualifications.[116]

Bloated and Blind? The State's Regulatory Power

The clientelist bureaucracy, although growing quickly, remained generally inward-looking. Even where more concerted efforts at planning were made, usually with the involvement of expatriate advisors, their scope was limited by the absence of data on the Saudi population and its economic activities. The state had grown, but its information-gathering capacities had not expanded accordingly. Contradicting Kiren Chaudhry's account, but in line with conventional rentier theory wisdom, the Saudi bureaucrats of the 1960s simply seemed to be uninterested in systematic collection of information on activities outside of their agencies.

In 1963, estimates of the population of Riyadh ranged from 170,000 to one million, and there were few statistics on basic subjects such as trade or industrial production.[117] Lacking birth, death, marriage, and divorce records, the Ministry of Social Affairs in 1964 struggled to check the soundness of welfare applications.[118]

The Ford Foundation reported in 1965 that "unfortunately, price and wage data are seriously lacking for the Kingdom at the present time."[119] The fifth annual SAMA report issued the same year conceded problems with project budgeting due to the "lack of proper surveys and advance planning."[120] Data on business activities—a prerequisite for any consistent regulatory effort—was similarly deficient. In 1965, several thousand firms were apparently unregistered, prompting expatriate advisors to propose re-registration to gain basic administrative control over businesses.[121] Even those businesses that might have been willing to share

---

115. "Under King Feisal...If corruption has not disappeared, it is circumscribed." Morris to Stewart, Internal Scene, January 1969. Under Faisal, at least occasional punishments occurred; Biweekly summary (Dhahran), 29 April 1964, NARA RG 59, 250, 5–7, box 2644, folder POL 2–3 Politico-Economic Reports SAUD (1/1/64). He had clearly reduced royal waste; Faisal's Character (n.d.), Mulligan Papers, box 1, folder 70.
116. Saudi Arabia: Government Officials, PRO/FCO 8/1168.
117. Report on visit of BBME manager, 6 January 1963, PRO/FO 371/168868 (BS 1015/3).
118. *Taqrir 'an wizarat al-'amal wal-shu'un al-ijtima'iyya* (1383): 78, IPA, folder 181 (Report on the Ministry of Labor and Social Affairs, 1963).
119. Central Planning Organization, An economic report, 1965: 144, FF/IPA, Box 30.
120. *MEED*, 28 October 1966: 532.
121. Management analysis and reorganization of the Ministry of Commerce and Industry, September 1965: 21–22, FF/IPA, Box 4.

information either kept no books or maintained accounts that were too primitive to allow systematic data-gathering.[122]

The few statistics that were available were not sufficiently used, and Saudi bureaucrats seemed unaware of the utility of company registration for purposes of data-gathering.[123] There was no meaningful statistical activity in the Ministry of Commerce and Industry.[124] While the construction sector expanded, the absence of a building permits system prevented the collection of meaningful housing data.[125] *Contra* Chaudhry, taxation does not seem to have played a major role in data-gathering. Indeed, Saudi citizens and companies were not subject to income taxation.[126] Foreign observers regarded the Saudi administration as incapable of raising domestic taxes even if it had tried.[127]

Indeterminate and inconsistent official rules would further undermine any regulatory effort. The absence of a proper company and commercial code baffled foreign investors. More generally, royal and other decrees were often incomplete, not published, or superseded by new decrees but not formally abrogated, all of which led to a steady accumulation of rules that compounded bureaucratic opacity.[128] Most agencies kept no systematic tabs on their own regulations, driving into despair foreign consultants who wanted to take stock.[129] Administrators often interpreted existing regulations idiosyncratically.

---

122. According to official statistics, 285 industrial plants and 18,549 small businesses were registered between 1956 and 1969. However, in the same period, a mere 113 licenses for "free professionals" were issued. This included 31 accountants/auditors, 35 barristers, and 38 engineering consultants; *MEED*, 13 March 1970: 325. A national economy with 31 accountants is difficult to penetrate for the best of regulatory bureaucracies.

123. Management analysis: 21–22, FF/IPA.

124. Ibid, p. 49.

125. A foreign advisor recommended introducing building permits in order to improve national accounts; cf. C. Hastings, Observations on and recommendations concerning national accounts work in Saudi Arabia, November 1970: 7, FF/IPA, Box 5.

126. Ramon Knauerhase, A Study of the Revenue Structure of the Kingdom of Saudi Arabia: Report No. 4 (May 1970), FF/IPA, Box 4. Even a tobacco tax was "almost impossible to enforce"; ibid.: 14. An income-based "road tax" was only levied on government employees, while *Zakat* collection was inefficient. Interview with former minister, Riyadh, December 2005.

127. Improving the Budgetary System, 27 October 1969, NARA RG 59, 150, 64–65, box 816 (FN), folder FN 15 SAUD (1/1/67). Even after Faisal had been king for six years, most of these weaknesses persisted, and the lack of statistics on most sectors remained a heavy constraint on planning; cf. Edens and Snavely 1970. Among other things, current production data on agriculture was absent in 1970, and advisors recommended that annual reports from banks should be collected (with the implication that this had not happened); cf. Hastings, Observations, FF/IPA.

128. Spiekerman, final report, FF/IPA; David J. Keogh, Advisor to Ministry of Agriculture and Water, Summary report of the activities March 1969–January 1971, FF/IPA, Box 13; Al-Awaji 1971: 172.

129. A task force in the Ministry of Health in the 1960s "soon became convinced that it was impossible to compile all current laws and regulations. The Ministry's legal files were disorganized and incomplete." Report on health administration: 7, FF/IPA.

State-Society Relations: Swallowing and Reshaping Social Groups

The state's low regulatory capacity is all the more striking when one considers how well Faisal's regime was capable of controlling society in political terms.[130] Politics remained driven by a handful of players at the exclusion of independent social groups. Instead of being politically engaged, non-state social elites and the middle class were gradually co-opted by the state, whose distributional politics would slowly extend beyond the regime elites and their clients. Increasingly, this happened not only through personal patronage but also through expansion of the anonymous bureaucracy, even if frequently shot through with personal interests. The "buying off" of potential opposition, broadly predicted by rentier theory, had structural consequences for Saudi state and society that went far beyond just keeping the place politically quiescent.

Continuing a trend that Abdulaziz had started with pre-bureaucratic means, previously semi-autonomous groups and centers of social power would either be marginalized or become clients of the state, which expanded in several diverse directions at once. In the most extreme cases, social clienteles themselves became part of the expanding formal state, losing their political independence and coherence but gaining somewhat amorphous control over segments of the bureaucracy. Bureaucratic mobility and opportunities to gain local status in the state apparatus offered avenues of advancement, especially to urban citizens. Rapidly growing revenues made it possible to expand the state laterally and "grant" institutions to specific groups, making them dependent while also fragmenting the state institutionally.

One pivotal group that Faisal progressively "bureaucratized" were the ulama.[131] The king granted them extensive administrative power over education, judiciary, and the morality police as a quid pro quo in which they agreed to tolerate institutional and legal modernization in other areas,[132] trends they had often tried to prevent.[133] While this helped to restrict their influence, it also led to an increasing disconnect among different parts of state.

---

130. A 1965 diplomatic dispatch summarizes the paradox: "Saudi Arabia remains in many spheres an ill organized Kingdom where appalling inefficiencies permeate every branch of government yet arouse no ground swell of resentment." Power Structure, NARA.

131. Piscatori 1983: 60–61.

132. July 1968 monthly review, NARA RG 59, 150, 64–65, box 2470, folder POL 2 SAUD (1/1/68).

133. A leading Saudi lawyer in 1968, for example, deplored that the ulama had "succeeded in 'freezing' the commercial courts," which were supposed to operate outside of religious control; cf. Well-connected Saudi Expressed Views Regarding Recent Defense Contracts, Reform, and Royal Family Succession, 9 June 1968, NARA RG 59, 150, 64–65, box 2470, folder POL 2 SAUD (1/1/68). Commercial courts would in fact be created—at least on paper—only in 2007.

In 1964, Faisal managed to institutionalize the politically touchy field of education for girls by attaching it to a new, separate institution under religious control.[134] Two years before, Faisal had decreed the creation of the Ministry of Justice (MoJ), with the aim of bringing *sharia* courts into the state apparatus.[135] It wasn't until 1970, however, that the MoJ was in fact set up and a minister appointed; this was probably because the powerful Grand Mufti, who died in 1969, had previously vetoed the project.[136] The creation of the MoJ, the Council of Senior Ulama and of Dar Al-Ifta (the fatwa administration) in 1970 and 1971[137] meant that the religious establishment was being fully centralized and technically brought under the king's control—whereas the Mufti had been an independent institution until then.[138] The king had become the only formal link between the religious bureaucracy and the rest of the system. Until today, Dar Al-Ifta and Council of Senior Ulama report directly to him, as do the morality police.[139] Otherwise, religious institutions are completely autonomous from the state apparatus.

Ulama were granted a circumscribed stake within specific segments of the state, provided that they would accept their subordinate and functionally restricted role.[140] The most intransigent ulama were ousted, while others were given salaries and institutions to fill, including the MoJ, religious universities, charities, parts of the Ministry of Education, and the expanding morality police, in which otherwise unemployable graduates of religious institutions could hope for a reasonably remunerative career with high social prestige.[141] Students in the separate system of Islamic schools under the Grand Mufti enjoyed considerable privileges.[142] But religious forces also exerted control over most other areas of education, where "the traditional system with its spirit, methods, and even curriculum survived in the modern Saudi system as nowhere else in the world."[143]

At the same time, religious authorities drifted away from some core functions of administering social order they had once fulfilled, including the supervision of markets and taxation. Since the 1960s, manifold quasi-judicial bodies have

---

134. Al-Yassini 1985: 113; Al-Fahad 2000.
135. Creation of Supreme Judicial Board, 27 December 1967, NARA RG 59, 150, 64–65, box 2472, folder POL 15–1 SAUD (1/1/67).
136. Dhahran weekly summary, 24 November 1964, NARA RG 59, 250, 5–7, box 2643, folder POL 2 Saud (8/26/64).
137. *Lamha tarikhiyya*: 16.
138. Teitelbaum 2000: 18.
139. Al-Yassini 1985: 70–71; Phebe Marr, Public Morality Committees, 17 June 1961, Mulligan Papers, box 3, folder 7.
140. Bligh (1985: 43) mentions a "permission to run their own systems."
141. Al-Rasheed 2002: 124–25.
142. Training program for sharia graduates (Proposal, July 1967), FF/IPA, Box 9.
143. A.L. Tibawi, cited in Nyrop 1977: 99.

arisen in technocratic ministries to avoid commercial matters from falling into the hands of the ulama, progressively fragmenting jurisdictions.[144] Nonetheless, that the core of the judicial system has been the preserve of religious forces has greatly complicated the business environment in Saudi Arabia, as ulama-controlled courts reject, among other things, interest payments and mortgage contracts.[145] At the same time, the religious dominance of school curricula has made many Saudis unemployable on the private labor market, leaving the state their employer of last resort. The wholesale inclusion of ulama into the bureaucracy has imposed heavy costs on Saudi development—costs that probably only a rapidly expanding rentier state would find politically expedient, or even possible, to bear.

Whereas the ulama were bureaucratized, the tribes further faded away as cohesive political actors. They had already been much weakened under Abdulaziz, and as mid-level tribal leaders became clients of princes or started receiving state salaries by virtue of their local administrative functions, their autonomy decreased still further.[146] By the mid-1960s, the Ministry of Interior under Fahd formally administered tribal stipends.[147] Tribes were clientelized.[148] Although tribal identity persisted, Faisal's modernization deprived tribes as collective agents of much of their political clout.[149] Beyond some rumors of small-scale unrest, little was heard of collective tribal demands in the 1950s or 1960s. The vestiges of the pastoral economy, which could have given them some economic autonomy, were largely destroyed through the importation of foreign products. Many tribesmen settled in the periphery of cities, and significant parts of major tribes were employed as members of the National Guard. Apart from the latter channel of mobility, the nomadic sections of tribes have been among the losers of modernization in the Kingdom, ill-equipped to join either the urban economies or the higher levels of bureaucracy.[150]

As indicated in chapter 2, the growing merchant class in various regions of the Kingdom occupied a luckier spot in the growing Saudi system. Due to the liberal economic orientation of the regime and the traditional role of merchants as an

---

144. Interim report on establishment of an integrated public works agency for the Government of Saudi Arabia, May 1965, FF/IPA, Box 17. These bodies have remained ad hoc and not subject to any regulation coordinating their activity; Al-Saleh 1994: 105–6.

145. In 1981, at the height of the boom and just before the bust, the Ministry of Justice ruled that mortgages were not to be used as security for loans, depriving many Saudis of the chance to monetize their tangible assets in times of scarcity; *MEED*, 7 May 1982: 39.

146. Al-Rasheed 2002: 127.

147. Survival Prospects, NARA.

148. Al-Fiar 1977: 140; El-Farra 1973: 74, 200; Al-Seflan 1980: 151.

149. The significant subsidies to tribes in the budgets of the late 1950s already indicated dependence and clientage rather than bargaining power. The 1958 annual budget allocated 192 million out of 1375 million SR of total expenditure to the tribes; cf. *MEED*, 14 March 1958.

150. High-status bureaucrats were from urban social groups. Sons of traditional officialdom were advantaged in particular thanks to family connections and (relatively) better education. Only 2 percent of 271 "top officials" in a late 1960s survey were born to tribesmen; cf. Al-Awaji 1971: 169–76.

elite coalition partner, they escaped the fate of their peers in other Arab countries who were crushed or dispossessed. They instead profited greatly from the growth of state budgets. By gradually increasing national privileges through investment and trading restrictions on foreigners, the state created a loyal clientele of local merchants.[151] By the same token, however, their orientation gravitated ever more toward the state and its elites, as this was the center of distribution. Merchant families carefully cultivated their contacts in the state bureaucracy. Their dependence on the state and the Al Saud naturally grew, with their individual orientation toward the center further undermining their corporate coherence. While the growth of the system provided space for new entrants, those who could not find favor among the rulers would disappear as significant actors.[152]

The shifting balance between state and business is vividly reflected in the increasing provision of public goods by the state that had before been financed by the merchant community, including local electricity generation, education, and sports clubs.[153] Previously autonomous sectors of local artisanal production were crowded out by imports from more developed countries.[154] Most economic activities remained small scale: a 1968 establishment survey showed a total of 44,481 establishments with 104,698 workers—barely more than two workers per company.[155] The role of the merchant communities was limited to quiet political support and acting as intermediaries and subordinate providers of contracted services, while the state led development and policy-making. When new regional chambers of commerce were gradually set up, this happened under the aegis of the government.[156]

While swallowing or marginalizing old collective interests, King Faisal's state allowed no new autonomous interest groups to emerge. Disaffection in society never was channeled into organized, broad-based demands.[157] The Saudi security apparatus crushed the marginal oppositionist parties of the 1960s, while imports of foreign labor prevented the emergence of a national labor class. No unions,

---

151. Cf. the 1962 "commercial agency" rules limiting trade representations to Saudis; royal decree m/11; 20/2/1382; *Munjazat wizarat al-tijara fi ahd khadim al-haramain ash-sharifain 1402–1422* (Riyadh, Ministry of Commerce 2002): 75, folder 1061 (380), IPA (Accomplishments of the Ministry of Commerce in the period of the Custodian of the two Holy Mosques, 1982–2002).
152. Field 1984: 105.
153. *MEED*, 30 March 1967: 260; Pledge et al. 1998: 138.
154. Niblock 1982: 77.
155. *Nushra ihsa'iyya*, IPA.
156. This started with the first chamber in Jeddah in 1946; cf. *Lamha tarikhiyya*: 11.
157. Even during the 1958 crisis, in which the elite were deeply split, disaffection seems to not have been widespread or organized enough to pose a direct threat to the regime; cf. 1958 annual review, PRO/FO 371/140356; on the absence of ideological challenges to the Al Saud, cf. Piscatori 1983: 60; Impact of Youth and the US National Interest—Phase II, 17 December 1970, NARA RG 59, 150, 66–67, box 2584, folder POL 7 SAUD (1/1/70).

parties, or syndicates were allowed.[158] Formal civil society life was limited to small, state-dependent welfare clubs.[159] After Faisal had consolidated his power, he also discontinued the government's experiments with municipal elections. More generally, the Al Saud discouraged the aggregation of larger political interests and recognized petitions only if they were brought forward by individuals.[160] Traditional sectoral interest groups like the guilds in the Western region persisted, but they received no formal role in administration or decision-making, guaranteeing their marginalization to this day.[161]

Growing bureaucratic employment and formal education gradually created new social classes. This, however, did not result in the emergence of a cohesive "new middle class," as some observers have hoped.[162] Bureaucrats, usually recruited from the urban middle class, were politically dependent, tied up in individual networks of clientage, and fragmented into a disparate set of highly hierarchical institutions.[163] While individual bureaucratic mobility remained relatively high, no coherent political voice ever emerged from the ranks of the administration.[164] Like older social classes, the urban middle class was thoroughly co-opted. The military was perceived as the most dangerous section of the new bureaucratic stratum, but the peril of a takeover was contained through institutional and geographic fragmentation of the armed forces, which were moreover shot through with dozens of princely officers. A coup attempt in 1969 did not get far.[165]

The formal political system that had emerged from the elite conflicts of the late 1950s and early 1960s was extremely dominated by the executive. Saudi Arabia had become a bureaucratic state, with no formal forum for policy making, policy negotiation, or interest aggregation outside of the bureaucracy. The only nonbureaucratic politics worthy of the name happened in informal structures around the court, involving advisors, confidants, and representatives of a few

---

158. Vasiliev 2000: 368–72; Internal Cohesion, NARA.

159. The total expenditure of the seventeen registered societies in 1971 was only about 1.5 million SR, and 48 percent of their board members were government employees; cf. *Bahth ihsa'i 'an al-jami'at al-khairiyya al-ahliyya bil-mamlaka* (Ministry of Labor and Social Affairs, 1391/92): 1, 40, 59, folder 777 (361), IPA (Statistical study on the private welfare societies in the kingdom).

160. Discussions with senior advisor to Saudi government, Riyadh 2003 and 2004.

161. Chaudhry 1997: 71–76, 89–90.

162. Abir 1993; Seznec 2002.

163. A survey from the late 1960s shows that among 271 cases of high-status bureaucrats, only 17 had laborers, farmers, and tribesmen as fathers. Civil servant fathers made up 27 percent, shopkeepers 31 percent, businessmen 16 percent. The strong presence of shopkeepers seems to indicate considerable mobility for the urban middle class at an early stage; cf. Al-Awaji 1971: 177.

164. Anyone with an advanced degree, for example, had a wide choice of jobs; cf. Supporting study 1: personnel management: 9, FF/IPA, Box 1.

165. Lacey 1981: 381–82; Vasiliev 2000: 371–72. The military lacked credibility and was seen as corrupt; cf. The Power Structure, NARA.

senior families. The new Saudi state had marginalized the few public, collective, and consultative corporate structures that had existed in Saudi history. Contacts and entitlements toward the state were individualized rather than group-based, as is typical for patron-client systems.[166]

The lack of independent interest aggregation mechanisms and spaces for collective bargaining has been a defining feature of the Saudi political economy. It has become a hub-and-spoke system centered around the royal family and its bureaucracy, with many entitlements and claims for patronage in society but little integration or communication between the various entitled individuals and groups. While social interests had penetrated the state on the individual level, they remained deeply fragmented on the meso-level.

This system would become generalized and solidified through the 1970s oil boom. Whereas the administration had still been relatively small in the early 1970s and significant social groups still had little contact with it, by the early 1980s it had managed to wrap itself around the bulk of the Saudi populace. Public services and resources grew extremely rapidly during the 1970s, and through the sheer scope of expansion, both meritocracy and clientelism were boosted tremendously and simultaneously. While state structures were further formalized, new strata of middlemen emerged, making brokerage of state resources a ubiquitous pastime. Virtually all of society acquired direct stakes of some form in the state. Chapter 4 addresses the further differentiation and segmentation of the state and its clients in this crucial period, at the end of which the system would encounter the limits of growth, mobility, and institutional engineering, finally congealing into a more permanent setup: a large and surprisingly immobile conglomerate of huge clienteles reaching out into all of society, suddenly tied down by its numerous obligations. By the mid-1980s, full control over society had created full immobility of the state.

---

166. Cf. Al-Awaji's (1971) and Krimly's (1993) lively accounts of individual patronage and rent-seeking in the bureaucracy.

Chapter 4

The 1970s Boom

Bloating the State and Clientelizing Society

If a Saudi bureaucrat from the year 1970 were to use a time machine and travel to 1980, he could probably continue in his job in the same ministry in almost the same fashion. The internal functioning of state agencies and the micro-mechanisms surrounding them had not changed much, as hierarchies were still steep, horizontal coordination weak, agencies overstaffed with bureaucratic clients, the bureaucracy heavy, and the state's regulatory capacity limited. Yet, on stepping out of the ministerial door, our bureaucrat would quickly realize that the Kingdom had gone through a rapid, dizzying transformation on both the meso- and the macro-scale. A large and modern national infrastructure had come into being, administered by a plethora of new institutions providing money, employment, housing, education, and other services for large parts of Saudi society, creating new entitlements and whole new levels of middlemen between the state and the populace as well as between the state and local and international business. The state and its relations to society had grown greatly in complexity; stakes in and dependence on it multiplied. In the 1970s, the modern Saudi bureaucracy had acquired its definite shape.

On the administrative level, the age of rapid expansion started in 1970 with a five-year development plan, the first of many documents produced by the Central Planning Organization aiming to coordinate—with mixed success, as we will see—the large scale deployment of public services and industrial policies. The quantum leap from the early 1970s to the 1980s is quickly illustrated with a few figures. The total allocation of expenditure in the second five-year development plan for 1975–79 was nine times that of the first plan (1970–74), almost

500 billion SR.¹ The Saudi GDP, almost completely oil-driven, expanded from an estimated 148 billion SR in 1970 to 540 billion SR in 1981 in real terms, with the share of government consumption and capital formation strongly increasing.² The state of 97,000 civil Saudi employees in 1970 expanded to 175,000 employees in 1980 (with some 90,000 budgeted jobs remaining unfilled in addition), to reach 344,000 by 1988, complemented by between 150,000 and 200,000 publicly employed expatriates.³ The figures exclude extensive employment in the security forces and other special bodies.

Although the greatly increased oil income created some pressure toward increasing expenditure and it was politically expedient to share the wealth, the pace of growth and the developmental ambition were still determined by the leadership. Fahd and a few dedicated technocrats were the main force behind the development drive, backed up by an aging Faisal.⁴ When he was second deputy prime minister and minister of interior before 1975, Fahd already headed supreme commissions on planning and investment, pushing for higher development expenditure.⁵ When Faisal was assassinated, Fahd became first deputy prime minister and crown prince, playing a dominant role in economic matters thanks to King Khalid's muted interest in technocratic questions.⁶

High Politics: Expanding Fiefs, Creating Islands of Efficiency

The ambition of Fahd and his advisors was to provide state-of-the-art infrastructure, free and comprehensive public services, employment guarantees, and the conditions for industrial growth as quickly as possible. To this end, the formal state was greatly expanded and deepened.

With more functions, the state was split up into more agencies. The new cabinet in 1975 saw six new ministries: Industry and Electricity; Public Works and Housing; Higher Education (split off from Education); Post, Telegraph, and

---

1. Holden and Johns 1982: 396.
2. Based on constant prices of 1999. In current prices, the increase was almost thirty-fold. SAMA Annual Report 2003: 431ff.
3. Krimly 1993: 231; Huyette 1985: 106; Al-Ammaj 1993: 146.
4. Interview with former Saudi minister Riyadh, December 2005; interview with Michael Gillibrand, former advisor to the Saudi government, London, November 2005.
5. Already in 1970, Fahd was behind a series of articles demanding, among other things, more development expenditure and political modernization (accompanied by populist calls for lower electricity and water prices as well as protection from exploitation by merchants); cf. Activities of Prince Fahd, July 1970, PRO/FCO 8/1508.
6. Holden and Johns 1982: 386–87; Lacey 1981: 429–25; biographical sketch Khalid, June 1968, Mulligan Papers, box 1, folder 70.

Telephone (split off from Communications); Municipal and Rural Affairs (devolved from the Ministry of Interior); and Planning (an upgrade of the Central Planning Organization to ministerial status).[7] Several development funds (real estate, agriculture, industrial development) were either newly created or freshly endowed with great amounts of capital.[8]

This sprawl provided space for new talent to be recruited into the public service and gave a bigger role to commoners in several areas of policy execution. However, it also created new problems of coordination and, in its rapidity, also led to the isolated, parallel growth of agencies, which were—if at all—only accountable to the top and developed their own vested interests in the course of their day-to-day operations. With the few senior royals who had some technocratic interest severely overburdened, there was much space for the bureaucracy at large to develop according to its own logic.

Institutional Sprawl: The Royal Fiefs

Several senior princes jumped at the newly available surpluses to build or expand their own empires. Under Fahd, the name of the game was to distribute, employ, and develop sector-specific services as quickly as possible. This left much scope for the growth of autonomous, smaller states within the state. In this process, the core constellation of the state remained the same, however, as the core organizations and major players controlling them saw no change. The renewed phase of accelerated state growth was much more orderly than what the Kingdom had seen in the 1950s. Major fiefs had already been solidified through bureaucratic growth and formalization in the 1960s.

At the same time, state growth provided space for new, smaller "empires" within the bureaucratic apparatus. Institutional change was mostly achieved through adding new bodies. Two new princely departments emerged from the cabinet expansion in 1975: Public Works and Housing under Mit'eb and Municipal and Rural Affairs (MoMRA) under Majid.[9] Both were resource-intensive: The second five-year plan envisaged 53 billion SR for municipal spending alone, more than 10 percent of the total plan allocations.[10] In the same year, Mit'eb's ministry was given the authority to move ahead with a plan for one hundred thousand

---

7. Samore 1983: 348; Al-Seflan 1980: 88.

8. *Lamha tarikhiyya mawjaza 'an tatawwur al-ajhiza al-hukumiyya (idarat al-buhuth wal-istisharat)* (Concise historical overview of the government bodies' development; Studies and Counseling Bureau, n.p., n.d.): 25ff., IPA.

9. In a rather muddled process, MoMRA took over parts of the responsibilities of the MoI; cf. Al-Sabban 1990: 101.

10. *MEED*, 22 August 1975: 5ff.

houses.[11] Whereas these new institutions had a clear functional role, the fiefdom nature of the Presidency of Sports and Youth Affairs, controlled by sons of Fahd, was more obvious. The Presidency "spent like crazy"[12] on massive infrastructure projects, with little control by the Ministry of Finance (MoF).[13] Its budgeted expenditures increased from 143 million SR in 1975 to almost 3 billion SR annually in the early 1980s.[14]

MODA: THE BIGGEST FIEFDOM AND ITS CLIENTS

More important in scale, however, the existing ministries of senior princes were extended, most notably those controlled by full brothers of Fahd ("Al Fahd") who had gained administrative turf in the fight against Saud. It is worth analyzing this expansion in some detail, as it solidified the institutional setup and power balance that has shaped Saudi policy-making ever since.

Security continued to be the largest item in Saudi budgets, consuming about 40 percent of the total.[15] The Ministry of Defense and Civil Aviation under Sultan consistently received the largest allocations. Saudi Arabia overtook the Iranian Shah in military procurement in the late 1970s, and Saudi military spending per capita in the early 1980s was ten times higher than that of NATO countries.[16]

The MoDA projects in the 1970s aimed at creating a military establishment with an internal infrastructure largely divorced from the rest of the Kingdom. Several large "military cities" were built in peripheral regions, the biggest of which was estimated to cost 10 billion US$ in 1979 and designed to house seventy thousand people.[17] Separate large-scale housing, education, and health infrastructures were set up under MoDA control. The U.S. Army Corps of Engineers was tasked with managing this expansion.

As MoDA was organized around Sultan (and to a lesser extent his full brothers Turki and Abdalrahman, his successive deputies), all of its organizational units were oriented toward him as central authority.[18] Despite increasing formalization and bureaucratic deepening, armed services and the civilian employees in the

---

11. *MEED*, 5 December 1975: 22.
12. Gillibrand interview.
13. The Presidency was given an independent budget in 1974; *Lamha tarikhiyya*: 23.
14. *Statistical Yearbook*, ch. 10, various editions.
15. In 1970, another 12 percent reportedly went into foreign subsidies—another expenditure pattern hard to imagine in a nonrentier state; cf. NARA RG 59, 150, 66–67, box 2587, folder POL 1 SAUD-US (3/30/70).
16. Pelletière 2004: 83; Vasiliev 2000: 443.
17. *MEED*, 1 June 1979: 3; *MEED*, 29 July 1977.
18. Summary presentation: report on the organization and administration of the Ministry of Defense and Aviation (n.d.), FF/IPA, Box 12. Sultan had a personalized "shaykh-bodyguard" type of relationship

ministry itself cooperated little.[19] The MoDA bureaucracy itself was fragmented; at one point, there were six separate budgets within the ministry.[20]

Sultan is said to have profited considerably from his position and today is possibly the richest prince in the Kingdom. His older full brother Fahd reportedly took a 40 percent cut on his military deals.[21] The size of the MoDA budget did not only serve Sultan's immediate material ambitions, however, but also allowed the distribution of privileged services to growing numbers of military employees. As inefficient as the armed forces turned out to be,[22] they functioned well as a provider of contracts, employment, and privileged services to both officers and their relatives.[23] They also offered posts for more princes below the most senior level. Control over the national airline Saudia—attached to Defense as a result of the Talal-Mish'al rivalry in the 1950s—gave additional patronage in terms of procurement, employment, and contingents of tickets that could be informally distributed or resold.[24] Needless to say, responsibility for civil aviation never left MoDA, despite forty years of consultancy to the contrary.[25]

The following of brokers around the minister expanded and gained in importance, the most prominent being Adnan Khashoggi. With the explosion of the defense budget, which quadrupled from 1974 to 1975 alone,[26] there was always a demand for "fixers" to arrange procurement deals, to act as gatekeepers to senior figures, and to take care of contract details on a commission basis.[27] Senior military figures are also known to have profited from the opaque procurement methods. The large-scale distribution machinery of Sultan's MoDA generated both extensive formal and informal clientage.

---

with his officers; Jidda to State, telegram 200645Z, July 1970, NARA RG 59, 150, 66–67, box 1792, folder DEF SAUD (1/1/70).

19. Political Situation in Saudi Arabia (Form-At-A-Glance—June 1973), PRO/FCO 8/2105.

20. Report on the organization and administration of the Ministry of Defense and Civil Aviation (n.d.): 22ff., FF/IPA, Box 18.

21. Holden and Johns 1982: 324.

22. Defense Attaché Reports for Saudi Arabia, PRO/FCO 8/1915.

23. The British assessment in the late 1960s was that loyalty rather than efficiency counted for promotion, which was why many senior officers had a Najdi background; Craig to Stewart, internal security in Saudi Arabia, 14 August 1969, PRO/FO 8/1165.

24. Interview with former British diplomat, London, June 2005.

25. Report on Ministry of Defense and Civil Aviation: 22–23, FF/IPA, Box 18; discussions with international transport experts in Riyadh, 2003 and 2004.

26. U.S. equipment sales agreements were valued at 459 million US$ in 1973/74 and at 1,993 million US$ one year later; cf. Holden and Johns 1982: 359.

27. Several cases have been well documented through congressional investigations and lawsuits in the United States; cf. Holden and Johns 1982: 411–14; Lacey 1981: 464–71.

ABDALLAH'S NATIONAL GUARD

The National Guard under Prince Abdallah was smaller in scale but similarly constructed around him as the overarching figure. Unlike MoDA, its main socioeconomic function has been to provide employment and services to tribesmen, and the level of education in the Guard has been lower than among MoDA personnel. Abdallah has been reported to defend the institutions as his own preserve, preventing other senior princes from getting involved in Guard matters and trying to bring the Guard on a par with MoDA in its capacities and weaponry.[28]

In the course of the boom, the National Guard developed its own "cities" as well as its own housing, education, and health systems, which also provided services for relatives of guardsmen—again extending the clientele of the institution beyond the immediate circle of its employees.[29] Like MoDA, it was strongly dependent on foreign expertise and training.

Foreign specialists assessed the Guard's chains of command as ill-defined and overcentralized. As the Guard was also a government agency, the presence of civilians in its headquarters decreased its administrative coherence. Abdallah was the figure around which everything was concentrically organized.[30] Separate units of the Guard often had parallel functions.[31]

Although not associated with such big deals and cuts as MoDA, National Guard procurement was hardly more transparent than that of MoDA. Specific senior advisors around Abdallah, especially from the Najdi Tuwaijri family, have functioned as important gatekeepers to the prince.[32] Master broker Khashoggi, however, patently had no access to Abdallah, working mostly as client of Sultan.[33]

THE MINISTRY OF INTERIOR

The Ministry of Interior (MoI), like other "sovereignty agencies," has also been controlled by a maternal faction of the Al Saud. After Fahd became crown prince in 1975, his full brother Naif, until then his deputy, took over the ministry; his deputy, in turn, has been his full brother Ahmad. It is different from both MoDA and the National Guard, however, in that it plays a bigger role in domestic policy, including economic regulation. It therefore features the most prominently

---

28. Saudi Arabia: Defence: National Guard, 1969, PRO/FCO 8/1198.
29. Lippman 2004: 291–92.
30. At least as late as 1972 it also lacked an operations center and adequate maps; Defence Attaché Reports, Report on the Saudi Arabian National Guard, 11 October 1972, PRO/FCO 8/1915.
31. Management survey report National Guard, March 1966: 28–29, FF/IPA, Box 23.
32. One figure apparently peddling his services as contract go-between was Abdallah's brother-in law from the Lebanese Fustuk family (called "Bostick" in a U.S. telegram); cf. 20 March 1973, NARA RG 59, 150, 66–67, box 1793, folder DEF 6–4 SAUD.
33. Holden and Johns 1982: 364.

in the case studies in the chapters to come. Through its opacity and powerful, autonomous position, the MoI bureaucracy has been able to thwart or seriously delay numerous domestic policy initiatives, whether deliberately or not, whether through open vetoes or through gradual obstruction.

A seasoned British diplomat has dubbed the MoI a "rule unto itself"; its responsibilities are extensive.[34] The MoI has controlled immigration and hence workforce matters, and through its expansive interpretation of security and public order, it has developed much leverage over economic regulation. Until 1975, the MoI had full control over municipal matters, and even after the creation of MoMRA, it has remained the institution to which the regional governors report. This means that many senior princes are—in principle—accountable to the minister of interior.

Unlike other ministries, the MoI can recruit and control its extensive personnel independent of the rules of the Civil Service Bureau.[35] Like other Al Saud–controlled agencies, the MoI has its own expansive housing program.[36] The domestic security tasks of its various security agencies overlap with those of the National Guard. Procurement budgets are substantial. As will be explained in more detail in the case study on Saudization, a large network of brokers has evolved not only around MoI contracts but also around the labor import permits that were issued by the MoI until 2004.

INTER–FIEF POLITICS

MoDA, the National Guard, and the MoI engaged in repeated competition over budget allocations in the 1970s, betokening the status ambitions of their heads.[37] When Fahd was still in charge of the MoI, he reportedly embarked on developing a coastal guard and frontier force as a third power to counterbalance MoDA and the National Guard.[38] Faisal himself seems to have tried to balance the interests of his ambitious younger brothers. In one case he asked a senior British advisor in the National Guard to stay in Abdallah's trust but thwart his ambitions of augmenting his forces beyond acceptable limits.[39]

---

34. Phone interview with former British ambassador.
35. Similar exemptions exist for regional governors; cf. Al-Ammaj 1993: 171ff. Although controlled by royals and growing regional budgets, the governorates have generally lost importance relative to the central state, with the exception of Salman's all-important Riyadh governorate.
36. *MEED* mentions a ten-year program to develop four hundred thousand housing units in 1981; cf. *MEED*, 1981 Saudi Arabia yearly report: 50.
37. Some observers thought that fights over security budgets repeatedly brought the family close to the brink of conflict. Interview with former U.S. diplomat, New York, April 2007.
38. Craig to Stewart, internal security, 14 August 1969, PRO/FO 8/1165.
39. Saudi Arabia: Defence: National Guard, 1969, PRO/FCO 8/1198.

The high degree of internal autonomy of the princely institutional conglomerates has tended to thwart policy coordination. In a situation of temporary fiscal strain, MoF under Mohammad Abalkhail in the autumn of 1978 issued a blanket order that all agency expenditure above 70 percent of the allocated budgets and all contracts above 100 million SR would have to be approved by MoF.[40] As might be expected, different agencies interpreted the 70 percent rule differently.[41] Abdallah and Sultan, however, ignored the rules altogether.[42]

More generally, the different procurement and command-and-control structures of the various security agencies prevented interoperability and impeded a coherent national security policy. Different intelligence agencies attached to different princes did not coordinate, operating with different lists of people they wanted to arrest.[43] Attempts to rationalize health and housing policies foundered, not least due to the fiefdom character of MoDA, the National Guard, and the MoI.

Although a committed modernizer, within the family Fahd was a less forceful personality than Faisal.[44] In line with the segmented core institutions of the regime, the Al Saud under cautious King Khalid increasingly developed an oligarchical style of leadership, with a tendency to produce stalemates and redundant policies. Senior figures like Fahd and Sultan were represented on numerous "supreme councils" intended to focus decision-making, but these were little more than mini-cabinets staffed with fewer cabinet members or deputy ministers, reproducing existing cleavages.

Although MoDA, the National Guard, and the MoI were the most salient fiefdoms, many smaller organizations or departments within them were entrusted to princes with varying degrees of public service orientation and willingness to communicate. Such figures could leverage their family status and were even harder to dislodge than commoner bureaucrats. Remarkably, however, most of the commoner minister posts staked out by Faisal as such remained under commoner control, even if deputies or assistant deputies often came from the Al Saud family and its collateral branches.

---

40. *MEED*, 17 November 1978: 3.
41. Some paid only 70 percent of their contract dues, others scaled down their projects or stopped them; cf. *MEED*, 1 December 1978: 41.
42. Holden and Johns: 508; *MEED*, 1 June 1979: 25.
43. List of Aramco employees imprisoned, NARA RG 59, 150, 66–67, box 2586, folder POL 23 SAUD (1/1/70).
44. Fahd was considered not to have "the driving will or intellect to make inroads on the domains of other key figures such as Finance Minister Prince Musaad, Defense Minister Prince Sultan, or even Petroleum Minister Yamani and Planning Organization head, Hisham Nazer." Prince Fahd, the King, and the Inner Circle, 26 September 1973, NARA RG 59, 150, 66–67, box 2584, folder POL 2 SAUD (1/1/70).

Commoner Fiefdoms in the Oil Boom

One sector that has remained firmly in the hands of commoners, though of a particular provenance, is the field of religious organizations created under Faisal. The bureaucratic territory parceled out to the religious establishment expanded tremendously over the 1970s, a separate ideological recruitment ground alongside the secular bureaucracy and often linked to the rest of the state only through the person of the king. Budgets increased rapidly: the line item for religious endowments expanded from thirty-five million SR in 1969 to 1.7 billion SR in 1981.

The educational fiefs of the religious establishment also entered a phase of rapid expansion. The budget of the Islamic University in Medina, in existence since 1961, increased from 9 million SR in 1970 to 400 million SR by the late 1970s. Imam Muhammad University in Riyadh and the College of Sharia in Mecca gained full university status in 1974 and 1981, respectively, with budgets increasing accordingly (the College was renamed Umm al-Qura University). All of these universities awarded mostly religious degrees that imparted skills useful only for careers within the Saudi religious field, and they have been run largely autonomously from the rest of Saudi higher education.

The largest religiously controlled budget items were linked to school education. Female intermediate enrollment increased from 5,305 pupils in 1970 to 71,037 in 1980, while female primary school enrollment grew from 120,000 to 360,000.[45] The budget of the General Presidency for Girls' Education, under religious control since its inception, expanded from 95 million SR in 1970 to 3.8 billion SR in 1980. The budget of the Ministry of Education, responsible for boys' education, increased from 384 million to 9.6 billion SR in the same period. Though not exclusively under religious control, conservative forces had a strong presence in this institution too.

The expansion of the religious bureaucracy, however, was not what drew international attention to Saudi Arabia in the 1970s. More visible and more spectacular was the expansion of the secular bureaucracy and the services it provided, unencumbered by the parallel growth of the religious administrative machinery. In many ways, secular commoner bureaucrats were becoming more powerful than ever before, vigorously shaping the landscape of modern Saudi Arabia—and in the process building veritable fiefdoms of their own.

To be sure, technocratic commoners, although sometimes selected from venerable urban families, differed from princes in that they had little independent political standing and were easily dismissed if the Al Saud so decided, sometimes

---

45. SAMA, annual reports.

through a simple radio announcement. Most of the new members of the 1975 cabinet were clients of Fahd and accordingly beholden to him.[46]

Dependent on royal goodwill, technocrats have been oriented toward the few princes above them rather than their commoner colleagues in the cabinet. Different ministers could have different patrons within the Al Saud,[47] which often undermined the functioning of the cabinet as collegiate body, the assembled expertise notwithstanding.[48] Some ministers, such as Hisham Nazer, at times managed to balance between different royal patrons, but they too were necessarily oriented upward rather than laterally.[49] Individual ministers have generally tried to influence the king in private, not in public, making communication vertical and exclusive.[50] The "parallel government," in which the most important decisions were still made by the Al Saud and their advisors—most ministers not among them—continued to undermine the cohesion of the formal government.[51]

Due to the increasing complexity of administrative tasks and developmental ambitions, however, Fahd and his peers inevitably had to delegate technocratic powers when it came to the implementation of projects and policies by individual agencies.[52] Although often notoriously reluctant to take initiatives, commoners, once a task was under way, had great leeway in fleshing out the details, even in weaker agencies such as the Ministries of Commerce or Labor and Social Affairs. Several ministries were reportedly given wider powers to help the implementation of the second five-year plan, which envisaged projects on an unprecedented scale.[53]

---

46. Interview with former U.S. diplomat, New York, April 2007.

47. This patronage could be arranged in cross-cutting ways, balancing influence within the royal family. Whereas a minister would be Fahd's man, his deputy might be a client of Faisal; cf. Tatham to Dandi, Hunt, Wright, 10 September 1973, PRO/FCO 8/2105. The powerful Ministers of State without portfolio in 1970s cabinets were perceived to be Khalid's representatives. Gillibrand interview.

48. Both Fahd and Faisal have insisted on having all major decisions presented to them. That Fahd has reportedly devoted less attention to policy detail does not imply—as sometimes is erroneously assumed—that he was more inclined to delegate substantive powers. All ministers have been obediently oriented towards the king and his senior peers. Interview with former deputy minister, Riyadh, May 2004.

49. Anecdotes about the formal seniority of even small-time royals over commoners abound. Oil minister Zaki Yamani once was not able to take his seat in a plane because an "acned, young prince" happened to occupy it; cf. Holden and Johns 1982: 401.

50. Huyette 1985: 100–101.

51. Holden and Johns 1982: 390; phone interview with former British ambassador.

52. After 1975 and before the oil price crash in the 1980s, ministers tended to get money whenever they asked for it, and Fahd was not much involved in details of implementation. Interview with former Saudi minister, Riyadh, December 2005.

53. *MEED*, 5 December 1975: 22.

Within their strictly hierarchical institutions, senior individual commoners could be dominant figures.[54] Although the number of bureaucrats with Western education increased exponentially in the 1970s, by the early 1980s the layer of competent people with some experience was still pretty thin.[55] Saudi government departments, like Saudi businesses, were usually dominated by one or two individuals who made most of the decisions, usually those in good stead with senior royals.[56]

Such senior figures had large leeway over the internal operations of their rapidly growing institutions.[57] On the 16-level bureaucratic seniority scale, ministers could appoint employees up to level 13 of seniority, which included most bureaucrats (on levels 14, 15 and the highest, "excellence," the cabinet formally decided).[58] In the absence of a strong civil service tradition, this gave ministers considerable powers of patronage. Even if bureaucrats were difficult to dismiss, incompetent individuals could be sidelined or, at higher levels, in effect be promoted out of active service and given general "advisor" posts. Regionalism and kinship often played a role, but due to the scale of state growth, not automatically so.[59]

Ministers could give civil servants smaller land grants, an important resource of patronage in times of booming real estate prices.[60] Heads of agencies would control the budgets of their institutions and also specific services their agencies provided, such as housing. Housing compounds used by agencies were often owned by ministers and their relatives.[61] The growth of such resources was especially pronounced in some of the ulama-controlled parts of the administration, such as education.

The most important commoners in domestic economic policy-making were imposing figures like Minister of Planning Hisham Nazer, Minister of Oil Zaki Yamani, and Minister of Finance Mohammad Abalkhail. The latter was a descendant of a Najdi notable family and a close confidant of the Al Fahd, who in the absence of detailed interagency coordination used his control over budgets to exert a crude influence over line agencies run by other commoners, often much

---

54. Overcentralization was a constant concern of the Civil Service Bureau; cf. *At-taqrir ath-thanawi li'diwan al-muwadhdhafin al'am1391–93* (Biannual report of the Civil Service Bureau, 1971–73), IPA.
55. Interview with British diplomat who lived in the Kingdom in the early 1980s.
56. Montagu 1985: 66.
57. Phone interview with former British ambassador.
58. Decree 2, 18/8/1397, Civil Service Council, printed in: *Majlis al-khidma al-madaniyya: dalil qararat majlis al-khidma al-madaniyya* (Civil Service Council: guide of decrees), IPA.
59. Al-Hamoud 1991: 89.
60. Armitage to LJR Dandi, 6 July 1974, PRO/FCO 8/2332.
61. Phone interview with former British ambassador.

to their chagrin.[62] In charge of increasingly opaque national budgets, he was a crucial broker (and bottleneck) between the royals who held the purse strings and controlled the commoner ministers. Reportedly Abalkhail clashed repeatedly with Yamani, a client of Faisal.[63] Yamani also had serious conflicts over jurisdiction with Nazer, who once had been his deputy, but had moved closer to Fahd in the 1970s.[64]

POLICY-MAKING: PARALLEL FIEFS AND PLANS

Within the commoner bureaucracy at large, steep hierarchies combined with rapid proliferation of agencies and tasks made coordination difficult, increasing competence at the top notwithstanding. Agencies were created and staffed in a rapid top-down fashion based on locally delegated authority, with little balancing and supervising mechanisms available to integrate their operations.

Although there were formal mechanisms of coordination—not only the cabinet but also numerous interministerial committees—they were "badly serviced."[65] They either involved too junior personnel[66] or the same senior players who represented entrenched institutional interests.[67] The government itself recognized the persistent problems of miscommunication, overlapping jurisdictions, redundancy, and overcentralization but never took decisive action against it.[68] In the segmented and hierarchical Saudi setting, an independent, powerful institution for clearing interests, be it a parliament, a party, or a supreme planning commission with independent staff, was lacking and indeed difficult to imagine. The system was organized concentrically around the Al Saud, themselves increasingly fragmented. As long as the budget kept growing, however, redundancy and parallel structures were not a pressing problem—instead, new problems would often be "solved" by creating new institutions.

At a time of project-based growth, the political game revolved around maximizing individual agency budgets as granted by the king. According to expatriate

---

62. Abalkhail reportedly was happy about the decrease in government borrowing in the early 1990s, as this would result in ministries not being able to pay their bills, making them more dependent on him; Montagu 1994: 49. He reportedly came into government as a client of Prince Sultan. Interview with former U.S. diplomat, New York, April 2007.

63. Interview with former deputy minister, Riyadh, November 2005; interview with former minister, December 2005; Robinson 1988: 399–400.

64. Hertog 2008.

65. Phone interview with former British ambassador.

66. Senany 1990: 282.

67. Committee members in general tended to not have the authority to approve measures, the biggest problem of committee work, according to 74 percent of respondents in a 1980s survey; cf. Al-Shalan 1991: 224.

68. *At-taqrir ath-thanawi.*

advisors, many ministries therefore submitted budget requests anywhere from three to nine times their likely allocations in the preparation of the third five-year plan in 1979, including massive tenders for office buildings.[69] Rational coordination suffered accordingly. In the forefront of the crucial second development plan, Crown Prince Fahd, Hisham Nazer, and Zaki Yamani gave greatly varying estimates of its volume, varying from 50 billion to 150 billion US$. The scale of industrial projects varied similarly.[70] Grandiose projects would be announced by one agency, only to be denied by another.[71] Although endowed with little capacity to craft proactive, independent policies, commoner ministries had considerable power to "say no" to policies that involved them.[72]

On the level of day-to-day administration, different government agencies interpreted decrees differently, most notably in the case of Abalkhail's various budget measures. They would also grant exemptions from tender regulations without being formally allowed to do so, resulting in potentially unenforceable titles.[73] Variation in rule interpretation and procurement standards was pervasive.[74]

A general penchant toward secrecy between institutions on similar levels of hierarchy did not make coordination any easier. The Saudi Industrial Development Fund (SIDF) suffered from lack of cooperation on the part of the Ministry of Industry and Electricity and the government-run Industrial Studies and Development Center, both of which were unwilling to share data.[75] Different agencies tended to issue widely varying statistics. Officials from the Central Planning Organization admitted that the audits from different ministries did not tally up but had little leverage to remedy the problem.[76]

With growing complexity and the increasing ambition of extending services to every Saudi, problems of fragmented governance in several sectors became more severe. There was little coordination of the various institutions involved in increasingly important agricultural policies, with none of them possessing a

---

69. *MEED*, 14 December 1979: 51.

70. *MEED*, 9 January 1975: 6.

71. One famous case involved the minister of agriculture's repeated denials of plans to tow icebergs to Saudi Arabia for fresh water supply, a rumor that had been bandied about by Prince Mohammad bin Faisal, head of the desalination administration; cf. *MEED*, 10 December 1976.

72. Phone interview with former British ambassador.

73. Montagu 1994: 106.

74. Phone interview with Barry Lello, representative of the UK Department of Trade in Saudi Arabia in the 1970s and early 1980s, April 2005.

75. Among other things, an evaluation of important hydrocarbon projects was not shared with the SIDF. A coordination body that was set up in August 1978 did not seem to improve things significantly, as turf battles between SIDF and Ministry continued. *MEED*, 9 June 1978: 14, 25 Aug 1978: 36; Gillibrand interview.

76. Largest Boom in History Saudi Minister Claims, 16 August 1972, NARA RG 59, 150, 66–67, box 779, folder E 5 SAUD (1/1/70).

meaningful lead agency function. The increasingly well-endowed Agricultural Bank, for example, was under the control of MoF, depriving the Ministry of Agriculture and Water of crucial market intervention mechanisms.[77] In the case of water policy, responsibilities were spread among the Ministry of Interior, the Ministry of Agriculture and Water, the Water Desalination Organization and the Ministry of Planning.[78] The Ministry of Communications had no control over air transport, which remained part of Sultan's fief—despite the paramount importance of domestic flights for personal travel.

Administrative segmentation, overlapping tasks, and unclear responsibilities also prevailed on the level of local administration, where the relationships linking MoI, governorates, MoMRA with its attached municipalities, and other line agency branch offices were often unclear and settled according to the informal local distribution of authority and status, if at all. Centralization of decisions did not help local coordination. A 1980s survey revealed that representatives of different institutions involved in local administration consistently interpreted their responsibilities and roles in fundamentally contradictory ways.[79] In Saudi cities, it often came to "battles of the bulldozers"—rival government agencies giving contracts for the development of the same areas.[80] Refusing to coordinate their plans even after the fact, it was sometimes their respective foreign contractors who fleshed out a compromise among themselves.

As the state and its ambitions of distribution grew, parallel institutions tended to acquire as many tasks as possible, giving them budget entitlements, patronage power, and relative autonomy from other bodies. Due to the underdevelopment of general infrastructure and public services and the overcentralization of decision-making, government agencies under conditions of fiscal abundance strove to create their own services—housing, transport, electricity generation, education, and so on—to be able to develop autonomously, if only to fulfill their ambitious plan targets.[81]

This resulted in increasingly fragmented health and housing sectors. By the late 1970s, at least eight agencies were involved in running clinics and hospitals,

---

77. Chaudhry 1997: 179–80.
78. Lackner 1978: 156.
79. When interviewed by a Saudi student, even a powerful regional governor, Majid of Makkah, was unsure about his formal relationship to the MoI; cf. Samman 1990. One contributing factor to the undefined nature of jurisdictions was probably the resistance of powerful *umara*; cf. Awaji 1971: 131. The situation in the 1950s had been similar; cf. Municipalities in Saudi Arabia (preliminary report), May 1956, Mulligan Papers, box 5, folder 18.
80. *Saudi Introspect* (monthly newsletter published from Staniford Street, Boston), October 1981, Middle East Documentation Unit, University of Durham.
81. Gillibrand interview. On a smaller scale, ministries had done this in the 1960s already; cf. Supporting study 2: agriculture Case: 3, FF/IPA, Box 1.

in predictably uncoordinated fashion.⁸² This led to a surplus of beds in Riyadh, where most agencies were based, and a deficit elsewhere. The Ministry of Health was responsible for a bit more than half of capacity, with other hospitals run by the Ministries of Education, Higher Education, Labor and Social Affairs, Pilgrimage and Religious Endowments as well as the MoI, MoDA, the National Guard, and the morality police, in addition to establishments owned by the Faisal Foundation (run by King Faisal's sons) and several private business groups.

Parallel recruiting efforts under conditions of manpower scarcity were wasteful, and the rival technocratic ambitions led to a dominance of overpriced megaprojects. When the government attempted to coordinate project planning in 1979, it had become hard to break up existing patterns of distribution.⁸³ Four years later, *Middle East Business Intelligence* (*MEED*) still concluded that "the kingdom's health care system is one of the most fragmented in the world, with 13 agencies providing medical services."⁸⁴

Similarly, different agencies had their own housing programs, including MoDA, the National Guard, and the MoI, but also unlikely candidates such as the Ministry of Foreign Affairs, the quickly growing Saudi universities, and the vocational training administration.⁸⁵ As might be expected, there was little coordination among them.⁸⁶ Again, the Ministry of Public Works and Housing, although controlled by a senior royal, could not act as an effective lead agency, and coordinating bodies, insofar as they existed, were inconsequential. In addition to housing and educational institutions, some ministries had unlikely attachments such as their own large-scale printing facilities, TV studios, and, in at least one case, a ministry-run gas station—located far outside Riyadh, forcing bureaucrats' drivers to take great detours.⁸⁷

Islands of Efficiency

Ministries operated as islands, accountable only to the top, and in political rather than technocratic terms. However, within this fragmented environment, individual agencies achieved degrees of efficiency which were unprecedented, and many institutions saw considerable improvements in skill and depth of expertise. A U.K. diplomat who returned to Saudi Arabia in the late 1970s after having served there

---

82. *MEED*, 9 February 1979: 6–7.
83. Ibid.
84. *MEED*, 4 February 1983: 41. A national health council had still not been formed, although it supposedly was a priority of the third plan, which had started in 1980; cf. *MEED*, 11 February 1983: 43.
85. *MEED*, 28 March 1980: 45; *MEED* Special Report, August 1981: 104.
86. *MEED*, special Saudi Arabia issue, August 1978: 58.
87. Interview with former deputy minister, Riyadh, November 2005.

ten years earlier vividly described to me how stunned he was by the degree of material progress achieved, the relative efficiency in provision of specific services, and the competence of leading technocrats.[88]

The Saudi system progressively benefited from technocratic education, which it could afford to buy in the West. In this context, the Al Saud, while using much of the state as patronage resource, proved capable of co-opting effective administrators for high-level posts or specific elite organizations. An expanding state offered great upward mobility for a growing class of young, Western-educated commoners.[89] The trajectory of several bright, Western-educated and left-oriented young Saudis illustrates the power of such cooption: Handpicked as technocrats-to-be in young adulthood, they quickly shed their fashionably radical chic and embarked on stellar bureaucratic careers. Their ranks include ministers and deputy ministers like Abdalrahman Al-Zamil, Ghazi Al-Gosaibi, Ibrahim Al-Awaji, and Soliman Suleim.[90] Fahd in particular seems to have reared a clientele of such development-oriented "young Turks."[91] Not much was heard of their leftist ideas after the early 1970s.[92]

Several centers of (relative) administrative excellence emerged in the 1970s. These functioned—not unlike other agencies—in relative isolation from the rest of bureaucracy. They also operated at arm's length from the inefficiencies and patterns of favoritism of other bodies, however, being accountable only to senior royals who had an interest in creating autonomous drivers of development and the resources to set up additional structures without compromising patronage in other fields. In addition to traditionally well-managed SAMA, IPA, and Petroleum College (which gained university status in 1975), some essentially service-oriented institutions managed to recruit and retain good personnel, establish efficient procedures, and achieve impressive outcomes.[93]

One of these institutions was the Royal Commission for the new industrial cities of Jubail and Yanbu, one of Fahd's ambitious pet projects, which he gave

---

88. Craig interview.
89. In the British view, the government, although anachronistic, offered more opportunities for young people than any other Arab government; cf. Armitage to FCO, Saudi Arabia—The Al Saud and Internal Affairs, December 1973, PRO/FCO 8/2105.
90. Interview with former deputy minister, Riyadh, February 2007.
91. Al-Awaji reportedly even attended the regional conference of the Baath party in Damascus in the early 1970s and was a Baathist leader of Arab students in the United States before joining the government. Discussion with Saudi businessman, London, July 2005. Fahd, a much softer ruler than Faisal, brought many of the young radicals back into the Saudi system that Faisal had had expelled or imprisoned. Discussion with Saudi political researcher, Riyadh, November 2005.
92. In a graphic illustration of princely patronage over budding technocrats, a Saudi Ph.D. who is otherwise very critical of the Saudi system and the civil service thanks Prince Salman profusely in the thesis preface for his help in getting a scholarship to study in the West; cf. Al-Ammaj 1993: iii.
93. On the college, see Albers 1989.

special powers to bypass the regular bureaucracy in terms of recruitment and infrastructure.[94] Its vice-chairman was Fahd confidant Hisham Nazer. The agency managed to set up impressive transport, utility, and housing facilities in a short time.[95] Although suffering from severe shortages of cash in the 1980s, the two cities with their heavy petrochemical and other industries are today rated as the greatest industrial successes in the Gulf. Interference by the rest of the bureaucracy was minimized through exclusive accountability to the top.

Following on the heels of the Royal Commission, the cabinet decided to form the Saudi Basic Industries Corporation (SABIC) in 1976 with a capital investment of 10 billion SR, with the aim of setting up the basic large-scale industries that the private sector was incapable of financing. Although until today the state has retained majority ownership, thanks to its relative political autonomy the company's extensive petrochemical ventures have been a large success, quite different from the heavy industries of most other oil exporters. With the minister of industry as chairman, an effective and insulated management has run the organization with limited political interference, paying higher salaries than other public sector jobs and recruiting high-quality local personnel.[96] Incorporated as a company, its senior functionaries have consisted mostly of handpicked engineers and planning experts from Najd who enjoyed the protection of Fahd and his young minister of industry, Ghazi Al-Gosaibi. SABIC has been shielded from predation ever since, run by a stable core team largely recruited from inside the company. Many of the senior executives in 2009 were already with the company in 1976. Today it is the Middle East's leading industrial concern by a wide margin.

SABIC stood in striking contrast to state refining and would-be national oil company Petromin, a company created in 1962 under the Ministry of Petroleum. Petromin had early on become a source of patronage and brokerage for royals and friends of its governor, Abdalhadi Taher, a client of oil minister Zaki Yamani, who in turn was one of Faisal's close advisors. A largely autonomous fiefdom, Petromin's investments were not coordinated with other state agencies and were mostly failures. SABIC and, after nationalization in the 1980s, Saudi Aramco gradually took over Petromin's activities—part of the decline of Zaki Yamani's fortunes

---

94. The Commission reported directly to Fahd and had an independent budget and separate accounts; cf. royal decree M/75, 16/9/1395; printed in *Annual report of Royal Commission for Jubail and Yanbu* (1977), IPA. A circular from the presidency of the Council of Ministers signed by Fahd gave the Commission the license to ignore other parts of the bureaucracy, making it "solely responsible" for implementing its infrastructure projects "without necessarily taking into consideration the jurisdiction or responsibilities of the various ministries or public corporations in regard to such projects." Decree 502, 8/1/1397.

95. The Royal Commission was deliberately set up as "new empire" to make it function independently; Gillibrand interview.

96. *MEED*, 10 September 1976: 22; Rothnie to Wright, 22 August 1974, PRO/FCO 8/2332.

and of the rise of the new Najdi clients of Fahd. Petromin would, however, take almost three decades to be fully dismantled after being stripped of its first industrial assets in 1976. Once an institution is created in Saudi Arabia, it can outlive its political death sentence for decades, even if it carries on as little more than a moribund shell.[97]

Another example of effectiveness within the state in the 1970s was the Ports Authority under Fayez Badr, who, invested with ministerial status and the authority to ignore other bureaucracies, managed to clear the monumental logistical mess in the Jeddah port within half a year, despite rapidly expanding trade and great infrastructural bottlenecks.[98] The Ports Authority was one of the first agencies that would undergo successful semi-privatization in the 1990s.[99]

The most sociologically interesting island of efficiency is arguably the Al-Riyadh Development Authority, which has effectively functioned as a regionally circumscribed super-ministry under the powerful governor of Riyadh, Prince Salman, a full brother of Fahd. The well-funded organization has managed the breakneck expansion of Riyadh significantly better than the more fragmented local authorities of the other major Saudi city, Jeddah, whose public services and infrastructure remained woefully inadequate despite the city's slower rate of expansion.[100] The top ranks of the Al-Riyadh Authority are heavily staffed by well-educated technocratic notables from the more secular branches of the vast Al Sheikh ulama family.

Higher salaries and other perks, exemptions from general civil service rules, and, in many cases, public corporation status made for much greater flexibility than in the rest of the bureaucracy.[101] From 1964 to 1974, the number of employees in "public institutions" outside of the regular bureaucracy increased from 479 to 10,497.[102]

However, all of the new islands of efficiency took on clearly delimited tasks, often geographically circumscribed. No institution of central importance such as SAMA emerged anymore at this stage, as the new bodies were additions to rather than full substitutes for the general bureaucracy, which was growing in parallel.

---

97. For details, see Hertog 2008.
98. Holden and Johns 1982: 404ff. The authority was set up in 1976; *MEED*, 24 September 1976: 39.
99. Bakr 2001.
100. In early 2008, Riyadh was awarded several UN-sponsored urban management awards in London; cf. *Saudi Gazette,* 2 January 2008. The Riyadh Authority, however, was powerless against royal land grabs in Riyadh, showing the limits of insulated technocracy in the hierarchical Saudi system. Interview with Caroline Montagu, Saudi business analyst, London, July 2005.
101. Report on public corporations, 1978: 8ff., FF/IPA, Box 18. The Civil Service Council codified further exceptions in a 1979 decree (141, 27/5/1399), printed in *La'ihat al-mu'ayyanin 'ala bund al-ujur fi al-khidma al-madaniyya* (Ministry of Civil Service 1423/2003), folder 627 (351,1), IPA.
102. *At-taqrir ath-thanawi:* 7–8.

Development through Projects

The moderate efficiency exhibited by a few new administrative bodies explains some, but not all, of the material accomplishments of the Kingdom by the early 1980s—a phenomenon with which, waste and corruption notwithstanding, many spectators at the time were deeply impressed. A good deal of the new infrastructure was rolled out under the aegis of older organizations such as the Ministry of Transport (formerly Communications) or the Ministry of Interior. How can we explain the successes of these relatively under-institutionalized and clientelist bodies? Much of the answer lies in the distributive and project-oriented nature of development in the 1970s, which was much easier to administer with existing institutional resources than "conventional" development based on private sector growth and concomitant business regulation would have been. At a time when various line agencies were tasked with providing as much services and infrastructure as possible, usually through large-scale contracting, lack of coordination and small-scale discretion were less problematic than they would become in subsequent decades. Project-based development suffers under bureaucratic segmentation, but it is still possible.

Lack of data and meaningful interagency processes made planning difficult. But MoF under Prince Musa'd (who headed the plan review board) and then Mohammad Abalkhail guaranteed basic coordination of macro-spending, even if the result was not always what planners wanted.[103] Broad allocation is easier to control centrally than administrative procedures and day-to-day regulation, as the case studies will show. Moreover, the completely state-driven development in the 1970s and 1980s required private sector participation only in a subordinate role. Project management rather than regulation of markets and economic actors was thus the main requirement of growth.

Almost all of the great Saudi achievements of the boom period were projects: airports, industrial cities, roads, hospitals, schools, housing, water and electricity provision, and so on. During the 1970s, the Saudi sections in regional trade journals expanded massively, and big tenders dominated the news. Capital expenditure amounted to extremely high proportions of 60 percent and more of national budgets, dwarfing current expenditure on wages and maintenance.[104]

Waste, duplication and rent-seeking by royals and others were considerable, but the quantitative development achievements were nonetheless impressive: the electricity generated in the Kingdom, for example, multiplied 25 times between 1969 and 1984 to 44.5 billion kilowatt-hours. The production of desalinated

---

103. Holden and Johns 1982: 395.
104. *MEED*, 20 April 1982: 16.

water grew from 4.4 million gallons/day in 1970 to 355.3 million gallons/day in 1984.[105]

Services involving more than the provision of physical infrastructure, such as education and health, lagged somewhat behind due to manpower shortages and the administrative complexity of the tasks involved, but they still improved significantly.[106] Again, project-oriented, locally delimited developments yielded by far the most impressive results, including some regionally renowned high-tech hospitals.

Project implementation and follow-up involved relatively small teams of bureaucrats. These roles were often taken by relatively young, motivated Western-educated Saudis who had smartened up a great deal over their predecessors in the 1960s, quickly acquired skills of project management, and were capable of holding their expatriate contractors closely accountable for results.[107] Crucially, projects, once agreed upon, did not have a cross-cutting nature but could be managed by one organization.

Contracting of services, moreover, meant that by definition many tasks would be locally delegated, with the administration not necessarily involved in day-to-day operations, although a certain penchant to micro-management was already observable. Contractors functioned as local operational islands, from the mid-1970s on even being obliged to import all their labor from abroad by themselves, taking it back with them after the job was finished.[108] Frequently, external players dominated not only large parts of project implementation but also parts of the planning processes themselves, leading to a sharp increase in foreign consultants in the 1970s. Shadow teams of expatriates were installed in most major agencies.[109]

The State of the State in the 1970s: Regularization and Red Tape

The role of consultants meant that much of the regular mid-level bureaucracy was bypassed in policy analysis and information-gathering.[110] However, Saudi mid-level administrators, though often cut off from the planning of large projects, eventually played a pivotal role in the daily administration of most services and distribution.

---

105. Ministry of Planning, *Achievements of the Development Plans*, 2003: 136, 144.
106. Ibid.: ch. 6; Niblock 2007: 71–72.
107. Interview with former senior British diplomat, London, June 2005; Gillibrand interview.
108. *MEED*, 26 March 1976.
109. Gillibrand interview.
110. A UK ambassador present at the time called the government a "shell" for distributing jobs and money, with much of the research done by foreigners. Interview with former British ambassador.

Although otherwise described as rigid in his old days, Faisal had in 1971 managed to establish one crucial precondition for gradual administrative professionalization: he finally signed the comprehensive civil service law that had been promised many years before, the first law with a comprehensive job classification plan.[111] Systematic files on employees were introduced.[112]

The impact of the new rules on seniority, promotion, supervision, and other issues was, again, only gradual. Exam-based promotions often existed only on paper as favoritism continued.[113] Performance reports through assigned supervisors were not taken seriously. Most employees in a late 1970s survey claimed that such reports reflected kinship, friendship, or similar factors. Some supervisors were said to discriminate according to race and province of origin.[114] A good share of bureaucrats still did work that had little to do with their titles.[115]

But administrative reform was a process rather than an event, and most sources agree that a discernible regularization of procedures gradually took place. Faisal had been preventing the grossest abuses on the highest level since taking power, and during the later years of his reign, some institutions and capable senior commoners worked on recruiting, educating, and controlling the day-to-day bureaucracy.[116] Among them were the IPA and the Civil Service Bureau, which grew in influence under Abdulaziz Al-Quraishi, governor of SAMA after 1974.[117] The increasing complexity of administrative functions in the 1970s forced further regularization and differentiation of bureaucratic roles in public organizations,[118] although this did not always proceed smoothly: the civil service administration had to battle other line agencies, and new rules were frequently ignored.[119]

---

111. Morris to McCarthy, 12 February 1969, PRO/FCO 8/1165; Morris to McCarthy, Inside King Feisal, 2 April 1969, PRO/FCO 8/1169; Al-Hamoud 1991: 268.

112. *At-taqrir as-sanawi li'diwan al-muwadhdhafin al'am* 1/7/1397 to 30/6/1398, IPA (Annual report of the Civil Service Bureau, 1977/78): ch. 2.

113. Binsaleh 1982: 70ff., 118.

114. Ibid.: 136.

115. Tawati 1976: 131ff., 266; *At-taqrir as-sanawi*.

116. In his final report to the MoF, a Ford Foundation representative highlighted the Central Personnel Bureau, the IPA, and the Central O&M department as special achievements of reform in an otherwise modest record; Spiekerman, final report, FF/IPA, Box 1.

117. Some public officials had even been tried on misdemeanor charges; Armitage to McGregor, 31 July 1971, PRO/FCO 8/1733; Armitage: Saudi Arabia—The Al Saud and Internal Affairs, December 1973, PRO/FCO 8/2105.

118. *At-taqrir ath-thanawi*.

119. The noncooperation of different agencies with the civil service bureau led to the creation of an interministerial civil service board in 1977, a decision of the Al Saud-headed Supreme Committee for Administrative Reform; Binsaleh 1982: 33–34. On lower levels, supervisors often stuck to old civil service rules even after a new 1977 code had been introduced; Binsaleh 1982: 137. The Civil Service Bureau complained that some government departments continued to follow old procedures; cf. *At-taqrir ath-thanawi*: 20–21; *At-Taqrir as-sanawi*.

In the course of bureaucratic deepening, new layers of hierarchy were added, as reflected in the doubling of seniority levels in the new civil service code of 1971 and the differentiation of job descriptions in a subsequent 1977 reform.[120] Public employment expanded rapidly from year to year, with growth rates between 5 and 10 percent, sometimes above the already ambitious plan targets.[121]

Though much of the rapid recruitment into the state apparatus was still indiscriminate or based on patronage, the bureaucratic growth of the 1970s also allowed for a considerable degree of meritocracy, with ambitious and qualified young Saudis advancing rapidly to senior positions. Although there was little mobility from the very bottom—nomads and the urban poor—impressive bureaucratic careers were possible for representatives of the urban middle class.[122]

State-sponsored education in the United States created a whole new class of administrators who, if not automatically pro-Western, often developed a strong liking for Western approaches of policy and planning. Whereas 851 Saudi students were reported to be in the United States in the summer of 1970, the number had risen to some 11,000 by late 1979, many of them destined for civil service.[123] The regime was rearing its own bureaucratic class—or having it reared abroad.[124]

Formalization and education did not remove all problems, and some administrative inefficiencies indeed grew worse. While the pool of talent expanded quickly from a small base, tasks also multiplied. Decisional bottlenecks in many cases grew worse as most senior players remained reluctant to delegate substantial authority and a deep aversion to taking initiatives permeated the hastily recruited lower rungs of the bureaucracy.[125] There, as before, administrators frequently perceived inactivity as the securest and most convenient way to preserve one's guaranteed income. Senior administrators complained that nothing got done without their personal attention.[126]

---

120. Law issued by royal decree 49, 10/7/1397, printed in *Nidham al-khidma al-madaniyya wa lawa'ih at-tanfidhia bil-mamlaka al'arabiyya as-sa'udiyya* (Civil Service Bureau 1406) (Civil service law and executive bylaws, 1985); cf. also Al-Hamoud 1991: 268ff., 355ff.; Al-Ammaj 1993: 141–42, 168; Koshak 2001: 104–5.

121. During the 1978/79 fiscal year—although a year of relative fiscal restraint—the public sector grew by 23 percent, instead of the planned 12.9 percent; *MEED*, 13 April 1979: 40.

122. Of 252 middle managers sampled in a mid-1970s survey, only 3 had a nomadic background, whereas 95 were from small villages. Similar to the late 1960s survey by Al-Awaji 1971 (see note 150 in chapter 3), 101 (i.e. 40.1 percent) of their fathers had been shopkeepers, 53 officials, 74 farmers, 16 businessmen, and only 2 private sector employees, reflecting the tiny size of the corporate private sector; cf. Madi 1975: 62–63.

123. *MEED*, 17 July 1970, 14 December 1979: 51.

124. The IPA played a special role in sending Saudis abroad for degrees; *At-taqrir ath-thanawi*: 17–18.

125. It was only in 1981, toward the end of the boom, that the requirement of full cabinet approval for the creation of joint stock companies, which had held up private sector development, was abolished; cf. *MEED* Special Report, August 1981: 33.

126. Holden and Johns 1982: 459.

Trust toward subordinates generally remained low. Unpunctuality, patronage, and the shirking of responsibilities were among the main topics of a 1978 administrative workshop.[127] According to an official report, it was normal for some employees to arrive at work two or three hours late.[128] Despite their considerable formal powers, senior administrators found it hard in practice to dislodge inactive subordinates, who were seen as entitled to their jobs. In any case, it would have been difficult to find educated replacements. "Lack of administrators with power and initiative" was generally diagnosed as a main problem in the late 1970s, and business consultants urged foreign companies to avoid contract clauses that would later on require many bureaucratic signatures, as this would frequently result in great delays.[129] The Al Saud had used employment as a deliberate instrument of wealth-sharing and distribution, creating formal bureaucratic clientage; in the harsh judgment of one diplomat, the government provided "totally phony jobs" for any Saudi who wanted them.[130] At the same time, the government strongly raised salaries several times, by 70 percent at one point.[131]

Whereas only slightly more than fifteen thousand pupils attended the Kingdom's secondary schools in 1970, that number had swelled to one hundred thousand by 1980, broadening the recruitment basis for the bureaucracy.[132] But as recruitment virtually exploded, a large share of new recruits was still undereducated.[133] In 1975, more than a third of the employees in the Deputy Ministry of Social Affairs were illiterate.[134] Moreover, higher or Western education was not a guarantee of performance. A great deal of "deadwood" with Western degrees still permeates the Saudi state to this day. In some cases, ignorance gave way to deliberate obstruction or ineffectual cooperation of bureaucrats with personal agendas.[135]

While being formalized, bureaucracy generally grew heavier over the 1970s.[136] With public servants producing more and more paper, complaints over stalled

---

127. *MEED*, 7 April 1978: 37.
128. *At-Taqrir as-sanawi*.
129. *MEED*, 1 June 1979: 25.
130. Interview with British diplomat who lived in the Kingdom in the early 1980s; cf. Vasiliev 2000: 462.
131. Cf. decrees in *Majlis al-khidma al-madaniyya*.
132. SAMA annual reports.
133. The quality of local university output is subject to debate. In any case, in 1974, a mere 2,100 Saudis graduated from local universities, a total that increased to 5,100 by 1979; Ministry of Planning, *Achievements of the Development Plans*, 2002: 226.
134. That is, 417 of 1,196 male ministry employees; cf. *Al-kitab al-ihsa'i lil-wikala al-ijtima'iyya* (1395/96): 13 (Statistical yearbook of the Deputy Ministry of Social Affairs), IPA, Folder 769 (320).
135. Armitage to FCO, Saudi Arabia—The Al Saud and Internal Affairs, December 1973, PRO/FCO 8/2105
136. Lello interview.

decisions and lackadaisical effort were frequently heard, exacerbated by inconsistencies between agencies.[137] Formal complaints usually did not travel far in the administration.[138] Whereas personal intercession or "consultancy fees" might have helped a concern to be addressed in the relatively small bureaucracy of the 1960s, the state in the early 1980s was perhaps more regularized and predictable but often too complex to be easily maneuvered by any but the most powerful individuals.[139] Bureaucratic procedures had always been cumbersome. But now the bureaucracy had become harder to avoid, playing a much bigger role in business and people's daily lives, while trust in formal interaction remained low. The bureaucracy, one article complained, was "run by people with no power other than the power to obstruct."[140]

The size of the system and its regulatory ambition had grown swiftly by the 1980s, resulting in new layers of rules for business. Companies had to contend with inspections by MoI representatives, increasingly complex labor rules, municipal regulations, and, in the case of contractors, a cumbersome company classification scheme and complicated tender specifications at the Ministry of Public Works and Housing. The Saudi Arabian Standards Organization (SASO), responsible for product standards, had a staff of about twenty in 1977, with a goal of employing up to seven hundred people by 1985.[141]

More Bloated and Still Blind? Regulatory Failures

All of this made state-business interaction more complex, without necessarily adding much to the effective regulatory power of the state. Although availability of basic data had improved in several sectors thanks to a number of surveys since the late 1960s (some carried out by hired consultants), the government's information on the growing economy was by no means comprehensive. Relative to its size, the Saudi economy must have been one of the most poorly documented in the world. Due to lack of data and effective broader research, the first five-year plan was little more than a list of projects and loose targets, repeatedly stressing the need for information-gathering.[142]

---

137. Complaints came from U.S. government advisors, UK businessmen, and the Saudi private sector itself. *MEED,* 6 October 1978: 10, 3 June 1977: 10; Montagu 1987: 16, 115. The civil service bureau recognized the length of procedures as problem; cf. *At-taqrir as-sanawi.*
138. Similarly, formal complaints from within the lower levels of the administration had to be submitted to the person immediately above (instead of a higher or independent authority), meaning that they often did not make it far; cf. Binsaleh 1982: 136.
139. Gillibrand interview.
140. *MEED,* 23 July 1976: 3.
141. *MEED,* 26 March 1982: 34; Saudi Arabia Monitor, July/August 1987: 6.
142. Central Planning Organization, *Development Plan* (1390 AH).

The bureaucracy was so busy with administering the expanding services and managing its own growth that it had little in the way of resources to control in any systematic way what happened with its goods once they had been distributed. Formal interaction was largely a one-way street. Even the SIDF, a relatively efficient agency with high-quality expatriate manpower, lacked data on the projects it had commissioned. Many licenses never led to any business being set up, and conversely, much production was unlicensed—including large plants built right next to smaller, licensed ones.[143] Companies were not required to issue production figures, and the SIDF did not even know whether the phone numbers submitted by licensees were accurate.[144]

Data on specific economic sectors were similarly scarce.[145] Unobserved economic activities are by definition difficult to regulate, so bureaucratic control attempts often amounted to heavy-handed (though not always intentional) obstruction rather than coherent regulation. Information-gathering would have required much more patient bureaucratic effort and formal cooperation with counterparts in society than was possible in the heady days of the boom.

The Aramco index on contracting activity was still the main indicator of the business cycle for many years. American consultants from the U.S.-Saudi Joint Economic Commission were the "only people with a coherent stream of basic statistics."[146] Although the Kingdom had conducted a census—with disputed results—in the early 1970s, it again lacked up-to-date demographic, mortality, and health statistics by the end of the decade.[147]

State-Society Relations: Co-opting Everyone

The contrast between low regulatory control and omnipresent distribution was even more glaring during the boom years than in Faisal's calmer 1960s. While still struggling to regulate basic economic activities, the state's distributional efforts had managed to penetrate most of society in little under a decade.

One important channel of distribution was state employment. But the paternal ambitions of the regime went far beyond providing jobs. A quickly expanding service infrastructure benefited more and more Saudis: electricity and desalinated

---

143. *MEED*, 9 June 1978: 14.
144. In the mid-1980s, it was still very difficult to produce an accurate picture of the Saudi industry; cf. Montagu 1987: 38.
145. There were few numbers on the all-important construction materials business for example; cf. *MEED*, 30 June 1978.
146. Interview with British diplomat who lived in the Kingdom in the early 1980s.
147. *MEED*, 9 February 1979: 6–7.

water were provided to small consumers at strongly subsidized prices. Domestic consumption of—strongly subsidized—refined products increased from 26.1 million barrels in 1970 to 395.2 million barrels in 1984. The number of operating telephones multiplied more than tenfold between 1970 and 1984 to about nine hundred thousand.[148] Hospital beds increased from 9,000 in 1970 to 26,800 in 1984, with medical services provided free of charge. Medical and paramedic personnel grew from 6,200 to 55,500. The number of individuals covered by social insurance grew from zero in the late 1960s to 3.6 million by 1984.[149]

In 1970, almost half of the economically active Saudi population worked in the agricultural and nomadic sectors. Large parts of the population had experienced little change from oil at that point.[150] By the early 1980s, Saudi Arabia had become one of the most urbanized countries in the Middle East. Simultaneously, not least due to a rapidly growing transport infrastructure, the state's services reached deep into rural communities. Unearned income in rural populations in the early 1980s was so high that it often removed the incentive to work. State loans and grants to farmers were often frittered away, and welfare payments were common.[151]

Individual dependence on the state and its subsidized services had grown to a level surpassing most other political systems in the world, save a few small and even wealthier rentier states. While the paternal state rapidly expanded its cooptation and distribution mechanisms, most of Saudi society remained apolitical.[152] By and large, Saudis preferred individual, small-scale networks to pursue their material interests—which was, under circumstances of affluence and a large, authoritarian, but generous state, quite a rational strategy. Cooptation seemed mutually beneficial.

A Society of Brokers and Fixers

The role of these networks within and around the state grew to unprecedented proportions. The formal state was rich and omnipresent but at the same time rigid and somewhat blind. This created a setting in which huge numbers of intermediaries flourished whose main business was, in one form or another, to make the abundant state resources accessible to the manifold interested parties who otherwise could not "work" the complex system of steep and segmented hierarchies.

---

148. Ministry of Planning, *Achievements of the Development Plans,* 2003: 128, 143.
149. Ibid., 2002: 223, 231, 235.
150. Birks and Sinclair 1982: 200–201. Agriculture, though stagnating, was still by far the largest employer in 1970; cf. Nyrop 1977: 237.
151. Birks and Sinclair: 204.
152. The 1970s were probably the post–World War II decade with the least organized oppositional activity.

In the absence of other, meso-level channels of collective interest aggregation and negotiation, individual brokerage flourished.

Intermediaries between state and society came into being both on formal and informal levels, and the two overlapped frequently. Personal mediation of state resources had of course been a pervasive trait of the soft Saudi state ever since its inception, as documented in chapters 2 and 3. But the scope and complexity of intermediation, the many new horizontal layers and vertical segments of brokerage made Saudi society, without much hyperbole, a society of intermediaries.

HIGH-LEVEL BROKERAGE

Personal brokerage flourished most prominently on the level of large business deals but also through the regular bureaucracy. Legendary figures like Khashoggi, Pharaon, and Adham used their access to senior princes to put together huge deals for international contractors who did not have a direct channel to the overtaxed and cautious royals. Khashoggi and his ilk were brokers for external actors by virtue of their client status with one or more senior patrons. Their access was often due to the traditional position of their family in court—frequently a historical accident—and sometimes thanks to kinship links to senior royals.[153]

The scope for intermediation was huge. Even for professionally managed large projects in the industrial cities or for Aramco, specific Saudis were needed to make an introduction.[154] In the face of extreme centralization and limited technocratic expertise among the Al Saud, there was a great need for trusted lieutenants to flesh out the details of big deals. Below the level of the big, well-known brokers, princes availed themselves of *wakeels* (representatives), "the often illiterate but millionaire messengers who ironed out minor problems for their masters." Selling their access to decision-makers, wakeels "could be paid by several parties at once."[155]

In the course of the 1970s, more and more intermediaries would be princes themselves, as Fahd restrained his relatives' business ambitions less strictly than Faisal had. Royals would peddle their access to more senior princes or, more frequently, peddle their political leverage toward the distributive bureaucracy. This they would sometimes, but not always, do through holding an official post—usually below the ministerial level—in which they tended to punch above their official weight.[156]

---

153. Lippman 2004: 164; Morris to McCarthy, Inside King Feisal, 2 April 1969, PRO/FCO 8/1169; Well-connected Saudi Expressed Views Regarding Recent Defense Contracts, Reform, and Royal Family Succession, 9 June 1968, NARA RG 59, 150, 64–65, box 2470, folder POL 2 SAUD (1/1/68).
154. *MEED* Special Report, July 1980: 31.
155. Field 1984: 115; Fallon 1976: 40.
156. Interview with British diplomat who lived in the Kingdom in the early 1980s.

In a rigid bureaucratic system, the name of the right prince, deployed at the right time, could help business greatly. Princes' functions as intermediaries varied widely, from taking commissions to holding trade agencies for foreign companies, from functioning as joint venture partners to engaging in oil barter deals to speculating in land.[157] Land was usually granted by the king and then resold to a restricted set of commoner clients, who would parcel it out further for higher prices, creating a veritable cascade of patron-client relationships. Individual princely largesse and intercession—for medical trips abroad, education for one's children, and so on—gained in importance for common Saudis.[158] As the royal family grew and its branches became more differentiated, the Al Saud also saw increasing internal clientelism, with smaller princes patronizing commoners while remaining under the patronage of senior players.[159]

BROKERAGE ALL OVER SOCIETY

Brokerage of state resources and services did not stop with the more salient senior "wheeler-dealers" and royal players but pervaded all of society, concentrically organized around the authoritarian but soft distributive state. The bureaucracy became gatekeeper to so many benefits in daily life that engaging with it became impossible to avoid for most Saudis, and certainly for most businesspeople.[160] Informal brokerage played a large role in making the state's resources accessible to broader classes of society, and a leadership bent on sharing the wealth with an underorganized and still undereducated society often saw fit to tolerate it.

As the great complexity of the state had become less easily negotiable for single individuals or companies, bureaucrats and figures with administrative connections would act as intermediaries. Often profiting from access to senior-level figures, they would make jobs, benefits, aid, state credit, licenses, smaller contracts, or scarce information available to friends, relatives, and friends of relatives or, sometimes, to the highest bidders.

The leeway for discriminatory access was large. At least as important as open manipulation of rules, administrative regulations, accumulated in a haphazard way, were often so ambiguous in their details that they left much scope for interpretation or, at a minimum, for stalling their enactment.[161] The fragmentation of

---

157. Montagu interview; Field 1984: 99, 114; Holden and Johns 1982: 411ff., 509; Montagu 1987: 24; Montagu 1994: 10.
158. Interview with former deputy minister II, Riyadh, December 2005.
159. Smaller princes were often dependent on the distributional goodwill of the king and other seniors; cf. Field 1984: 101.
160. Al-Rasheed 2002: 126.
161. Numerous decrees and memos were carried over from previous years and left in an ambiguous state; cf. *MEED,* 26 August 1983: 39–40; *MEED,* 21 November 1980.

the administration on lower levels added to such imponderables. The "harsh, rather arbitrary, business climate" in the Kingdom usually required expatriate businesses to hire experienced (and expensive) lawyers who would know about legal and administrative practice.[162] But even this was no guarantee against arbitrary decisions—or an absence of decisions.

In such an environment, the privilege an administrative position could confer to its holder and his contacts is obvious. Unsurprisingly, many Saudi civil servants took to moonlighting in business.[163] In a recurring pattern, bureaucrats would suggest that businessmen seeking contracts with their agencies do business with them.[164] Moonlighters usually engaged in the lines of business overseen by their ministries.[165]

Reflecting the segmented nature and limited reach of different parts of the administration, brokerage links on lower levels would only have local utility. Smaller businessmen might have good contacts within MoF but not the Ministries of Justice or Agriculture.[166] Similarly, the patronage of a smaller prince might help a commoner client and supporter to obtain services from a specific agency or act as go-between in a government department, again offering locally specific access to the system. Switching to a new line of business often required developing links to new administrative agents and intermediaries.[167]

Absent other forms of organization, networks would often be based on kinship or, more loosely, on common regional origins. As a result of geographical and educational mobility and growing bureaucratic complexity, simple links of acquaintance and friendship also played a more important role. As opposed to the early 1970s, family connections often were no longer sufficient to access the state, so businessmen increasingly used networks based on past business relations.[168] Even in these cases, brokerage remained personalized and access exclusive and stratified. The increasing complexity of the administration, however, also led to a commercialization of brokerage in some areas, as intermediaries made themselves more generally available for payment—either informally, through retainers or one-off payments, or formally as "agents," consultants, joint venture partners, or contractors.

The growth of such intermediation cannot be reduced to nepotism and corruption. Much of the commercialized brokerage has instead come about as a

---

162. *MEED*, 15 June 1979: 15; Montagu 1987: 126.
163. To comply with (changing) formal incompatibility rules, administrators often opened businesses in the names of relatives, including sons who were still in school.
164. Binsaleh 1982: 142.
165. Though not everyone did so, it was a fairly frequent phenomenon; Lello interview.
166. Awaji 1971: 177–78; for similar linkages in Jordan, see Cunningham and Sarayra 1993: 11.
167. Lello interview.
168. Ibid.

result of deliberate, formal state policies intended to create intermediaries. Although informal relations were undoubtedly important, the Saudi system was not just the tangled web of favoritism, whether personalized or commercialized, that some observers have described,[169] but functioned in more regular, institutionalized ways than many other oil states.

A number of government decisions created new, formal intermediary functions for Saudi nationals and businessmen in particular. The aim, achieved with remarkable success, was to improve locals' chances to share in the boom and to create a larger indigenous business class. Saudi Arabia's economy and public services had been heavily dependent on foreign imports and foreign businesses in most respects ever since first oil exports had started, offering ample opportunity for local intermediation. As early as 1962, Faisal as crown prince issued a decree stipulating that all commercial agencies were to be held by nationals, requiring every foreign producer to sell goods and services through Saudi partners.[170] With the import boom of the 1970s, the stratum of agents expanded rapidly, now with many Najdis in its ranks.

In the course of boom-induced diversification, the government created further opportunities for intermediation. Industrial projects would only qualify for some of the government's generous support measures if there was a share of Saudi ownership. In some professions, like the all-important engineering consultancy, Saudis were increasingly privileged. In public tenders, increasing provisions were made favoring the involvement of local businesses and local procurement. A Saudi "sponsor" was needed for foreign firms to operate on Saudi ground. Finally, and most directly, from 1977 on, Saudis—and only Saudis—as agents were by law entitled to receive commissions of up to 5 percent on contracts their principals won. This 5 percent rule was "apparently being widely interpreted as a 5 per cent floor," that is, as the very minimum cut Saudis would accept.[171]

Even if the official role was as equal business partner, the primary function of Saudi businesses usually was to act as brokers in the local environment, as they usually lacked production and management expertise of their own. In industrial joint ventures, the better-connected Saudis arranged access to land, subsidies, and loans, whereas the plant would be run by the foreigners.[172] Whatever their function, the expansion of industry led to a great increase in the number of Saudis active in the sector. By the end of 1971, 188 industries had been licensed under

---

169. Islami and Kavoussi 1984.
170. *MEED*, 27 July 1962: 344.
171. *MEED*, 5 January 1979: 3; Business International 1985: 67.
172. In 1976, good Saudi joint venture partners were still "extremely rare"; many took 50 percent of the profits and did little more than a bit of paperwork; *MEED*, 23 July 1976: 3. The situation in the mid-1980s was not much different; cf. Montagu 1985: 65.

the National Industries Ordinance.[173] In 1984, 1,196 factories were recorded to be in operation.[174]

The state's attitude toward domestic business was paternal, as it was government that decided which sectors to move into, whom to support, and how. Technocrats dominated policy-making, literally deciding which new industrial sectors to create. The private sector had few objections to such interventionism, however, as it benefited greatly from the state-provided opportunities and its privileged linkages to the bureaucracy. At the time, bureaucratic meddling tended to be beneficial.

Contracting is the one sector in which the state's hand was even more visible. Through its humongous public works programs, the government created a large new class of local contractors. Construction accounted for more than 10 percent of GDP during most of the 1970s.[175] It was an area in which the ambiguity of the many new formal regulations combined most clearly with informal connections to create a new class of privilege. In contracting, rushed national privilege rules were perhaps the most confusing and created the greatest need for interpretation. Regulations on local procurement, preference, or exclusive access for local bidders or joint ventures changed from year to year. They quickly became more restrictive for foreigners[176] and hence created expansive spaces for Saudi partners—who would, thanks to the ambiguity of the rules and the various interpretations of different agencies, usually be well-connected players who knew how to work around both.[177] As Saudi contractors could often not fulfill their official tasks due to lack of experience and capacity, many of them subcontracted implementation to foreigners again, limiting their role to informal brokerage (and in the process inflating project costs).[178] In many cases, Saudi businessmen would subcontract

---

173. Knauerhase 1977: 141.
174. Ministry of Planning, *Achievements of the Development Plans,* 2003: 150.
175. *MEED* Special Report, August 1982: 6.
176. New branches, agencies, or representative offices of foreign companies could by 1982 only be opened with the permission of the Ministry of Commerce. Practically all foreign economic activity by then was licensed through the government; cf. *MEED* special construction & contracting issue, March 1982: 81ff.
177. The 1977 tender rules gave general priority to Saudi bidders but were unclear on crucial points, for instance the duties of foreign actors and local agents and the scope of tax privileges. The deputy minister of commerce, conceding its ambiguous language, explained that the decree should not be seen "too legalistically." *MEED,* 5 January 1979: 3. Similarly, the implications of the 1983 rule that 30 percent of business within projects should be subcontracted to local businesses was "a constant source of uncertainty." Montagu 1985: 50; for further examples of ambiguous rules, see *MEED* Special Saudi Arabia reports 1980: 33, 1981: 50; *MEED,* 16 September 1983: 55. Saudi businesses would often avoid the 30 percent rule by posing as fully Saudi although they effectively had foreign partners through management contracts; cf. Business International 1985. On the use of local partners to work around rules, see Moon 1986.
178. Lello interview.

to Saudi subcontractors, who would in turn subcontract. Four to five layers of intermediation were not uncommon.[179]

The scope for favoritism and brokerage more broadly was immense, as the government was keen to create indigenous business players, had unprecedented resources at its disposal, and was difficult for most foreigners to deal with. In many cases registered Saudi companies ran no actual business operations at all. Some would use their administrative connections to import large quantities of foreign labor as their official employers or "sponsors" and then resell it to other companies in actual need of labor.[180] A massive unregulated market for imported labor was created, serving state-generated demand for cheap services. The "sponsorship" system institutionalized the dependence of foreign laborers and the role of employers as brokers of their fate, whether actively trading with them or not. Similarly, Saudis would use their status as nationals to register smaller trade and service companies that were actually run and financed by foreigners, usually Arab expatriates. This "cover-up" business, thriving on Saudis' formal privilege as national investors, was a popular pursuit of even lower-class Saudis. In the context of labor and "cover-up" brokerage, formal and informal privileges and brokerage blended. The occasional Ministry of Interior threats against this new class of intermediaries never led to a broad crackdown.[181]

A Society Atomized through Clientelism

Social change in Saudi Arabia accelerated in the 1970s, and the Kingdom witnessed modernization along many of the standard criteria: higher incomes, vastly improved education, rapid urbanization, a larger middle class. The most dramatic change was the greatly enhanced role of the state in people's lives. Nevertheless, this process did not lead to much greater integration of society or the formation of broader, coherent social groups along nontraditional lines. With few exceptions, there were no signs of political mobilization and little demand for formal politics.[182] After the coup and party formation attempts of the 1960s, Saudi Arabia saw few coherent attempts to organize class interests.

The importation of a foreign labor class contributed to this. Even more important, perhaps, the clientelist orientation of various groups in society toward

---

179. Interview with former deputy minister, Riyadh, May 2004.
180. Montagu interview.
181. Interview with senior advisor to the Saudi government, Riyadh, November 2005. The MoI issued a decree against cover-up practices in 1979 (decree 3S/1177, 29/1/1400).
182. The occupation of the *haram* in Mecca in 1979 involved a group that was millenarian rather than political; cf. Hegghammer and Lacroix 2006. The harshly suppressed Shiite unrest in the Eastern Province in the same year was a regionally circumscribed phenomenon.

the state and of individuals toward specific agencies and patrons, meant that there was little space for horizontal class formation. Brokers almost by definition had a local, specific function, usually tied into exclusive hierarchical networks. If there were any collective interests to defend, they were related to protecting specific state-related privileges. Moreover, below the highest levels of society, structures of collective interest aggregation were highly deficient.

This does not mean that individual Saudis had no access to the state: formal distribution and material entitlements through jobs were extensive. More important, individual patronage could make the otherwise rigid bureaucracy locally responsive, and patronage and brokerage, as we have seen, could have considerable distributive effects for those lacking a formal job or formal entitlements. Wasta, the practical utility of personal contacts, was still an effective means for a large number of Saudis to pursue their individual material interests.

This means, however, that access was usually local, organized on the microlevel, and not channeled through broader institutions—certainly not ones independent of regime and bureaucracy. As Saudi Arabia had not seen a "national moment" of collective social struggle, the social intermediation of the state's resources happened through small-scale networks, reproducing social fragmentation.[183] Although mobility and education allowed for new social ties on a smaller scale, the split between tribal and nontribal identities persisted in many regards,[184] as did regional divisions, sometimes reproduced through patronage and networks within state institutions.[185] Within the administration, individuals were tied up in the hierarchies and networks of their agencies, and no broader bureaucratic ethos emerged. The bureaucracy never coalesced as a collective lobby group. While social interests penetrated the state on the micro-level on a much larger scale than in the 1960s, society remained just as fragmented on the meso-level.

Oil wealth would allow for extreme luxuries of social segmentation. Complete public segregation of sexes, the creation of all-female educational and administrative institutions, and the prohibition of female driving were expensive policies that would arguably not have been possible in a functioning "production state" that needs to rely on the economic dynamism of national business and an integrated labor market. Similarly, only a state with access to rapidly growing external resources could afford to build a large educational and juridical bureaucracy for a religious establishment that was dead set against social modernization, secular bureaucracy, and fundamental principles of modern economics. In the

---

183. Chaudhry 1997: 192.
184. Al-Rasheed 2002: 128.
185. Such relations would indeed often be boosted and revived through the growing state and its resources; interview with former deputy minister II, Riyadh, December 2005.

short run, the political returns on these policies outweighed the costs; there was enough money to expand the more functional parts of the state at the same time. In the long run, after budgets had stopped expanding, both policies came to haunt the Kingdom's leadership.

While the state kept growing and the Saudi economy was based on distribution, conflicts between different segments of the state were limited. The same was true of society. Political orientations and material expectations were centered around the state and, in the last analysis, the Al Saud family, creating segmented and hierarchical state-society relations in a hub-and-spoke structure, with little communication between the segments. Within each segment, brokerage and clientelism ruled supreme. Unlike traditional, dyadic patron-client relations, Saudi rentier clientelism could be multilayered, cascaded and intertwined with formal institutions; players could move in different circles. Mobility had increased, the state had become complex, its resources vast, its channels of access variegated. All the same, most access was specific, usually individual, and tied up in hierarchical structures, and there was little incentive or space to organize outside of the state and its resources, which were centrally disbursed and locally redistributed.

Up to the mid-1980s, however unpalatable to some Western observers, the growing system provided great mobility to its clients and servants. It was the last time that great numbers of "Young Turks" in their thirties and forties in high ministerial positions were tasked with building new programs and institutions. It was also the last time that big new names entered the business arena, be it Najdi contractors and traders or the Lebanese-Saudi businessman Rafiq Hariri, head of mega-contractor Saudi Oger, who rapidly emerged from obscurity and was "made" by Fahd over a few dizzying years in the late 1970s.[186] Even then, however, most of the big business names profiting from the boom went back to earlier decades, indicating the advantage of early privilege in the rapidly growing Saudi system.[187]

The growth of the Saudi state accelerated exponentially in the 1970s, making ever larger sections of society its clients through employment and the provision of state services. Despite a slow process of bureaucratic maturation, the increasingly complex state apparatus remained overcentralized, segmented, and opaque, disposing of limited regulatory powers. This resulted in the growth of informal networks of patrons and brokers on various levels who would make state resources

---

186. Saudi Oger's contracts were usually won in direct negotiations with the MoF, not on open tenders, as most other contracts were. This is not to say that Oger was not qualified—Hariri had reportedly first impressed Fahd by finishing a hotel contract in Taif in 1977 before his deadline. Without the king's patronage, however, even the most capable contractor would not have reached Oger's privileged position; on Oger, see *MEED*, 5 March 1982: 64; *MEED*, 11 February 1977: 11, 27.

187. Business International 1985: 192–93.

available to merchants, foreign partners, or the population at large. At the same time, the growing resources available in the 1970s also allowed for the creation of several efficient bureaucratic islands, further accentuating the internal heterogeneity of the Saudi state. Since the main target of state-building at the time was distribution of resources, state fragmentation did not undermine its functionality.

Interlude: Two Stagnant Decades

Would the reengineering and differentiation of the system and the generation of new actors, institutions, and distributional policies ever stop? Indeed it would, and quite suddenly at that. The post-1983 oil bust brought institutional change in Saudi Arabia to a halt and, with it, state-provided mobility. Stasis set in on many levels.

The bust, to be sure, was spectacular and created great losses to numerous players. It did not, however, reengineer the Saudi system but rather froze it in place. Since the early 1980s, the Saudi bureaucracy has matured in many respects, but the changes have been gradual, involving largely the same actors and relationships between them.

Bust and Budgetary Retrenchment

The story of the oil crash has been told in detail elsewhere.[188] Hardly a state was hit as severely as Saudi Arabia, which had taken on the role of "swing producer" to balance global oil markets. Official oil income decreased from 329 billion Saudi riyals in 1981 to 42 billion in 1986. Total official state income decreased from 368 to 76 billion SR. Until the early 2000s, state revenues never fully recovered, hovering between 130 and 200 billion SR in the 1990s (see fig. 4.1).

The government slashed state expenditure by more than half between 1981 and 1986 but nonetheless incurred deficits every year from 1983 to 2000. In this atmosphere of gloom, hard decisions had to be made for the first time in decades, and the endurance of entitlements was put to a test.

The Limits to Austerity: Sticky Entitlements

Entitlements proved remarkably enduring. As in other Gulf states, project budgets saw the heaviest cuts, by far. This was discretionary spending that did not require

---

188. Yergin 1991; Abir 1988, 1993; Robinson 1988.

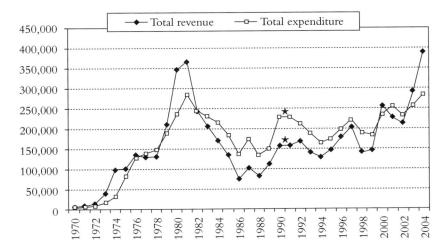

Fig. 4.1 Saudi fiscal history (million SR)
Source: SAMA annual reports, nominal figures.
★ Data for the Gulf War years (1990/91) is only available in aggregate; for convenience of exposition, the figures have been divided into equal halves for both years.

the rescinding of broad distributional programs or state employment. Over time, government capital spending, which had reached about half of the national budget, declined even below the replacement requirements of Saudi infrastructure. Budget cuts curtailed the leeway for commoner ministers to initiate and see through projects. Austerity stalled Saudi development programs, as many projects were put on hold or cancelled. It took until the 1990s until the private sector would be able to pick up some of the slack in national capital formation.

The collapse in project spending affected the relatively large class of Saudi contractors, already hit through increasingly common payment delays, which had reduced trust between the bureaucracy and business.[189] But distribution to contractors had never been institutionalized, and many of them had been little more than paper firms involved in bogus subcontracting. By focusing on capital expenditure, the impact of spending cuts on the system at large was limited.

Unlike what Chaudhry has argued, it appears that the Al Saud under Fahd went out of their way to sustain the service and income entitlements of regular Saudis, that is, of society at large as it had been enveloped by the distributive state. While it is true that various utility tariffs were raised on several occasions, the government retained the tiered system for domestic water and electricity, which

---

189. One third of contractors went bankrupt or were in dire financial straits; cf. Vasiliev 2000: 454.

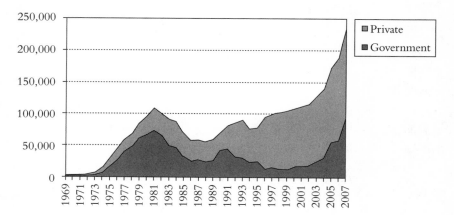

Fig. 4.2 Gross fixed capital formation (million SR)
Source: SAMA.

effectively cross-subsidized smaller consumers.[190] In 1984 the government imposed a surcharge of 0.05 SR/kwh for electricity consumption exceeding 2000 kwh per month, putting the onus on larger customers.[191] Smaller consumers continued to receive electricity and desalinated water at prices far below production costs. In most other countries, by contrast, marginal utility prices are either constant or decrease with larger consumption.

On several occasions, subsidy cuts were reversed, forcing large losses on the public utilities. In late 1985, Fahd decided to cut electricity prices, barely a year after increasing them. After increases in the water tariff scale in February 1985, prices were cut again by 50 percent in March 1986.[192]

Subsidies to business were hit more strongly than more general subsidies. For instance, SIDF disbursements declined more rapidly than those to the Real Estate Development Fund, which supported individual housing, and social security subsidies declined much less rapidly than agricultural subsidies.[193] Costly agricultural price guarantees, Chaudhry's main example of elite privilege in bust times, were cut back strongly in the early 1990s, resulting in strongly reduced production.[194]

190. Montagu 1985: 3; Malik 1999: 104.
191. Talal Bakr, Electricity Crisis: Diagnose Before Prescribing Medication, *Arab News,* 2 October 2006.
192. *MEED* Special Report, Saudi Arabia, May 1986: 2.
193. Both facts are demonstrated by Chaudhry's (1997: 150, 152) own tables; cf. SAMA annual report 2003: 371.
194. Krimly 1999: 265. As a result, wheat and barley production dropped steeply; cf. Ministry of Planning, *Fifth five year development plan* (ch. 9), *Sixth five year development plan* (ch. 8); Central Department of Statistics, *Statistical yearbook,* various issues.

When technocrats appealed to Fahd for across-the-board subsidy decreases and higher service fees, he repeatedly declined.[195] Students in higher education continued to receive a 300 US$ monthly stipend.[196] The regime also intervened if universities' admission policies got too strict, to ensure a place for the swelling ranks of young Saudis.[197] At the same time, stipends for overseas studies, crucial for building the Saudi managerial elite in the 1970s, were reduced to a small trickle.[198]

Reflecting the regime's concern for broad-based distribution, reductions in education and health expenditure throughout the 1980s were far below those seen in other fields. Both increased their relative shares in shrinking budgets, while development-oriented expenditure collapsed (see fig. 4.3). Only royally controlled defense was comparably successful at defending its turf. Budgetary retrenchment did not lead to the fiscal rationalization that political economists might expect. Instead, spending cuts hit economically vital but politically less sensitive targets, while the broader distributional decisions of the 1970s proved sticky. Rating a prince as stingy, as *bakhil,* is a strong condemnation in Saudi Arabia, one that Fahd was not willing to risk.

International consultants decried (to little effect) the Saudi combination of minimal capital spending with continuing wide-scale distribution. They also fretted over the welfare function of large-scale state employment, but the share of public wages in government expenditure rose continuously over the 1980s and 1990s (see fig. 4.4).[199] Despite gross overemployment and shrinking overall budgets, the number of public employees grew every single year. Paternal distribution of wealth seemed to trump the necessities of public sector reform.[200]

A significant part of the expansion is explained with the decrease in unoccupied positions after 1984. However, the number of positions called for in the budget also increased every single year until today, albeit at a slower pace after

---

195. Craig interview; interview with Western diplomat III, Riyadh, April 2003. In December 1988, at the height of austerity, Fahd reportedly instructed all ministries not to cut benefits to citizens; cf. *Saudi Arabia Monitor,* January/February 1989: 8.

196. Al-Rasheed 2002: 151.

197. Abir 1993: 101.

198. According to Institute of International Education data, the number of Saudi students in the United States declined from about 10,000 to 4,000 in 1989/90; available at http://opendoors.iienetwork.org. An increasing share of the latter were privately funded.

199. The numbers in the graph only include official employees registered by the Ministry of Civil Service. Unofficial estimates of total state employment are much higher; one 2007 estimate is of 1.8 million people; Saudi-British Bank, Giving a Boost, SABB Notes, 7 February 2008: 5.

200. In 1988, a trade journal commented that while the Saudi fiscal shortage was nothing special, the scope of hangers-on dependent on the state's distribution was indeed extraordinary. *Saudi Arabia Monitor,* June 1988: 10.

122 Chapter 4

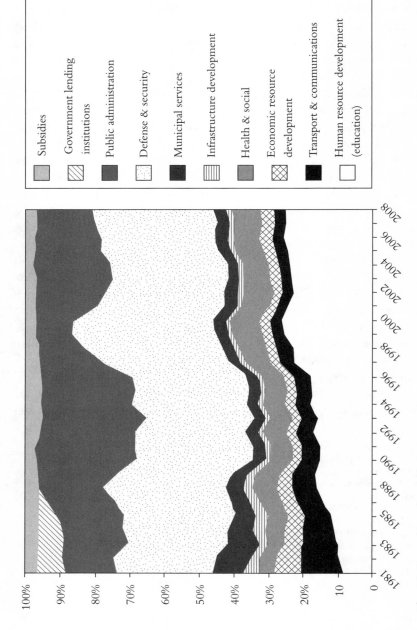

Fig. 4.3 Composition of annual budgets
Source: SAMA.

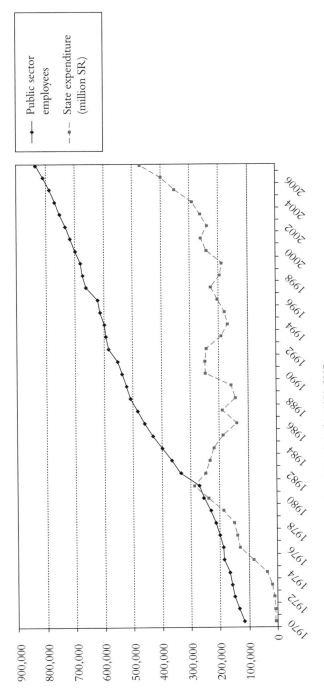

Fig. 4.4 State employment vs. state expenditure in Saudi Arabia, 1970–2007
Source: SAMA, Ministry of Civil Service.

1985.[201] Expenses on wages, salaries, and allowances grew from 46,788 million SR in 1986 to 70,071 million in 1990/91.[202] As proportion of the budget, salaries went up from 13 to 49 percent between 1981 and 1990.[203] By 1999, the share had reached 60 percent, representing a full 23 percent of GDP, a very high proportion by international comparison.[204]

For new entrants, however, the recruiting state was a completely different entity than it had been in the 1970s. First of all, wages were frozen from 1983 on.[205] In 1985, Council of Ministers Resolution 101 limited overtime pay and working weekends and cancelled numerous allowances, which for some bureaucrats amounted to as much as 60 percent of annual income.[206] While not decreasing core wages, this greatly reduced opportunities to manipulate civil service rules in order to maximize effective income.

More important in the long run, most employees were forced to stay on the same level of seniority for many years.[207] The times of rapid bureaucratic mobility for young, educated Saudis had largely passed, which had a demoralizing effect on latecomers.[208] From the mid-1980s on, a large share of bureaucrats got stuck on the low- to mid-level of the civil service.[209] Rapid mobility was followed by great immobility.

The regime strove to maintain broad distribution, but by the same token, its real value eroded. The sharp decrease in mobility and the decline in real wages within the bureaucracy after the mid-1980s budget crunch further contributed to the low motivation of low-level administrators. Incentives for performance were lacking.[210] Inactivity and misdemeanor were seldom punished,[211] and although the senior bureaucracy gradually expanded and acquired experience, the bureaucratic atmosphere and the level of obstructionism, if anything, became worse.[212]

201. Al-Ammaj 1993: 149–50.
202. IMF 1991: 19.
203. Krimly 1993: 233.
204. Diwan and Girgis 2002: 12.
205. *MEED* Special Report, July 1983: 4. The last major increase had come in 1981, with public wages reaching from 1,500 to 20,750 SR; cf. Civil Service Bureau, *Al-khidma al-madaniyya fi mi'at 'am*: 387 (100 years of civil service), IPA.
206. Al-Hamoud 1991: 189; *MEED* Special Report, May 1986: 2.
207. Al-Hamoud 1991: 357. Promotions were strongly reduced: while there were about 17,000 in 1981, only about 5,500 bureaucrats were promoted by the mid-1980s, while the total number of employees kept rising; cf. *Al-khidma al-madaniyya bil-arqam*: 23.
208. Gillibrand interview.
209. Interview with former deputy minister, November 2005.
210. *Saudi Arabia Monitor,* June 1987.
211. According to one survey, the time Saudi bureaucrats spent on actual work in the late 1980s was about 49 percent of official working time; cf. Al-Hamoud 1991: 86.
212. Interviews with businessmen, consultants, and diplomats living in Saudi Arabia in the 1980s.

The bureaucracy of the late 1990s at first glance looked much like the same body two decades earlier. After the push toward more formalized structures in the 1970s, little further regularization happened. Personal networks still played an important role within the inherited formal framework and the functioning of departments still depended a great deal on who their directors were.[213] Nonetheless, the bureaucracy had seen some gradual, "subterranean" change in its composition and role.[214] On the one hand, it included more experienced figures with good economic expertise.[215] On the other hand, additional small rules and procedures in state-business interaction had accumulated, and red tape had further grown.[216] A heavy-handed bureaucracy with increasingly less to distribute imposed growing burdens on business. In a related development, "old hands" generally agree that petty corruption in the lower levels of administration worsened.[217] Gone were the boom-time days when bribery was limited to large high-level deals. And personalized networks of nepotism in some organizations—often not corrupt in a strict sense—gave way to a more anonymous market corruption in which specific bureaucratic services simply became available for anyone willing to pay the price.[218] Corruption in the Kingdom was still much less pervasive than in poorer Arab states. Still, in Saudi terms, it seems that the bureaucracy had considerably changed its function from a driver of state-led development to a retarding factor.

Royal Expansion into a Stagnating System

Chaudhry's claim that the state was unable to curtail elite privileges in the late 1980s is true in two cases: royals and clergy.[219] Members of the royal family largely managed to defend their special status. While many contractors and industrialists suffered from a collapse in project expenditure, princes often expanded their business turf.

Within the government, the five-year plans were frequently ignored, with preference given to select projects favored by royal fiefs such as MoDA and the larger governorates.[220] Select ventures under the patronage of senior princes—such as an overpriced, gigantic underground oil storage facility, the badly planned Petromin

---

213. Interview with former minister, Riyadh, December 2005.
214. Thelen 2003: 233.
215. Interviews with former senior British diplomat living in the Kingdom in the 1980s.
216. Interview with former Saudi minister, December 2005; Gillibrand interview.
217. Montagu, former British diplomat interviews; interview with Michael Field, Gulf business consultant, London, July 2005; *Arab News,* 4 September 2005.
218. Lello interview.
219. Chaudhry 1997: 296.
220. Koshak 2002: 105; *Saudi Arabia Monitor,* June 1988: 29.

refinery in Rabigh, and a number of large military procurement deals—all went ahead while other capital spending collapsed.[221] Many of the pet projects turned out to be white elephants, but they did produce sizeable kickbacks. From 1984 on, procurement based on oil barter agreements allowed for increasing involvement of princes as senior middlemen.[222] More broadly, princes leveraged their senior position to move into commerce and act as agents for foreign companies, often using commoner frontmen.[223] Able to undermine the business of less powerful commoners, mid-level princes gradually enhanced their presence, bringing along an expansion of middlemen and brokers.[224] Frustration among the business community was high, but it proved incapable of opposing princely predation in an organized way.

The Ulama's Tenacity

Religious organizations were the other type of fiefdoms that successfully defended their distributional stakes throughout the 1980s. After the seizure of the Grand Mosque in 1979, Fahd ceded further ground to religious forces. Many observers have remarked that Faisal, with his better religious credentials, was much more capable of standing up to religious forces than Fahd, known as a bon vivant. The latter gave the morality police enhanced power and resources to monitor public behavior and allowed the ulama to control school curricula.

All budget items related to the religious establishment remained constant or expanded from 1981 to 1985, while the overall budget shrank by a third. From the mid-1980s on, religiously dominated Imam Muhammad bin Saud University received more money for projects than older and larger King Saud University, which had previously received much more funding. While the budget item for religious endowments was 1 billion SR in 1979, it remained close to 2 billion SR all the way through the 1980s.[225]

In 1984, the 130 million US$ King Fahd Holy Koran Printing Complex in Medina went into operation to spread the word of Islam internationally. The government allocated up to 200,000 barrels of oil per day to an Islamic missionary fund, which would reportedly generate more than 27 billion US$ over the years. By 1986, an estimated 16,000 of the Kingdom's 100,000 university students were pursuing religious studies.[226]

---

221. The gigantic Yamamah armament project was further expanded in 1988, one of the years with the worst fiscal results.
222. *Saudi Arabia Monitor,* May 1987: 9.
223. Montagu interview; *Saudi Arabia Monitor,* March 1990.
224. Interview with James Greenberg, chairman, Devcorp, Riyadh, December 2005.
225. Statistical yearbooks, various years.
226. Lacey 2008: ch. 10.

The other area in which capital spending was not frozen by the late 1980s, apart from defense spending, was the expansion of the holy sites in Mecca and Medina.[227] In 1985, when the regime slashed all other budgets, Fahd initiated an eight-year, multi-billion-dollar project to expand the mosques, catering to his core clientele in the religious establishment.[228]

While such expenditure was fiscally problematic, the part of the religious state that imposed the heaviest burden on post-slump economic development more broadly speaking was probably the court system, built up under religious control in the 1970s. Suddenly the economic costs of a fragmented state with incompatible components became salient. The post-1983 bust led to numerous business disputes. The difficulties of enforcing debts in sharia courts severely hamstrung their resolution, as borrowers cited the prohibition on usury—read interest—to stall on repayment. International contracts were seldom recognized, and neither was, in most cases, international arbitration.[229]

Western bankers encountered the Saudi reading of the sharia as a pervasive problem. A 1981 fatwa had declared mortgages Islamically questionable, and the religiously controlled notary public offices refused to register them in the names of banks, leaving Saudis dependent on increasingly scarce government financing for houses.[230] The Ministry of Commerce had to find ways around the notary's refusal to deal with banking transactions, dating by the Gregorian calendar, and other accoutrements of modern life.

Even well-functioning institutions such as SAMA found it difficult to regulate their sectors insofar as they had to rely on other parts of the state to back them up. A special SAMA board to settle the numerous banking disputes of the 1980s had no executive power and could force neither courts nor the Ministry of Interior to enforce its rulings.[231] SAMA also ran into jurisdictional conflicts with the (secular) Ministry of Commerce.[232]

Those islands of efficiency that were less reliant on the rest of the state, however, usually continued to perform their tasks well throughout the seventeen-year slump. The stasis of the highly centralized state apparatus (and the paternal guarantee of the various stakes in it) also served to protect institutions that in other countries might have been politicized and preyed upon by rent-seekers in a time of crisis.[233]

---

227. *MEED* Special Report, 31 March 1989.
228. Coll 2008: 280.
229. Al-Fahad 2000.
230. Ibid.; *MEED* Special Report, May 1986: 20.
231. *MEED* Special Report, March 1989: 14; *Saudi Arabia Monitor*, September 1989: 7.
232. *Saudi Arabia Monitor*, July/August 1987.
233. Cf. Geddes 1990.

While the national economy contracted, fledgling heavy industry champion SABIC was rated as a "runaway success."[234] It turned profitable as soon as its plants went on stream in the mid-1980s and has not seen a single year of losses since. Relying on both state loans and international project finance, it systematically expanded its operations through international joint ventures, replenishing its management through internal recruitment and drawing on the Kingdom's best engineering graduates, often from the well-regarded University (formerly College) of Petroleum and Minerals. SABIC's role in its joint projects, in management as well as equity terms, grew progressively until it started launching ventures with exclusively local ownership.

The most significant addition to the Kingdom's pockets of efficiency in the 1980s was Aramco, whose ownership the Saudi state took over in 1980 and which it formally converted into a local company in 1988. Deciding to keep Aramco and Petromin separate, King Fahd essentially grandfathered the existing corporate and social culture of Aramco, which continued to be run by the same rules as it was under American ownership—maintaining a strong role for U.S. consultants but with Saudi "Aramcons" constituting the senior management and much of the field staff. Aramco is the one national oil company in OPEC whose managerial structures were not revamped in the course of nationalization, and it is today ranked as one of the world's best NOCs.[235]

Saudi Aramco, as it is now called, is directly accountable to the king through the Supreme Petroleum Council and protected from all but the most senior predators. With more administrative independence than any other secular fief, it constitutes a little state of its own: women can drive in its Western-style compounds, genders mix in the workplace, and discipline and recruitment standards are tight. It is in many regards the most Western enclave anywhere in the Middle East—located in the region's outwardly most conservative country. Saudi Aramco's separation from the rest of the state at such a late stage in state-building, however, was only possible because an autonomous, foreign-dominated team had managed it as a separate enclave since before the inception of modern Saudi Arabia.

Policy Stagnation after the Bust

While some senior players expanded their fiefs and most others defended their territory successfully, the overall institutional setup remained frozen. The state after the mid-1980s saw little change and few new policy initiatives. Even Petromin, which at the boom's end in 1983 had effectively been abandoned as

234. *MEED* Special Report, March 1989: 4.
235. Hertog 2008.

national oil company in favor of Aramco, took more than two decades to be fully dissolved and folded into other agencies.[236] An attempt to sort out the system of regions through a new ordinance in 1993 changed little about the persisting confusion of jurisdictions among governors, the MoI, MoMRA, and other line agencies.[237] Reigning in princely fiefs on the governorate would quite probably have been too difficult. The anticipated and administratively sensible split of the Riyadh governorate and the merger of smaller northern provinces did not come to pass.[238]

Policy areas such as education, health, and infrastructure remained largely fragmented. Ministers remained in office for decades, and the average age of the cabinet increased gradually. Changes in civil service regulations after 1981 were limited to spelling out details.[239] The system of judicial tribunals saw some revisions but without reversing the basic fragmentation of quasi-judicial bureaucratic bodies and sharia courts.[240] Although privatization of some public services such as PTT was discussed, no action followed.[241] Existing institutions and entitlements proved difficult to change.

Government agencies incurred mounting debts with each other, and in tandem with growing mandatory current expenditure, the government increasingly lost maneuverability. Debts to contractors were often only serviced if there was any likelihood of their gaining access to the king to complain personally.[242] A 1988 attempt to reintroduce an old income tax on foreign residents was revoked within forty-eight hours, after the business community, fearing higher employment costs, protested vigorously.[243]

In the words of a former deputy minister, the government in the 1980s and 1990s mostly busied itself with "putting out fires" from crisis to crisis, while little substantial change was initiated.[244] The 1990/91 Gulf War was another deep shock to the Saudi system, but it initiated little economic policy or administrative change. The state incurred further massive debt, while subsequent years saw an indecisive back-and-forth of subsidy decreases and increases, starting with a

---

236. *Arab News,* 25 October 2005.
237. Interview with former deputy minister II.
238. Business International 1985: 91.
239. Al-Ammaj 1993: 154–55.
240. Business International 1985: 18.
241. *Saudi Arabia Monitor,* October 1988: 12.
242. Montagu 1994: 49.
243. *Saudi Arabia Monitor,* February 1988: 11.
244. Interview with former deputy minister, Riyadh, November 2005. The 1985–2000 phase saw few new initiatives but was instead dominated by measures to cope with "changing conditions." Niblock 2007: 30.

TABLE 4.1.
Major laws governing economic life in Saudi Arabia up to 2000

Law	Year
Income tax law	1950
Commercial agencies decree	1962
Regulation of companies decree	1965
Banking control law	1966
Labor law	1969
Law of the judiciary	1975
Government procurement law	1977
Foreign capital investment law	1979
Board of grievances law	1982
Arbitration law	1983
Copyright law	1989

Note: Based on a survey of online collections of Saudi commercial law.

decrease of utility prices to counter domestic political unrest after the war.[245] The temporary spike of promotions in the civil service after the war probably occurred for the same reason.[246] As a reaction to the Islamist mobilization following the Iraq War, the regime also started to bankroll avowedly apolitical Islamist groups.[247] Insofar as service prices later increased again, they usually targeted higher consumer brackets.[248] As before, the regime tried to throw money at problems, while shunning any challenge to vested interests. The only (timid) institutional changes happened on the political level, in the form of a revamped law on regional governance, a "basic law" that essentially enshrined the Kingdom's autocratic monarchical character,[249] and a reconstituted consultative council staffed mostly with technocrats that had been promised repeatedly since the 1960s.

An overview of the major economic laws up to the year 2000 shows that most of them stemmed from the state expansion phase of the 1960s to the early 1980s (see table 4.1). Unlike what comparative political economists might expect, the

---

245. Directly after the war, the regime attempted to boost its legitimacy through a variety of fee reductions. Commitments to broad-based consumer subsidies remained substantial; Luciani 1995: 142–43. Growing debt led to fee and price increases in 1995 (Reuters, 29 December 1995) and 1998. The 1995 increases were again focused mostly on larger consumers; cf. *MEED*, 10 March 95: 25–26. One important austerity measure in 1998 was a secret decree in May that, again, halted expenditure on new projects (Reuters, 14 October 1998).

246. Civil Service Bureau: *Al-khidma al-madaniyya bil-arqam* (1416/1417): 22 (The civil service in numbers), IPA, folder 630 (301,1).

247. Lacroix 2008.

248. Niblock 2007: 111.

249. Al-Fahad 2005.

deep economic crisis of the 1980s did not lead to any significant economic or regulatory reforms.

Only one major piece of legislation, the 1989 copyright law, emerged after the mid-1980s oil price collapse. Other commercial and judicial regulations from the period were concerned with secondary administrative issues.[250] The 1990s saw not one major piece of legislation. Only after 2000, under new leadership, the government would again issue significant economic laws.

After the mid-1980s, the expansion of the Saudi state stopped, and with it change and mobility within the system. The institutions that had grown since the early 1950s congealed into a permanent setup—through bureaucratic formalization and resource constraints, but also through the sheer size and scope of the state, which had created numerous vested interests, brokers, and distributional expectations on various levels. As society had grown into the state through various forms of clientelism, now it tied it down. The great oil-based autonomy of the leadership to reengineer state structures was greatly diminished. Yet it was precisely after the oil price crash when a strong hand would have been needed to clear the institutional thicket of the Saudi state. With much less money to go around, business was expected to pick up the slack of economic development. But a costly and deeply fragmented state, once so successful at rolling out its services, now proved incapable of regulating business in a coherent fashion. The case studies will elucidate in greater detail to what extent intra-state politics has hampered policy-making, regulation, and state-business interaction until today.

Conclusion to the History Chapters

The three preceding chapters have told the story of the Saudi bureaucracy since the early 1950s, covering three different but closely linked periods. Chapter 2 analyzed the first steps of building a national bureaucracy, a highly fluid, fragmented, and personalized process mostly happening above Saudi society at large. Chapter 3 showed how the bureaucracy and its cleavages stabilized under Faisal's centralized regime in the 1960s, and how it served the increasing cooptation of social elites. The fourth chapter examined how the segmented Saudi state accelerated its expansion during the boom decade, with cooptation becoming a nationwide process and brokerage of state resources a national pastime despite further bureaucratic formalization. It concluded with an account of how the end of state growth led to sudden and pervasive institutional stagnation after the mid-1980s.

---

250. These include the Law of Commercial Books of 1989 and the Law of Commercial Register of 1995.

The Path Dependency of the Institutional Landscape

Saudi Arabia has seen the creation of a vast state apparatus telescoped into a few decades of more or less continuous growth. In a path-dependent fashion, early decisions have had large-scale consequences further down the road, and all institutions strongly reflect the conditions of their creation. These have varied greatly, resulting in a spectrum ranging from opaque bureaucratic sinecures to professional pockets of efficiency.

Contingent individual decisions and elite conflicts have left indelible imprints on various parts of the state, reflecting the great leeway that political elites enjoyed for several decades in using the unprecedented oil income. The state was predisposed to grow in a rapid and uncoordinated fashion, as revenue at any point between 1950 and 1980 greatly exceeded inherited administrative capabilities.

In the absence of a civil service tradition, rapidly expanding new bureaucratic bodies tended to become insular, being accountable only to their royal patrons in a strictly hierarchical system—a process that could also make for local insulation from nepotism. As the Saudi system changed mostly through the accretion of new organizations, institutions in the Saudi hub-and-spoke system proved remarkably resilient once established from the top, especially after bureaucratization set in under Faisal and generous employment policies cemented bureaucratic job entitlements. Already during the boom decade, institutional innovation seemed increasingly limited to the periphery of the state apparatus.

Numerous early contingencies have been etched into the Saudi institutional landscape and often magnified many times through state growth: civil aviation's attachment to the Ministry of Defense, the Vice Ministry of Finance, the stepchild status of the Ministry of Economy, the ubiquitous role of the MoI, the subordinate role of the Royal Guard as compared to the National Guard, the fusion of the posts of king and prime minister, the absence of royals from technocratic ministerial posts, and the privileged positions of most large Saudi business families.

More than any other event, the 1962 cabinet deal shaped the face of Saudi politics for the half-century to come, but it also appears somewhat contingent in retrospect. For example, there were reported offers by Saud to make Fahd his prime minister, which the latter rejected.[251] Had Fahd accepted, Saudi Arabia and its institutions of governance would look quite different today. Earlier in the 1950s, there had been rumors that Mohammad bin Abdulaziz, an irascible older brother of Fahd's, could be made minister of finance, which again would have changed the balance of power—and quite likely, institutions—tremendously.[252]

---

251. Yizraeli 1997: 89.
252. Enclosure No. 1 to Despatch No. 111 (from Basra), 4 May 1950, USRSA, Vol. 2.

Contrary to what has been argued, the modern Saudi state was in many regards shaped decisively and irrevocably in its first two decades and often as much by personality as by structural forces.

Clientelism and Cooptation

The clientelist nature of bureaucratic employment—whether personalized or based on anonymous large-scale recruitment—contributed to the resilience of organizations, once set on a path of growth. Swallowing chunks of society through employment and comprehensive service provision transformed the state itself. Abundance of resources allowed for the luxury of setting up different state segments to accommodate a variety of social groups, including tribes, ulama, and—more diffusely—merchants and other urban classes. A non-oil regime might have attempted to subdue the ulama and their capacity to set social rules in order to create a coherently regulated market. The Saudi state could swallow them and hence fence them in—but at the cost of ceding bureaucratic and regulatory territory to them.

Michael Herb has analyzed bureaucratic cooptation as a mechanism to overcome conflicts between factions within ruling families.[253] Arguably, the logic of cooptation—with all its tradeoffs—also applies to the leadership's relationship to other elite groups. Through reorientation of social networks towards the state, its organizations, and its hierarchies, society was kept fragmented and dependent. The distributional state could grow while avoiding the functional, political, and social integration of society. No state is completely homogeneous, but in few systems are the different components so independent of each other, so radically different socio-culturally, and based on incompatible basic assumptions about the state's very nature. In the Saudi process of modern state-building, no elite group was ever vanquished—instead they were all sucked into the orbit of an increasingly rich state. The sheer heterogeneity of groups that the Al Saud simultaneously co-opted—or even created—is now reflected in the state itself. And it exerts a price.

Shifts in State Autonomy and "Layering" of Institutions

Oil gave to the political elite great leeway in earlier phases of state creation, as rentier theory makes us expect. Yet, the bureaucratic behemoth that emerged at the end of the boom in the 1980s proved highly resilient to change. Once growth stopped, state fragmentation imposed increasing costs, but the vast administrative machineries of the Saudi state also became difficult to reengineer or curtail.

253. Herb 1999.

This is true of fiefdoms on the senior level. The major political shifts and crises after 1979 did not result in any of the most senior princes losing his job—the contrast to the flux of the 1950s and early 1960s is stark. Their massive fiefdoms, increasingly bureaucratized on lower levels, had become practically coterminous with their respective names and gave them large, autonomous resources of power and patronage.

Changing the bureaucracy below the royal level also became much more difficult. Some individuals could be sanctioned or sidelined, but the diffuse mass of civil servants were difficult to control. Despite blatant overemployment and two fiscal crises, the Al Saud never dared to curtail the payroll, and in a context of general immobility, performance incentives were strongly diminished. Regulatory capacity remained accordingly limited.

Top-down change in civil service structures had largely been limited to boom times. The last of a regular succession of civil service reforms happened in 1977. All of the seminars and discussions about public sector performance and motivation never led to broader reform in the 1980s or 1990s. Clientelist employment meant low control over day-to-day behavior in the bureaucracy but large obligations for the regime. Creation and discretionary staffing of new agencies had ceased to be an easy option for reengineering the system. Flexibility rapidly decreased, while individual interests and small-scale social networks of brokerage continued to undermine the coherence even of individual agencies and their policies. Whereas local redistribution of jurisdictions among the commoner bureaucracy happened on a few occasions, the internal structures of agencies were almost impossible to change. Unflappable bureaucratic passivity appeared to be a greater development burden than corruption or open rent-seeking. After the state had conquered all larger structures in society, small-scale social interests conquered it in turn.

Cumulative bureaucratic growth especially after 1962 means that the timing of institutional decisions became crucial. Once an agency had established its presence, it usually managed to defend its turf in distributive and regulatory terms. The new islands of bureaucratic efficiency emerging in the 1970s hence had only delimited reach and influence. Growth through accretion automatically positioned administrative players in a historical hierarchy. SAMA, set up in the 1950s, and Aramco, a "state within the state" inherited from the Americans, are the only pockets of efficiency that have fundamental and essential functions within the state, as they go back further in history.

Continuity and Change over Half a Century

Historical-institutional analysis helps us to distinguish the ephemeral from the enduring and to explain why specific types of problems recur. Just as important, diachronic analysis allows us to understand how the uses and effects of enduring social institutions can change according to circumstances.

Hardly a society has seen as many deep changes as Saudi Arabia has since the early 1950s. And yet, many of today's patterns of behavior and organizational interaction, especially on a micro- and meso-level, seem to have had clear equivalents half a century ago: a highly opaque royal court as the center of decision-making; steep hierarchies; delegation of administrative functions on a strictly local and case-by-case basis; bureaucratic fragmentation and organizations as fiefdoms; a clientelist mode of exchange resulting in an impassive bureaucracy; islands of efficiency operating under direct royal patronage; uneven, often individualized access to state resources; and a generally top-down, state- and elite-driven mode of politics.

Then as now, ordered delegation of administrative functions on a larger scale appears difficult in the Saudi system. With great material resources, but limited administrative capacity, controlling the bureaucracy has remained a challenge. Similarly, despite all bureaucratic growth, regulatory penetration of society remains weak, increasing only slowly in an environment of low bureaucratic motivation, one-way distributional linkages to society, and persistently low trust toward formal administration. The islands of efficiency in the Saudi state are theoretically informative exceptions to these patterns, but their existence has little effect on the bulk of the state apparatus.

We should not expect any of these features of administrative behavior and state-society interaction to be easily removed by any reform—and if they are, then this might indicate a far more fundamental change than any "snapshot" analysis of politics can reveal.

But while many mechanisms within the administration remain the same, or at least very similar, we have to recognize that they function in a completely different macro-context, making for different conditions of policy-making and reform. While the personalization of forces at the very core of the state has persisted, much of the rest of the state has been thoroughly bureaucratized. The enormous growth of the system has robbed it of much of its suppleness. Institutional flexibility and state-induced social mobility are largely a thing of the past. It is true that significant professionalization has taken place—during the 1960s as well as during the bonanza of the 1970s—resulting in much greater bureaucratic expertise on the upper levels of the system and in its efficient islands. At the same time, the setting has become much more rigid and complex; even if red tape has not become worse on the level of individual offices, its reach and complexity throughout the large and fragmented system has grown tremendously.

While vertical exchange still predominates, clientelism has in many cases been formalized through large-scale service provision or employment. Here lies the most significant difference: today's state reaches deep into Saudi society. This implies a great presence of state agencies in people's lives, many more levels of brokerage (in parts formalized), and a large weight of distributional obligations which tie the state down as soon as it stops growing. Informal penetration of the bureaucracy by small-scale social interests might be less blatant today than under Saud and more hemmed in by formal rules, but it is a much larger-scale phenomenon with a dynamic of its own—one that is much harder to control.

PART II

POLICY-MAKING IN SEGMENTED CLIENTELISM

Crown Prince Abdallah and the Second Fiscal Crisis of the 1990s

By the mid-1990s, Saudi Arabia had endured a dozen years of fiscal retrenchment without policy reform. "During King Fahd's latter days the country at best stood still or moved backwards in both business methods and nepotism."[1] After the brief Gulf War spike, the oil price mostly lingered between 10 and 20 US$ per barrel. The government axed development expenditure further, state services gradually deteriorated, and contractors suffered from severely delayed payments. After a brief recovery in early 1997, oil prices plunged below 10 US$ in late 1998 as the Asian economic crisis caught OPEC off guard.

Public debt, although held domestically by banks and state social security organizations, veered towards 100 percent of GDP. The 1980s fiscal crisis had led to stalemate. The 1990s crisis, however, put the Kingdom so clearly on the path to fiscal disaster that the leadership seemed to realize, finally, that policy change was inevitable—change that would have to go beyond the subsidy cuts and fiscal tinkering of the previous fifteen years.[2]

It was a new ruler who would lead the charge. In December 1998, Crown Prince Abdallah publicly told the Saudis that the fat years were over, that they had to learn to live with less, and that it was time for a new lifestyle that did not rely

---

1. Montagu 2001: 6.
2. Malik 1999: 104, 107.

entirely on the state.³ After Fahd suffered from several strokes in the middle of the decade, the crown prince had emerged as de facto ruler of the Kingdom. He enjoyed a reputation of relative probity, piousness, and concern for regular Saudis. He was, however, constrained by the entrenched power of his senior brothers, as Fahd's twilight reign exacerbated the fragmentation of the regime.

Abdallah tried to rein in the royal family, but it is not clear how successful he was in curtailing princely privilege. He reportedly cut the stipends of newly born royals, but at the same time, many adult princes worked to expand their business activities while Fahd was incapacitated—especially if they had access to the frail king's signature.⁴ Princely patronage through real estate redistribution and privileged access to bureaucratic decisions, often replacing the direct distributional privileges of previous decades, remained widespread.

Abdallah nonetheless emerged as a leader on economic policy issues. The orthodox rentier state model, according to which states could forego economic policy, did not apply to Saudi Arabia any longer.⁵ With the population still growing at more than 2 percent a year, there was an urgent need to create employment and growth independent of state circulation of rents. The hopelessly overstaffed civil service had lost its capability of soaking up the growing numbers of new labor market entrants. Real income per capita had declined strongly since the early 1980s.

Private investment by domestic, but also foreign businesses would have to be stimulated, and international cooperation and technology transfer would have to be solicited. In the absence of international debt, international actors had no direct leverage over Saudi economic policy. Nonetheless, the diversification successes of more liberal neighbors such as the United Arab Emirates demonstrated the need for making Saudi markets more attractive, and the government was subject to a constant barrage of Washington Consensus–style policy advice by Article IV International Monetary Fund missions and World Bank advisors.⁶

Abdallah had only received a basic education, and his understanding of the nuances of economic policy is reported to be limited.⁷ Nonetheless, unlike the more urbane and Westernized Fahd, he started to tackle substantive legal reforms. By 2001, numerous regulations were under review, including the company law, sponsorship and agency rules, capital and other service market regulations as well as taxation, labor, mining, and competition laws.⁸ Further widely-touted reform

---

3. *Financial Times,* 31 December 1998
4. Montagu 200: 7; interview with former deputy minister, November 2005.
5. Luciani 1990.
6. Interview with senior Western lawyer, United Arab Emirates, February 2009; Niblock 2007: 96.
7. He reportedly learned to read and write only as an adult.
8. Montagu 2001: 35.

initiatives included the Saudi Gas Initiative, which was supposed to open the upstream energy sector to foreign investors for the first time since Aramco had been nationalized, and a comprehensive privatization strategy.[9]

Most regulatory reforms were packaged into three ambitious policy projects. Hoping to join the WTO, the government attempted to adjust its domestic regulation to WTO requirements. Hoping to attract foreign investment, it embarked on a comprehensive drive to reform foreign investment regulations. Hoping to increase the share of Saudis in national employment, it stepped up its efforts to "Saudiize" the labor market through various new regulatory mechanisms.

These three policy projects are the subject of the next three chapters. Regulatory reform put Saudi institutions to a different kind of test than the distributional expansion of the oil boom had. The new policies served as a test of how Saudi Arabia's historically grown bureaucratic clientelism would react to—and shape—top-down reform initiatives aimed at fundamentally revamping the role of the administration and giving business a much wider role in development.

Old and New Players in the Case Studies

In addition to senior princes such as Crown Prince and, since 2005, King Abdallah and Minister of Interior Prince Naif, the next three chapters will involve a whole raft of institutional players, including the Ministries of Finance, Commerce, Industry, Labor, and Justice as well as various subordinate units of the Ministry of Interior. By the late 1990s, these were usually headed by senior commoners who had been in the Saudi administration for decades.[10]

The bureaucracy was more experienced but also heavier and more predatory. Its regulations had fallen behind those of Saudi Arabia's liberalizing neighbors like Bahrain or the UAE. Whether the issue was antiquated rules and inflexibility or opacity and corruption, by the late 1990s it had become clear that regulatory reform was urgent.

The meso-structure of the Saudi state of 1999 was very similar to that of twenty years earlier. It had merely accrued two additional bodies of relevance to economic policy-making: the Supreme Economic Council (SEC) and the Majlis Ash-Shura, or consultative council. The SEC, created in August 1999, is essentially

---

9. The Saudi Gas Initiative makes for an interesting case study of bureaucratic politics outside of the regular domestic political economy, in which turf-conscious Aramco technocrats managed to significantly undermine the government's strategy of opening; cf. Robins 2004.

10. A comprehensive reshuffle of commoner ministers in 1995 had brought only a partial rejuvenation.

a mini-cabinet of the ministries concerned with economic matters, chaired by Abdallah. It also has an advisory body that includes a number of specialists from the public sector and a few select businessmen. Although it has little staff of its own, it has repeatedly served as a sounding board for new policy projects that could then be debated more thoroughly and with greater business participation than in past decades.

The Majlis Ash-Shura was created by King Fahd in 1993 as a reaction to internal unrest after the 1990/91 Gulf War. Staffed predominantly by educated technocrats, it has gradually carved out an important role in the policy debate through its specialized committees and hearings. Although it cannot overrule the cabinet, its voice is heard and taken into regard. Albeit the Majlis is, properly considered, part of the state rather than a true representative body, it has further contributed to the formalization of policy-making.

Both SEC and Majlis Ash-Shura have provided new access points for lobbying by the private sector, be it through the Chambers of Commerce and Industry or through the invitation of individual business representatives to debates. Of all the established players involved in policy-making in Saudi Arabia, the private sector has changed the most thoroughly since the early 1980s. Whereas the state has in many parts been atrophying, business has been slowly but gradually expanding, and while the state was continuously forced to distribute its income, businesses were able to accumulate capital.

Despite inevitable failed experiments, many business figures have managed to reinvest that capital productively in advanced contracting, service, and industrial ventures. With state contracts making for a significantly smaller share of business income, businesses have increasingly catered to the considerable private demand in Saudi Arabia and the Gulf Cooperation Council. The share of the private sector in Saudi GDP has increased from 36 percent in 1979 to 43 percent in 1999. More significantly, whereas in 1979 private gross fixed capital formation, that is investment in physical infrastructure, was less than half that of the government, by 1999 the private sector's capital outlays were about seven times as high as those of the government (see also figure 4.3 in chapter 4).[11] After government funding had dried up in the mid-1980s, business became increasingly active on the fledgling capital market.[12] Private manufacturing has matured considerably, and Saudi businesses have become the leading regional investors in the Middle East.[13] In most sectors, Saudi business has moved far beyond its initial role of mere brokerage and developed substantive managerial capacities.

11. Calculated in real terms acc. to SAMA annual reports, statistical appendixes.
12. Molyneux and Iqbal 2005: 117.
13. Luciani 2005.

At their core, the economic reforms of the last ten years have been about giving the private sector a role in development and job creation that the state cannot fulfill anymore. Business interests were thus bound to play a larger role in process of regulatory change than they did in our historical account of the Saudi state. Nonetheless, as we will see, the attempts to reengineer the business environment have exposed deep-seated continuities in the Saudi institutional structure, both within the state and the private sector. These continuities have meant that neither the leadership's will to reform nor the private sector's higher managerial capacities have smoothly translated into comprehensive reform. Like the history chapters, the case study chapters will first look at the meso-level politics surrounding specific reforms and then proceed to analyze implementation issues and bureaucratic practices on the micro-level.

Chapter 5

The Foreign Investment Act
Lost between Fiefdoms

"My Kingdom will survive only in so far as it remains a country of difficult access, where the foreigner will have no other aim with his task fulfilled, but to get out."
—Ibn Saud

Although built as an inclusive system, the political economy of the Kingdom is hard on latecomers. This is as true for regulations and institutions as it is for individuals. The foreign investment law of April 2000 and SAGIA, its adjunct administration, have been perhaps the most prominent victims of this rule in recent years.

Together with the Saudi gas initiative, the new foreign investment act was the bellwether project of the Saudi reform drive starting in 1999/2000. In the way it was promoted and negotiated as well as in the way in which its implementation faltered, it reflects both resilient, systemic continuities and possibilities for localized change in the Saudi political economy.

This chapter is broadly divided into two parts, the first part looking at how the foreign investment act (henceforth FIA) and related legislation were debated and decided upon, and the second part investigating how the act and the general drive for foreign direct investment (FDI) reverberated and, in many cases, came undone within the Saudi administration and its fragmented networks. We will see that despite a surprisingly large presence of the private sector in policy negotiations, much of the reform was still state-driven. More specifically, it was an elite project promoted by a particular faction of the leadership and its immediate supporters.

Epigraph: *Le Monde,* 27 March 1975, cited in Holden and Johns 1982: 384.

Against the background of a segmented and soft state, this is the main reason that the reform had only limited impact on the actual investment climate and FDI flows until at least 2005. On the meso-level of interagency politics, unaccountable bureaucratic organizations issued vetoes to dilute formal details of the reform after it had been agreed on in principle. On the micro-level of bureaucratic behavior, much implementation was thwarted, often inadvertently, by decision avoidance, ignorance, locally divergent interpretations of rules, and the persistence of preexisting brokerage networks—sustained by exclusively vertical structures of bureaucratic communication and accountability within organizations. While some of the commoner ministries adjusted to some SAGIA demands after a long time lag, deeper fiefdoms like the Ministries of Interior and Justice have thus far proved impenetrable. The FDI successes of the post-2003 boom have happened within the segmented clientelist framework, with either SAGIA acting as an ad hoc, case-by-case broker or larger projects implemented in partnership with old islands of efficiency such as SABIC or Aramco past the rest of government.

Background: Rules and Practices until 2000

Like many regulations, the pre-2000 Saudi foreign investment law was written in the dynamic boom years.[1] Although relatively liberal at the time, by the late 1990s the 1979 law had fallen behind the formal standards of openness in several of the neighboring Arab countries, including formerly socialist Egypt.[2] As it stood, it allowed foreigners to hold only minority stakes in local companies and forced them to acquire a Saudi "sponsor" to be able to operate in the Kingdom.[3] There was no possibility for foreigners to own real estate, and the corporate tax rate on foreigners was a high 45 percent. The law tasked the Foreign Capital Investment Committee in the Ministry of Industry and Electricity (MoIE) with giving licenses to applicant investors.[4]

The licensing process was often arbitrary and took between six and twenty-four months.[5] Based on a positive list of possible investment activities, it required

---

1. Royal decree M/4, 2/2/1399 (1 January 1979), printed in Jeddah Chamber of Commerce and Industry, *Nidham wa ijra'at al-istithmar* (1408), IPA, folder 300 (332, 6) (Law and procedures of investment).
2. Egypt permitted foreign landownership and abolished the requirement for a local partner in the late 1980s; Springborg 1993: 156. It has used the "negative list" system since 1991; cf. Handoussa 2002.
3. *Arab News*, 21 February 2000; the sponsor is not necessarily the actual business partner.
4. Montagu 1987: 155.
5. World Bank 2002: 57. Things moved faster when the government was a partner; cf. Niblock 2007: 188.

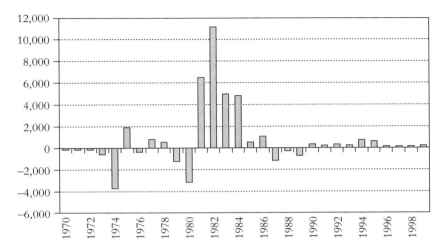

Fig. 5.1 Saudi FDI inflows until 1999 (million US$)
Source: UNCTAD.
Note: The 1974 and 1980 capital outflows are explained by Aramco's nationalization and do not constitute capital flight.

the individual screening of companies by the Saudi administration, which evaluated the desirability of any proposed project.[6] The committee's decision was final.[7] The bureaucracy enjoyed much discretion toward foreign supplicants.[8] Serious investors usually had to hire local help to construct the right kind of proposal. The main legal incentive provided for investors were tax holidays: ten years for investments in industry and agriculture, five years for other sectors.

By the late 1990s, a general sense prevailed among liberal technocrats and international advisors that the rules, like many others, were antiquated and no longer competitive and that potential investment was being lost.[9] Most FDI stock building in the Kingdom had indeed occurred between 1981 and 1984, the years of state-driven industrial expansion.

The limited post-1984 inflow mostly went into joint ventures with publicly owned SABIC, a successful heavy industry player in many ways insulated from the rest of state and business.[10]

6. Ministry of Industry and Electricity, Foreign Capital Investment Guide (n.d.), IPA.
7. World Bank 2002: 48.
8. Al-Saleh 1994: 45.
9. Interview with former deputy minister, May 2004.
10. Reuters, 7 November 1999.

Interests Involved, Public and Private

The MoIE under Hashim Yamani formally produced the first draft for the FIA. Internally, the World Bank and the International Monetary Fund (IMF) were strongly involved through studies submitted to the Ministry of Finance (MoF).[11] Although it reportedly enjoyed broad backing by the MoIE, the MoF, and the Ministry of Commerce (MoC), Crown Prince Abdallah played the supreme political role in driving the project, mainly through his role as head of the Supreme Economic Council (SEC).[12] Other senior royals were largely silent on the planned reforms in public.

The promotion of the FIA and related bylaws became even more of a "personality-driven initiative"[13] when Prince Abdallah bin Faisal bin Turki became the head of newly created SAGIA under the patronage of the crown prince.[14] Until his retirement in 2004, Abdallah bin Faisal, as the most consistently liberal and outspoken of all senior administrators in the Kingdom, has played a prominent role in the West as promoter of a modernizing Saudi Arabia.[15]

After its emergence from the Saudi administration, the FIA draft was, by Saudi standards, extensively debated, both in the Majlis Ash-Shura and the private sector.[16] The latter was a novelty: Business elites were not a driver of the FIA, but the crown prince and Abdallah bin Faisal allowed them to be part of the debate.[17] However, business never reached a unified standpoint on the overall project.

Although industrialists seemed less skeptical of the law than traders, SAGIA functionaries and other observers saw the private sector as not favorably disposed to the project.[18] Similar to the WTO case, which will be discussed in chapter 7, in many sectors there was a vague sense of something being taken away from Saudi business, of "losing a share of the cake."[19] The organized private sector also put a

---

11. Interview with former deputy minister, May 2004.
12. Interview with senior advisor to the Saudi government, June 2004.
13. Or a "royal initiative," as another interviewee had it. Interview with chief economist of Saudi bank II, Riyadh, May 2003; interview with Mario Sander, GTZ head of office, Riyadh, May 2003.
14. Although a nephew of King Fahd through his mother's side, the prince was generally perceived to be firmly within what is called the "reform camp" of (then) Crown Prince Abdallah; interview with Western diplomat II, Riyadh, May 2004.
15. Popular with Western business, the German embassy rated him a fortunate choice; cf. dispatch no. 91 to Foreign Office/Bonn, 12 April 2000.
16. Interview with Dr. Bandar Al-Hajjar, member of the Majlis Ash-Shura, Riyadh, May 2003.
17. The "private sector" here refers to larger actors, often organized in the chambers, who can be expected to have a certain interest in and be consulted about policy issues. This leaves out large parts of the Saudi small and medium enterprises.
18. "The private sector in general hates the law." Interview with SAGIA functionary I, Riyadh, May 2004; Sander interview.
19. Interview with SAGIA functionary II, Riyadh, April 2004.

more positive spin on its FDI-related views in official papers and declarations than its members did in private.[20]

One reason for unease was that significant parts of the business community were not well informed about the technical issues at stake.[21] The Chambers of Commerce and Industry (CCIs) themselves, although vocal and increasingly policy-oriented, still had only limited "brainpower" at their disposal.[22] Although they regularly organize symposia and discussion groups and contract studies on individual issues, they lacked the resources for policy research. While business should in principle have had a considerable interest in FDI policies, in practice a clear policy position seldom emerged.

Government and Majlis representatives complained that there was not enough private sector interest in regulatory issues, as it "does not spend time on it," and that the CCIs' leaders were impassive, possibly preferring personal shortcuts to government figures to organized lobbying if they had an important interest to voice.[23] The private sector was neither a driver of the FIA in general, nor did private sector groups put up strongly organized, open resistance at any point.

Although debated for quite a while, in comparison to the big issue of Saudization, the FIA was a relatively minor topic for large parts of Saudi business and was often perceived as a less immediate concern.[24] As we will see in the other cases, this fit the traditional mold of policy-making in the Saudi distributive state. Although individual businesses are much more advanced today than during the 1970s, the collective policy stances of the private sector often remain reactive, focused on conserving existing clientelist privileges over proactive policy-making.

Slow, Slow, Quick, Quick, Slow: The FIA's Origins

The government first announced its intention to revise FDI regulations in 1997.[25] This ushered in a dance of trial balloons and postponements in which the crown

---

20. Business actors reportedly prefer to be seen as modern and open: "The private sector sometimes lies about FDI in its papers." Interview with SAGIA functionary I, Riyadh, May 2004 (this person once worked with a CCI).
21. Hajjar interview, April 2004.
22. This was said specifically with regard to the FIA; Hajjar interview, April 2004. "The chambers are busy with themselves." Interview with former deputy minister, Riyadh, March 2004.
23. Interview with former deputy minister, May 2004; Hajjar interview, April 2004. In contrast to mid-sized businesspeople, top business families usually get to meet royals; cf. Malik 1999: 269.
24. Ibid. "The private sector did not care so much." Interview with former minister, Riyadh, March 2004.
25. *Arab News,* 21 February 2000. The crown prince reportedly had commissioned a review of laws from chambers and the Majlis Ash-Shura; cf. *Financial Post,* 31 March 1999.

prince repeatedly had to prod the sluggish bureaucracy to move forward. The inert policy-making stood in marked contrast to the forthright technocratic development thrust of the 1970s.

In November 1998 the crown prince stated that the Kingdom had no choice but to diversify, widen investment channels, and encourage international FDI. At the same time, officials promised new tax break incentives and measures to reduce red tape by the end of the year.[26]

These steps did not materialize; instead there were further announcements that new investment rules, as well as new company and agency laws, were in the making.[27] Other issues under debate were reformed sponsorship rules, real estate ownership and majority company ownership for foreigners, tax cuts, and a special body to review laws and propose incentives.[28] The CCIs were included in the deliberations.[29] During 1999 both Minister of Interior Prince Naif and King Fahd publicly mentioned plans to encourage investment and to move quickly on this matter.[30] According to Minister Yamani, King Fahd "had been made aware of obstacles that could discourage foreign investors," a motley list that included lack of transparency, labor regulations, weak dispute resolution mechanisms, lack of incentives, and lack of a body to follow up investment regulations.[31] A consensus, it seems, had emerged among the senior leadership. During the summer of 1999, there were more announcements of new regulations, to arrive before the end of the year.[32]

New Year's Eve came and went, and no such law appeared. On February 21, 2000, however, the press reported that Abdallah had issued a directive demanding that the study of the draft regulations be expedited.[33] At the same time the SEC forwarded the draft to the Majlis Ash-Shura.[34] Barely two weeks later, the Majlis had approved the law and forwarded it to the cabinet.[35] On April 10, the cabinet passed the FIA.[36] After more than two years of procrastination, this acceleration was remarkable, and the end result in parts differed considerably from the various

26. Reuters, 7 November 1998; *MEED*, 27 November 1998: 22.
27. *MEED*, 26 March 1999, 17 September 1999: 9.
28. *MEED*, 1 November 1999, 26 November 1999: 4, 3 November 1999: 8.
29. Reuters, 26 September 1999.
30. Even cautious Prince Naif publicly favored a measured acceleration of policy-making: "I have to say we are moving slowly.... We want to move faster...not offhandedly, but objectively." Reuters, 7 November 1999.
31. Ibid.
32. Reuters, 16 June 1999.
33. *Saudi Gazette,* 21 February 2000.
34. *Arab News,* 21 February 2000; *MEED,* 24 March 2000: 8.
35. *Arab News,* 8 March 2000.
36. U.S.-Saudi Business Council, *U.S.-Saudi business brief* 2/2000.

trial balloons of the preceding two years. It is worth looking at the main debates and the changes introduced, often during the final weeks. Despite the unusual inclusion of the private sector in preceding deliberations, the government cut it out of the picture wherever it was convenient. Business either remained indifferent or was defeated on specific contentious points.

The extent of foreign control over assets is at the core of any FDI regulation. In late 1999, non-Saudi equity of up to 75 percent was the number generally quoted for the draft law. Foreign investors would still remain eligible for benefits and be entitled to real estate ownership.[37] The 75 percent cap was in line with a view held among industrialists that at least for big projects, less than 100 percent foreign ownership was advisable in order to encourage technology transfer and local development.[38] The law that emerged in April 2000 surprisingly allowed for 100 percent foreign ownership, however, following a unilateral, last-minute government decision to follow Washington Consensus standards.

Just as suddenly, fiscal incentives related to national employment and regional development, which had also been discussed extensively and openly, disappeared from the draft during the last weeks. Decisions were again made within the extended state apparatus: After the Majlis Ash-Shura financial committee had removed the Saudization incentives in spring, it was the Ministry of Finance, interested in revenue maximization, that had "forced its opinion" on removing regional investment incentives.[39]

Indifferent on many other details, business actors did adopt a more coherent position in the debate about the general tax holidays for foreigners. During 1999, the discussion seemed to focus on extending or "improving" the existing tax breaks.[40] In addition to calling for a substantial decrease of the corporate tax on foreigners to 20 percent, the chambers proposed tax holidays of longer than ten years.[41] The most active business lobbyists again were mostly industrialists with foreign joint venture partners.[42]

In April, however, the tax breaks were abolished, despite private sector calls to keep them.[43] The holidays were substituted by an infinite loss carry forward. The private sector's one concrete attempt at lobbying was defeated. Only later on, in

37. *MEED*, 3 November 1999: 8.
38. Interview with Saudi industrialist, Riyadh, May 2004.
39. Saudization incentives were defeated in the financial committee; Hajjar interview, April 2004; interview with former deputy minister, March 2004.
40. *MEED*, 3 November 1999: 16; *Bfai Unternehmerbriefe*, 15 March 2000.
41. *MEED*, 1 November 1999. At a minimum, the sentiment was that the holidays should be kept. Interview with Saudi industrialist, May 2004.
42. Interview with senior advisor to the Saudi government, June 2004.
43. Investment predating the decision would go on enjoying the holidays, however. The decision was technically part of the reviews of the corporate tax law but was negotiated together with the FIA.

the context of general tax reforms, did Saudi business—mostly industrialists—succeed in lobbying for a lower corporate tax rate for foreigners.[44] Despite the 20 percent demand from the CCIs, the initial maximum tax rate on foreign business promulgated in 2000 was 30 percent. Thanks to subsequent business lobbying, after an interim reduction to 25 percent the rate was finally reduced to 20 percent in the comprehensive January 2004 tax law.[45] It constituted change on the margin.

Patterns of Consultation over FIA: Granted, not Claimed

The debate and consultation about the FIA was much more extensive than that seen on previous pieces of legislation.[46] Even these extensive deliberations took place completely on the government's terms, however. At one stage in the process, senior government personnel handed out copies of draft legislation to chamber representatives, only to collect them again at the end of the meeting.[47] Paternal attitudes and a penchant for secrecy and control in its relations with business as a client group persisted. Although the CCIs played a prominent role, they never enjoyed an institutional guarantee of being consulted.

The private sector's limited influence on the FIA was also due to its heterogeneity of views and limited interest in more complex issues. Quite a few of the bigger actors, especially industrialists, have been eager to attract foreign partners in order to pursue more advanced projects.[48] Even their interest did not extend to all FIA issues, however, and they did not command a clear majority in the business community. Those businessmen with a stake did not have much of an infrastructure for policy-making. Business remained a policy-taker, and policy-making retained its top-down character—the heightened economic autonomy of business and extensive consultation notwithstanding.[49]

---

44. Interview with SAGIA functionary I, May 2004. There have been some further demands for cuts since, but the government has not reacted.

45. Interview with former deputy minister, May 2004; interview with chief economist of Saudi bank I, Riyadh, December 2003; Ernst and Young 2004. The reduction supposedly was hotly debated in the SEC. Interview with Abdalrahman Al-Zamil, Saudi industrialist and member of the Majlis Ah-Shura, Riyadh, May 2004.

46. Interview with former minister, Riyadh, March 2004. At least at this early stage of the reform drive, the degree of openness amounted to "an exception." Interview with chief economist of Saudi bank I, December 2003.

47. Interview with SAGIA functionary II, Riyadh, April 2004; discussion with Prince Abdallah bin Faisal in Leiden, Netherlands, February 2004.

48. All industrialists with foreign partners reportedly were pro-opening. Interview with board member of Riyadh CCI, Riyadh, May 2004.

49. Montagu 2001: 60.

The government under the guidance of the crown prince and the SEC by and large drove the negotiation process. Much was decided during the hectic last weeks, when less transparent interagency politics came to the fore. Individual government actors like the minister of finance had a more important final say after the public consultations and could overrule business concerns and even previously publicized government plans.

Although not revolutionary, the eventual FIA was a clear commitment to openness to international business. In addition to the features already mentioned—100 percent foreign ownership of local companies, foreign real estate ownership, infinite loss carry forward, and lower corporate taxes—the FIA package provided for equal entitlement to subsidies and cheap loans from public funds, unhindered repatriation of profits, and the opportunity for companies to "sponsor themselves," potentially eliminating one type of national intermediation. The first projects under the new rules were announced in the summer of 2000, including Iranian, Syrian, and Indian investors in plastics and oil products.[50]

The FIA also created SAGIA as both policy and service agency in charge of foreign investment matters, broadly defined.[51] Nine ministries as well as SAMA were represented on the SAGIA board, in addition to two individuals chosen from the private sector.[52] SAGIA inherited some bureaucratic structures from the government-owned Saudi Consulting House. Its first head, Abdallah bin Faisal, had considerable leeway to bring along trusted administrators from his previous job as chairman of the Royal Commission for Jubail and Yanbu (RCJY), however, and was able to offer higher salaries due to SAGIA's status as an independent administrative body. Compared to most other bureaucratic organizations in the Kingdom, SAGIA was characterized by a remarkable degree of openness and mid-level initiative.[53] Its internal administrative structure remained somewhat undefined in its first years, as relationships revolved around the prince as central figure, but this did not undercut its dynamism.[54]

---

50. *MEED*, 15 September 2000: 26.
51. FIA and other SAGIA-related acts can be found on the SAGIA website, available at www.sagia.gov.sa.
52. These two do not necessarily represent the broad interests of the private sector, however. For example, Sulaiman Mandil, a former functionary in the MoF who was also a member of the RCCI board, generally harbors more liberal views than the bulk of Saudi businessmen.
53. All SAGIA interviewees were very helpful with supplying official documents and discussing political matters and problems with other agencies in a way unthinkable in other ministries. Beyond that individual impression, most foreign actors I met were impressed by SAGIA's willingness to discuss administrative problems.
54. Discussions with SAGIA employees; visits to SAGIA.

Negotiating the Fine Print

To some extent, the FIA was a symbolic act as much as a regulatory one, as in typical Saudi fashion, many of its features had yet to be defined in concrete terms through bylaws, which held up the formal implementation of the investment regime. Abdallah bin Faisal justified the delays with a deliberation process that aimed at transparency and involved consultation with many interested parties.[55]

The cabinet approved the first bit of legal follow-up, the property regulations, in July 2000.[56] Additional details were issued one month later per royal decree.[57] To own a private residence, a foreigner had to obtain a permit from the Ministry of Interior (MoI), whereas the responsible line agencies would give licenses for investment-related real estate. A minimum investment of 30 million Saudi riyals was required if the purpose of acquiring real estate was reselling or renting it out.[58]

The latter level was high and signified a measured withdrawal from the FIA's open-door approach. More significant backpedaling came with the general bylaw to the FIA, which was finally issued in August 2000. The bylaw introduced minimum investment levels to all types of FDI: 25 million SR in agriculture, 5 million SR in industry, and 2 million SR in other areas (various types of services).

The general opinion in the private sector was that even higher minimum investment levels were needed. This view prevailed not only among Saudi intermediaries who made a living from acting as the official owner of expat-run businesses but also among "real" smaller Saudi businessmen who had investments below the thresholds and little voice in policy deliberations. It also was predominant among bigger actors—although they had no direct material stake in them.[59]

The actual imposition of investment thresholds however was likely an intra-government affair, with the Ministry of Interior and the Ministry of Labor and Social Affairs as their main proponents—two agencies not especially involved in the early stages of the FIA.[60] Prince Naif specifically has developed an ambition of being a protector of "little guys,"[61] and the minimum investment levels corresponded with the general defensive mood in a country struggling to create employment and entrepreneurship among nationals.[62]

---

55. *Arab News*, 26 July 2000.
56. Royal decree M/15, 19 July 2000 (printed in SAGIA leaflet, June 2002).
57. *Arab News*, 12 July 2000. They had earlier been passed by the Majlis Ash-Shura; cf. *Arab News*, 11 July 2000.
58. *Arab News*, 14 August 2000.
59. Interview with Saudi industrialist, May 2004.
60. Interview with former deputy minister, Riyadh, May 2004.
61. Ibid.
62. The minister of interior, it is said, wanted to prevent small expatriate investors from becoming independent actors. Interview with senior expatriate lawyer, May 2004.

On the SAGIA board, only Prince Abdulaziz bin Salman from the Ministry of Petroleum and Minerals, Abdallah bin Faisal, and Sulaiman Mandil (a liberal former MoF functionary) opposed the imposition of an investment threshold.[63] The other board members, representing their respective sectoral agencies, supported the restrictions.

SAGIA functionaries complained that the threshold concept had no basis in the main FIA and hence should not be added post hoc through a bylaw. But that this constituted technically bad legislative practice seemed to faze no one.[64] Judging from SAGIA's later reports, the authority apparently felt excluded from the discussion about investment thresholds.[65]

The "Negative List" of Restricted Sectors

Minimum investment levels had found a clear and powerful majority. The "negative list" of activities prohibited for foreign investors was a trickier issue, as the stakes of the relevant players were higher and more complicated. The list, which was supposed to supplant the pre-FIA "positive list," was issued in early 2001 after repeated delays, reflecting its sensitive nature. It was the biggest issue for the private sector in the FIA context.[66]

Discussions over the negative list were initially conducted by the SAGIA board.[67] Its interdepartmental composition meant that the list was circulated around the senior levels of the administration at an early stage. This explains why the initial draft of the negative list was so long: despite a general agreement on liberalization among senior government ranks, each organization protected its own turf.

Business actors were also allowed to air their opinions, and representatives of several sectors voiced great concern about their national privileges.[68] Actors with retail and wholesale interests strongly demanded that commercial distribution should be kept off-limits.[69]

When the draft reached the Supreme Economic Council, the crown prince rejected it on account of its length.[70] The absence of an agreed list put SAGIA in

---

63. Interview with Saudi industrialist, Riyadh, May 2004. Abdallah bin Faisal is said to have "clashed" with the private sector on the minimum investment issue. Interview with SAGIA functionary I, Riyadh, May 2004.
64. "We did a bad job on this one." Ibid.
65. SAGIA 2003: 73.
66. Interview with senior advisor to the Saudi government, Riyadh, June 2004.
67. *Arab News,* 26 July 2000.
68. *MEED,* 1 December 2000: 32.
69. Interview with SAGIA functionary III, Riyadh, May 2004. On the historical role of "commercial agencies" for Saudi merchant families, see chapter 7.
70. *MEED,* 16 March 2001: 24.

the position of having to decide on the issuance of investment licenses without knowing which areas were actually available for investment.[71]

The first negative list was eventually published in February 2001, with the first yearly review scheduled for February 2002.[72] It included twenty-two areas, among them distribution and transmission of energy, telecoms, land and air transportation, fishing, health, education, retail and distribution, insurance, pipeline transport, explosives for nonmilitary use, real estate brokerage, printing and publishing, and radio and television.[73] Investment in gas exploration and production, power generation, water utility services, and mining were not on the list (a new sectoral law on mining was in the works).[74]

SAGIA had "strongly resisted pressure to include many other sectors on the list." In this context, the listing of telecoms, education, and insurance came as a surprise to local representatives of Western business, as "that [was] where they need[ed] investments."[75] According to the World Bank, the negative list was "excessively lengthy by international standards."[76] "Everything to make money with was kept on the list, including wholesale, retail, and insurance," stated one observer.[77]

Business interests, although in parts clearly articulated, reportedly played a subordinate role in the intra-state negotiations.[78] Instead, individual government agencies apparently got their way with the crown prince, successfully shielding their respective sectors. According to a former minister, when combined with the negative list, the FIA did not appear very different from the old investment regulations; indeed, in his opinion the old law was possibly better.[79]

In the runup to the next revision of the negative list, SAGIA lobbied strongly within the government for its reduction. In the summer of 2001, Abdallah bin Faisal publicly urged that the telecom sector be opened up.[80] Such recourse to

---

71. Interview with SAGIA functionary IV, Riyadh, May 2004.
72. *Arab News*, 14 February 2001, 10 April 2002.
73. *MEED*, 23 February 2001: 4, 16 March 2001: 24.
74. The law was only issued in September 2004; one reason for the hold-up supposedly was that two princes had interests in mining ventures that would be affected in opposed ways by the new law. Discussion with Western economic advisor III, Riyadh, February 2004.
75. *Arab News*, 14 February 2001.
76. World Bank 2002: xiii.
77. Interview with Western diplomat I, April 2004.
78. Interview with former deputy minister, November 2005. Private sector opinion reportedly favored opening insurance to FDI and otherwise was divided on many items. Interview with board member of Riyadh Chamber.
79. Interview with former minister, Riyadh, March 2004. This opinion was shared by a former head of SABIC and member of the Majlis Ash-Shura. Interview with Ibrahim bin Salamah, Riyadh, May 2004.
80. *Arab News*, 30 July 2001.

the public by a minister-level figure from outside the royal family core was unheard of.

The review of the negative list by a special interministerial committee was again delayed, this time dragging on until February 2003. In another unprecedented step, Abdallah bin Faisal openly contacted foreign missions to elicit comments.[81] Business interests again would participate in the deliberations through invitations to the council of advisors at the SEC, which under the crown prince's chairmanship had final say over the list.[82]

The main sectors eventually removed from the list included insurance (pending the new insurance law), parts of telecoms (where an international GSM license bid was under preparation), and electricity transmission and distribution.[83] All utilities were now open to FDI. Education and most health services were also removed from the list.

Although substantial, these still were the relatively "easy" sectors with less deeply vested interests. Electricity was a sector with long-identified investment needs, in which issues of profitability and legal security had long muted investors' interest.[84] Telecoms was a new area in which the Saudi private sector would require international partners; its gradual opening proceeded in tandem with the successful privatization of Saudi Telecom. Insurance was another new area exposed to WTO pressure in which international expertise was required. Education and health were sectors in which the government had reached its limits of provision. Actual opportunities for foreign engagement would in many cases still depend on the licensing policies of the responsible line agencies. Sectors with large, established domestic players were kept off-limits, most of all retail and wholesale.[85]

The WTO as External Driver of Change

The negative list negotiations followed the general Saudi rule of thumb that bylaws tend to be more rigid in their details than the main legislation, as local interests of the bureaucracy seep into the negotiation process and initial top-level momentum is diluted.

The crown prince's reform drive had to combine with external pressure to overcome the resistance of the more entrenched fiefs. Only in the course of

---

81. *Arab News,* 10 August 2002.
82. Interview with SAGIA functionary IV. The CCIs as such were not formally part of the process.
83. The new insurance law was issued in the summer of 2004; see chapter 7.
84. The main issue are tariffs set by the state that amount to a subsidy of smaller consumers (see chapter 4); interviews with Western diplomat II, Riyadh, May 2003, and David Butter, editor, Business Middle East, Economist Intelligence Unit, London, May 2005.
85. Wholesale in particular was perceived as a core area of Saudi business; cf. Montagu 2001: 40.

negotiations over WTO accession did the leadership manage to open up some of the core turf of Saudi business to foreigners, namely banking and commerce. Since 2003, when the WTO talks were reinvigorated and the two sectors emerged as core stumbling blocks, SAMA saw fit to issue a few licenses to foreign banks, both from the Gulf and from the West.[86] The director of NCB, the largest Saudi bank, reacted in nationalistic tones, predicting that foreign banks would not add much of value to the sector in their quest for short-term gain, unlike "sons of the soil" like him.[87]

The opening of retail was similarly WTO-induced. After four years of foot-dragging, a plan to allow 75 percent foreign ownership by the end of 2005 started circulating in government in 2004, as the issue was raised in WTO talks.[88] Saudi negotiators managed to delay the 75 percent commitment to three years after the WTO accession, that is, 2008.[89] Still, the WTO had forced the government's hand.

Parts of the private sector appeared to be fighting rearguard battles against FDI liberalization through a variety of channels. The Riyadh CCI reportedly complained to the Riyadh governorate about the issuing of SAGIA licenses to investors who plan to set up foreign schools; similar complaints were heard on construction projects.[90] Business interests also attempted to exert direct pressure on SAGIA not to issue licenses in mature sectors like the retail and wholesale sectors and specific industries.[91] The effects were probably limited. The whole history of the negative list is one of turf defense by both parts of the private sector and government line agencies, who were gradually beaten into submission by either the crown prince or external pressures. At all stages of the game, business sectors and ministries have acted as meso-level veto players instead of forming broader policy coalitions.

Implementation: The Bureaucracy Strikes Back

So far, we have looked at patterns of political, that is, mostly meso-level, negotiations around the FIA. Although these were state-dominated, state interests themselves proved to be fragmented, as became increasingly clear in the course of spelling out specific rules. The story reflects a rapidly built state that consists of parallel islands of bureaucracy communicating mostly through the royal elite. But

---

86. *Arab News,* 1 June 2004, 20 July 2004.
87. *Saudi Gazette,* 10 August 2004.
88. Interview with SAGIA functionary III, May 2004.
89. See chapter 7.
90. Interview with SAGIA functionary I, May 2004.
91. Discussion with Western economic advisor I, Riyadh, April 2004; *Saudi Gazette,* 6 April 2004.

what about the fabric of the state itself, the low-level clientelist bureaucracy with its smaller-scale hierarchies and networks?

Their impact on regulatory reform becomes clearer only in an analysis of policy implementation, where our attention is drawn to the bureaucracy at large—where additional layers of veto players exist on a micro-level, and the state is further fragmented. Here, the relationship between state and society is quite different from and more even than that on the policy-making level.

Micro-links of authority and communication within and around the state make for additional, arguably more significant, dilution and unintended reversals of reform efforts. Excessive hierarchies of long standing and accumulated rules exacerbate segmentation and stalemate on a smaller scale, which can often be circumvented only by resort to brokers of various sorts—which make the bureaucracy supple but further undermine the coherence of cross-cutting policy efforts.

When the FIA was issued, elite expectations of change were great, and enormous projections of future FDI circulated. Some of this was deliberate public relations work, but among certain senior bureaucrats, there was also a real expectation of change through the FIA.[92] However, as one observer expressed it, the problem turned out to be other bureaucrats.[93] A World Bank report issued more than two years after the publication of the FIA provides a succinct summary:

> From the point when an investor wants to obtain a visa to enter Saudi Arabia to when his business enterprise is up and running, he is faced with an overly-bureaucratic, occasionally obstructionist, and generally unhelpful regime that is characterized by laws that are contradictory or not uniformly enforced, bureaucrats that are unresponsive and lack appropriate customer service skills, pressure to hire local workers regardless of their degree of enthusiasm and qualifications, procedures that are slow, and systems that do not work well.[94]

This might seem surprising, as conditions for reform seemed to be relatively benign on several accounts. Saudi Arabia is not too corrupt by Middle East comparison; its bureaucrats are comparatively qualified; SAGIA appeared to be a committed lead agency; and the law was more liberal than the regulations in any other Gulf state, including the freewheeling United Arab Emirates. As we will see, it was the historical timing of the reform combined with the micro-structures of authority in the bureaucracy that thwarted much of the FIA reforms for the first five

---

92. German embassy, dispatch no. 93 to Foreign Office/Bonn, 17 April 2000.
93. Interview with former deputy minister, May 2004.
94. World Bank 2002: viii.

years at least. FIA and new foreign investors collided with numerous preexisting rules, administrative power structures, and entitlements that were difficult to overcome in a hierarchical and segmented system.

Legal and Regulatory Fragmentation

SAGIA and the FIA were implanted into a system in which they were not embedded. As incompatible legal and administrative rules in other agencies often remained untouched, significant parts of the foreign investment regulations were "up in the air and not operationable,"[95] on occasion exacerbating existing problems of bureaucratic fragmentation.

The final article of the FIA postulates that the law supersedes all contradicting regulations. This clause, as well as SAGIA's claim to be the primary actor on all FDI-related matters, collided on a grand scale with resilient preexisting structures. The authority and many of the foreign investors it was tasked to look after had to face numerous local rules and rule-makers who remained ensconced in their own hierarchies, unfazed by SAGIA demands.

In some cases, organizations would ignore new legislation altogether. Line agencies operating under their old regulations disregarded superiority clauses in the FIA and related regulation. Ministries appeared "like holes" into which any new rules disappeared, as they simply refused to give licenses or would stipulate investment conditions that contradicted the FIA.[96] As often, agencies would stick with incompatible or counterproductive administrative procedures hardly addressed by the broad FIA legislation. Unwritten documentation requirements, interventions in managerial decisions, and refusal of SAGIA deeds often reduced FIA to a law in name only.

New investors attracted by SAGIA's promise would encounter inconsistencies and segmented information on the administrative level. Whereas FIA, SAGIA statute, and other SAGIA-related laws are freely available on the internet, other agencies often did not publish their regulations properly.[97] In the absence of a functioning public legal depository, many agencies were reluctant to provide copies of important rules, to which administrators often felt that only they should have access.[98] Changing internal procedures, based on ministerial memos issued frequently and without warning, have tended to be opaque.[99]

---

95. Sander interview.
96. Interview with SAGIA functionary II, April 2004; cf. SAGIA 2003: 43.
97. Interview with Norbert Lehner, Hoshanco factory manager in a Riyadh industrial city, Riyadh, March 2004; SAGIA 2003: xviii.
98. SAGIA 2003: 18.
99. Interview with SAGIA functionary IV; SAGIA 2003: 2.

Minister of Commerce Usama Faqih, widely criticized in the business community, reportedly made himself the butt of jokes with his changing and incoherent memos, against which there were no meaningful formal appeal mechanisms.[100] More generally, rules about daily government-business transactions in the Ministry seemed "to change each time there is a new head of department."[101] Even when memos were available, there were usually no non-Arabic translations for investors. Individual rulings effectively had the force of law and were not publicized.[102]

Despite SAGIA's mandate, there was no effective mechanism to ensure interoperability of rules between agencies. Agencies would make incompatible documentation demands, and businessmen often had to double their efforts or find themselves stuck between ministries due to mutually unrecognized procedures.[103]

The Ministry of Justice (MoJ) has proved particularly resilient to regulatory reform. Due to the distrust of all secular regulation, the rivalry between MoJ and other parts of the bureaucracy has been especially deep.[104] The judicial staff manning the Kingdom's courts are not systematically kept up to date about legal changes and communicate little with the rest of the bureaucracy.[105] Judges fail to keep appointments, work slowly, and issue decisions often perceived as arbitrary. A separate circuit of commercial courts to serve business has been announced many times over the years but did not materialize until 2009.[106]

The occasional discretionary interventions of more powerful agencies into the turf of other agencies further complicate the picture. One example is a memo sent by the MoI to the MoC demanding that the general manager of a newly established company has to be either Saudi or a registered resident. As a residency permit can only be acquired after a company's commercial registration, however, this meant that investors were not able to bring their own head of branch and install him with power of attorney, undermining standard investment practice.[107] Such decrees can emerge suddenly and often remain unpublicized.

---

100. Interview with former deputy minister, May 2004; he does not appear to have played a very important role on the FIA.
101. Interview with senior expatriate lawyer, May 2004.
102. Interview with SAGIA functionary III, May 2004; SAGIA 2003: 32.
103. World Bank 2002: 62, 90.
104. "They [MoJ] refuse to be pushed by anyone." Interview with former deputy minister, May 2004
105. World Bank 2002, p. 22.
106. One Saudi interviewee described the judiciary as "too independent" in Saudi Arabia, claiming that in their day-to-day operations, courts could basically operate as they see fit. Interview with senior Saudi banker, Riyadh, December 2003.
107. SAGIA 2003: 52. Some Westerners had the impression that through such rules, Prince Naif intended to obstruct FDI in the Kingdom. Interview with senior expatriate lawyer, May 2004.

Bureaucrats and Their Interests

When new bylaws or procedural rules are issued, they are often not sufficiently communicated, either downwards or laterally. New laws frequently create inconsistencies, and even bylaws sometimes contradict the main laws to which they pertain.[108] But even if formal rules are clear, public servants frequently have little incentive to act coherently and predictably. FIA or SAGIA changed little about the inability of civil servants in certain agencies to keep appointments.[109] Administrators also tend to stick to formalities and obsolete rules in order to avoid effort or risk or, in some cases, to deliberately inhibit transactions. As has been the case since the inception of the modern Saudi bureaucracy, indolence and obstruction are too frequent to be punished effectively, and superiors have few incentives or sanctions at their disposal.

Senior technocrats tend to impose rigid hierarchies to control their sprawling bureaucracies. In the MoCI, only the minister and his deputies have something amounting to official discretion. This can severely slow down the flow of paper, however, and the absence of specific individuals responsible for given areas (for example, foreign applications) can stall whole departments.[110] The pervasiveness of such structural rigidities can make them very resilient. Even if the leadership changes and develops a genuine reform orientation, as has arguably been the case in the Ministry of Commerce after its merger with Industry under Hashim Yamani in 2003, there is little exchange or rotation of personnel on the middle levels.[111]

The boundary between bureaucratic inertia and deliberate obstructionism is often blurred. Stories abound of unexplained or arbitrary demands for additional documentation, obstruction due to mistyped names in forms, tactical use of formality and inflexibility, arbitrary fees, and delays. Overlapping and unclear rules that have accumulated over decades exacerbate the situation and can give mid-level bureaucrats considerable de facto discretion once a decision has to be made.[112] Trust toward supplicants is low, and coordination with bureaucratic peers on the same level is often lacking. SAGIA functionaries have complained that arbitrary decisions not only are a result of bad coordination but also demonstrate the interest of bureaucrats in exerting power.[113]

---

108. World Bank 2002: 22; interview with SAGIA functionary III, May 2004.
109. World Bank 2002: 91.
110. Interview with senior expatriate lawyer, May 2004.
111. Interview with senior accountant, Riyadh, April 2004; interview with senior expatriate lawyer, May 2004.
112. SAGIA 2003: 43.
113. "They die for power." Interview with SAGIA functionary III, May 2004.

The Persistence of Bureaucratic Brokers

As indicated in chapter 1, there are further groups with an even clearer interest in maintaining secrecy and complexity in the system: intermediaries and brokers operating between the formal bureaucracy and business on various levels. In addition to specialized consultants, lawyers, small-scale paper brokers and influence peddlers, a sizeable part of Saudi business itself has been more or less directly engaged in "facilitating" interaction with the administration. The links of such brokers to the administration often go back many years and have been slowly established in a generally immobile system. The new layer of SAGIA rules left broker interests largely untouched, as they were attached to and dealing with other bodies. Whereas in the 1970s brokers were often used to get access to the state's ample material resources, in the FIA context after 2000 their main role was simply to get paperwork done and forestall administrative obstruction.

Brokers include lawyers and accountants who specialize in keeping track of the memos from various institutions and offer their exegetic services, often based on personal contacts and familiarity with the administration. Several institutions do not give access to foreigners, and a national needs to be authorized for administrative transactions.[114]

Larger foreign firms usually go through law firms to deal with the administration,[115] whereas smaller businessmen, both foreign and local, hire smaller brokers in different institutions to follow the progress of their applications and forms through the bureaucracy.[116] Hundreds of "expediter" offices are working around the Ministry of Commerce alone.

In the event of administrative "difficulties," a customer usually has to pay higher rates to the broker.[117] It is not always clear where the fees which intermediaries charge end up. As charges of corruption are common,[118] the shared interest of brokers and administrators to keep the lower levels of the bureaucracy opaque and complex becomes obvious.[119] In the absence of bribery opportunities, one often needs to leverage senior contacts to move things past administrative stalemate or obstruction. Such contacts—often princes—either can be deployed by

---

114. Interview with SAGIA functionary I, May 2004.
115. In the 1980s, many law firms were—unusually, but quite understandably—simply taken as joint venture partners; cf. Montagu 1985: 63.
116. Interview with SAGIA functionary I, May 2004.
117. Interview with senior GCC functionary, Riyadh, December 2005.
118. Corruption in the MoCI is said to be rather widespread. Interview with former deputy minister, May 2004.
119. A typical offer from a broker in case of difficulties would be that one can pay either the official fine or less than the fine—which makes only sense if a venal official is involved. Interview with GCC functionary.

a low-level broker or can act as direct brokers themselves. In a complex bureaucracy, only the better brokers are connected in several agencies; many specialize in one agency.[120]

In this system, many foreign businesses still chose to acquire a local "sponsor" and a local investment partner in order to establish themselves—even though there is no longer a formal obligation to have either. Visa acquisition and other administrative hurdles would be hard to handle without local support.[121] The legal and procedural framework for "self-sponsorship" remained unclear for a considerable time.[122] Certain mid-sized companies specialize in offering administrative services for foreigners to get entry to the Kingdom, with a view to offering a partnership after a few months.[123]

Real Estate for Foreigners: Yes, but No

Bureaucratic obstacles and brokerage have been particularly pronounced in real estate acquisition and commercial registration, two areas of specific importance for foreign investors.[124] Insofar as they have tried, many foreign businesses found it difficult to acquire or develop real estate under their own names, the new framework regulations of mid-2000 notwithstanding. They encountered either a raft of unchanged, complex, and interventionist rules or a simple lack of procedures catering to foreigners.[125]

Real estate acquisition usually involved interaction with between five and eight agencies. In the case of industrial cities, investors had to follow detailed Arabic instructions on how to prepare drawings for new buildings, and the Minister of Industry himself had to approve the minutes of the committee responsible for new applications. Administrators had leeway to get involved in the planning of private projects, for example, when deciding on the utility needs of new installations.[126] The civil defense department of the Ministry of Interior reviews all new physical structures. If they do not decide to acquire a local partner, foreign investors were well advised to hire local engineering and other consultants (until

---

120. Interview William Barillka, managing director, DevCorp, Riyadh, December 2005; interview with GCC functionary.
121. Interview Klaus Hermann, project manager, ACI-Aqua Project Consult, Riyadh, April 2004.
122. Sander interview.
123. Von Loebbecke interview.
124. Problems of visa acquisition, another major investment obstacle, are dealt with in chapter 6, on Saudization. It should be borne in mind that Saudi businesses can encounter many similar problems, although they tend to be better equipped to handle them.
125. SAGIA 2003: 48.
126. World Bank 2002: 71–82.

recently, consultancy was one of the fields in which no full foreign ownership was allowed).[127]

The most coveted industrial cities by the early 2000s were at full capacity; several of the related water authorities struggled to cover demand, and even the installation of a phone line could take considerable time.[128] Administrative decisions on land and service allocation therefore acquired great importance. With a tilted playing field—indicated by reports of administrative "irregularities"—local knowledge and access to the right networks became essential.[129] Opportunities for brokerage, for example through well-connected joint-venture partners, increased accordingly.

Whereas procedures in industrial cities were at least clear on paper, and whereas the RCJY at least has a respectable administrative track record, the rules for municipal land acquisition remained almost completely obscure. As late as the summer of 2002, the Riyadh municipality had no procedures for dealing with foreigners intending to buy real estate.[130] The apparently simple ownership transfer of a headquarters building of a larger German company took two years, as the necessary administrative procedures to implement the law were still lacking.[131] As with industrial property, the formal administration of municipal real estate was highly centralized: The minister of municipality and rural affairs, Prince Mit'eb bin Abdulaziz, had to approve all projects costing over 500,000 SR.[132] The MoI, another player that was required to agree to foreign realty ownership, seemed to "sabotage" realty applications for foreigners by offering no application procedures whatsoever and taking no action on individual requests.[133] By default, many foreign companies wisely stuck with local partners, who could acquire real estate much more quickly.

Commercial Registration: Getting Started, or Not

Whereas property issues can at least be left for a Saudi partner to handle, the acquisition of an official business registration is necessary for anyone doing formal business in the Kingdom. Since April 2000, the first step for foreign investors has

---

127. Von Loebbecke interview.
128. World Bank 2002: xiv.
129. In World Bank–speak, acquiring land in industrial cities could be "non-transparent" according to some investors, which would lead to "extra-legal financial inducements." World Bank 2002: 88. Allegations of corruption were also made in personal interviews.
130. World Bank 2002: 87.
131. German-Saudi Liaison Office for Economic Affairs, Gulf Economy 4/2002: 22.
132. World Bank 2002: 103.
133. Discussion with senior advisor to Saudi government II, November 2005.

been to apply for a foreign investment license with SAGIA. This application is subject to a "thirty-day rule," which means that the license is granted automatically after one month unless SAGIA specifically declines the application.

At least during the first few years, getting SAGIA licenses was easy. They are only a first step in the registration process, however. The SAGIA license by itself is of little help without a commercial registration (CR) issued by the MoCI, on the basis of largely unchanged rules that SAGIA has little control over. Without CR, an investor—whether Saudi or foreign—cannot lease or rent a facility, open a company bank account, or even commence any business activity. "The thirty-day rule is great, but then come six months at the Ministry of Commerce. And that's if you have a good lawyer," as one source put it.[134] Mid-level bureaucratic stalling happened especially during the project assessment phase.[135] Investors in new companies, moreover, were required to have forms certified by the notary public, which is part of the judiciary and has been described as a "nightmare," not keeping appointments and demanding changes in partnership agreements already accepted by other agencies.[136]

CRs have to be renewed regularly, which can lead to renewed complications. Tens of thousands of CRs lapse every year, which businessmen blame on the MoCI's sluggishness.[137] Despite its prominent role as an elusive "business ID" and the trouble that goes into acquiring it, the CR is not a powerful regulatory tool. The registration as such does not impart any information about a company's type and size.[138] Moreover, numerous CRs are held by "letterbox companies" companies that do not actually run any operations. The database of the MoCI is limited, and CRs are often used for purposes other than the one officially applied for.[139] This low level of regulatory penetration by CRs is closely related to the issue of minimum investment levels and "cover-up" businesses.

Minimum Investment Levels and Ways around Them: "Cover-Up" Businesses

So far, this section has discussed how the 2000 foreign investment reform has left large parts of the administrative setting untouched, never becoming digested by or embedded in existing bureaucratic rules, structures, and networks. In some cases, the FIA has arguably created additional administrative ambiguity. This was the

---

134. Interview with former deputy minister, March 2004.
135. Interview with SAGIA functionary III, May 2004
136. World Bank 2002: 67 (citing an anonymous investor); *Arab News*, 28 July 2003; interview with Saudi lawyer, Riyadh, January 2009.
137. Interview with Saudi industrialist, March 2004.
138. Discussion with senior Saudi bureaucrat, Riyadh, June 2004.
139. Cf. *Saudi Gazette*, 5 January 2004.

case with the minimum investment levels in the bylaw. By formally restricting access to markets while lacking the capacity to monitor them, the government created another layer of unimplementable rules and new levers of informal negotiation and manipulation.

The purpose and effects of the minimum levels merit a brief explanation. Their immediate impact, officially, is to keep smaller non-Western businesses out of Saudi markets. But many smaller establishments in Saudi Arabia are in fact financed and run by expatriates who have teamed up with Saudis whose sole function is to provide their name for registering the company; the share of small businesses actually run by Saudis dwindled during the boom. The expatriate usually pays a monthly fee to his Saudi partner. Minimum investment levels prevent the bulk of these "cover-up" businesses (called "*tasattur*" in Arabic) from shedding their Saudi intermediaries and registering under their real owners, and they prevent smaller investors from entering the country through official channels. The largest group of cover-up businesses is in small retail and services.

A study by a Saudi economist in 2004 estimated their number at 155,000, and a Ministry of Labor survey in 2000 found a large proportion of "dummy businesses" run by expatriates among local small and medium enterprises.[140] An "anti-proxy law" explicitly prohibits the practice,[141] and even the Higher Council of Ulama has issued a ban on acting as a rent-taking frontman.[142] The Ministry of Commerce employs special inspectors to investigate cover-up practices.[143] Yet little has changed about the practice, as crackdowns remain "very superficial."[144]

Many of the frontmen-brokers for larger companies reportedly are businessmen and public sector figures, even senior ones.[145] Moreover, frontmen often come from the retinue of princes of various levels of seniority, which facilitates administrative steps such as acquiring visas for the workers (which until recently was usually done through the Ministry of Interior).[146]

Official Saudi owners usually have no interest in controlling what is going on in an establishment as long as the regular fees are paid, which start at about

---

140. *Riyadh Daily*, 19 June 2003; *Arab News*, 7 March 2004. An expatriate interviewee puts their proportion at 80 percent of all small businesses; cf. *Saudi Gazette*, 12 August 2004. Even official reports deplore the "weak-minded" Saudis who sell their commercial registration and mention that "tens or hundreds of thousands" of foreigners are involved in such schemes; Daghistani 2004: 281.

141. Royal Decree M/22, 22 June 2004, The Combating Cover-up Law (unofficial translation by Jadaan Law Firm).

142. However, as one administrator complained, the judiciary "don't implement their rulings." Interview with SAGIA functionary I, May 2004.

143. Council of Ministers, Decree M/2, 1/1/1418.

144. Interview with senior advisor to the Saudi government, June 2004.

145. *Saudi Gazette*, 12 August 2004.

146. Interview with former deputy minister, Riyadh, May 2004.

300 SR per month for a small grocery shop but range considerably higher for larger companies. The official owner monetizes his or her position of privileged legal access and does little else. Due to the lack of formal property rights, it is also easy for Saudi patrons to betray and expel their partners.[147]

Some technocrats have called for cover-up establishments to be legitimized through scrapping the minimum investment levels, hence bestowing official status on probably the majority of businesses in the Kingdom.[148] On the one hand, they argue, this would reward the people who are actually doing the work, giving them a secure future in the Kingdom and, on the other hand, it would remove a class of "lazy, often illiterate" rent-takers.[149] This is a liberal minority demand, however.[150]

The introduction of minimum investment levels did offer certain opportunities for legitimizing businesses, but in a roundabout way that in some respects moved economic realities even further away from what is on the books of the MoCI, namely by inflating the assets of one's company in order to be able to embark on the official registration. "Virtual" capital can be created through a variety of means: some of the better-positioned small foreign businessmen started taking loans and collecting money to reach the required minimum investment level and then placed the capital into a local account, only to withdraw and pay back the loan later on. Another method is inflating the value of equipment.[151] The MoCI, which is responsible for evaluating the minimum investment, hardly has the capacity to trace what happens to money later on or to estimate assets of small companies. There are also credible stories of minimum investment certificates being sold by middlemen.[152]

Minimum investment rules did not end cover-up practices. Instead, they showed how ambitious regulations under conditions of weak regulatory power can create additional incentives for misrepresentation and influence peddling. The creation of minimum investment levels complicated the issue instead of legalizing it. Apart from occasional crackdowns, the government has little permanent control over small businesses, most of which do not pay taxes.[153] Developing some regulatory grip over them does not become easier if the official owner does not

---

147. *Saudi Gazette,* 12 August 2004.
148. Ibid.
149. Interview with former deputy minister, Riyadh, May 2004.
150. Predictably, one shared by the World Bank; cf. World Bank 2002: xvii.
151. Interview with former deputy minister, May 2004.
152. The problem here must be either with the MoCI or with the bank where the business capital account is held. Interview with SAGIA functionary I, May 2004.
153. *Saudi Gazette,* 12 August 2004.

know what his business is doing and the actual owner perpetually fears deportation from the country.

In the course of WTO negotiations, minimum investment levels have been removed in services apart from retail, where it was set at a high 20 million SR.[154] According to an informal agreement within the Saudi government, however, small foreign investors in other service trades will have to apply for investment licenses from their home countries. Most of the cover-up investors are unlikely to take the risk of leaving the Kingdom and of being denied access and control over their assets.[155] Cover-up businesses will, if at all, be eroded only gradually.

Fragmented administration, immovable bureaucrats, and cumbersome rules have created many brokerage opportunities for local business. It would be wrong, however, to generalize that all private sector players have a stake in administrative obstruction and complexity. The clear majority of active Saudi businessmen are in favor of administrative streamlining. The "Riyadh Economic Forum" business summit in the summer of 2003 asked for an extension of the thirty-day rule to procedures in all agencies.[156] As of this writing, obtaining visas for employees or business partners is still a daunting obstacle for Saudi businesses, and many companies have seen their CRs lapse due to administrative complications. Business groups have repeatedly demanded simpler rules. As operations become more complex and autonomous from immediate state assistance and trust towards the bureaucracy remains low, many businessmen would prefer leaner regulation.[157] This is probably the most significant area of overlap with SAGIA interests. It has, however, never led to a broad SAGIA-business coalition, as significant parts of business are suspicious of SAGIA's liberalizing agenda.

The General Investment Authority in the Battle

This is somewhat ironic, as SAGIA's very raison d'être is, at least in principle, to overcome segmentation, rigid hierarchies, and informal brokerage, even if foreign investors are supposed to be its main clientele. It has struggled valiantly to change the Saudi bureaucratic environment during its first decade. Its efforts make for a unique test of the resilience of segmented power structures, the limits that

---

154. SAMBA 2006.
155. Interview with SAGIA functionary I, December 2005.
156. Interview with former deputy minister, May 2004.
157. Confirmed in all interviews with private sector representatives.

a segmented system imposes on newly established actors, and the ability of new actors to "work the system."

SAGIA's objectives according to article 4 of its executive rules include the promotion of diversification, technology transfer, and competition to the benefit of consumers; export development; environmental protection; and the development of plans on training, customs exemptions, tax incentives, and criteria for granting concessions. Article 5 tasks SAGIA with proposing legal changes to improve the investment climate in addition to sundry economic monitoring tasks. Article 7 provides a mandate for developing multiple investment-related databases.

Considering the breadth of its portfolio, which overlaps with the jurisdictions of many other agencies, SAGIA is a small institution with modest headquarters in Riyadh and "investor service centers" in Riyadh, Medina, Jeddah, and Dammam. Its annual budget as of 2007 was less than 100 million SR.[158] Despite top-level commitment and a range of good administrators, it has struggled to establish its presence. Indeed, SAGIA found it hard even to establish its own licensing authority: the MoC, for example, would on occasion question the business activity as set down in a SAGIA license, which forced investors to shuttle back and forth between the two institutions. The RCJY, for its part, demanded no SAGIA license at all but other documentation from MoIE, for example, which seemed "to indicate that SAGIA's approval authority is not recognized and the Ministry has not ceded its own licensing authority in favor of SAGIA."[159]

The microcosm of the "investor service centers" vividly reflects SAGIA's travails in carving out its role. In 2000, the first such center in Riyadh was set up and promoted as a "one-stop shop" for foreign investors, where they could conduct all necessary bureaucratic business. It housed representatives from the MoI, MoFA, MoC, MoIE, MoF, MoLSA, the Ministry of Agriculture and Water, and the Ministry of Petroleum and Mineral Resources.[160]

Among foreign investors, the center soon acquired the nickname "the one-more-stop shop."[161] The various agencies' representatives, insofar as they were present,[162] did little more than collect the necessary papers, with subsequent administrative struggles, as outlined above, mostly occurring in the line agencies, as before.[163] Representatives referred all requests upwards and did not coordinate with their colleagues from other agencies at the "one-stop shop." The staff

---

158. Saudi-British Bank, Saudi Budget 2008, *SABB Notes,* 12 December 2007: 10.
159. World Bank 2002: 58, 90; cf. SAGIA 2003: 63.
160. World Bank 2002: 48.
161. Interviews chief economists of Saudi banks I and II, May and December 2003, respectively.
162. Many of the booths were unmanned until at least 2004. Interview with former British diplomat, London, June 2005.
163. Interview with SAGIA functionary I, May 2004.

in the center had little authority, and the institution effectively remained "not operational."[164] Foreign investors felt that there were far too many people involved in the approval procedures.[165]

Feeling investors' discontent, Abdallah bin Faisal actively tried to reduce other institutions' influence over the process.[166] He publicly mentioned that there were shortcomings in certain government departments.[167] On his insistence, the crown prince resolved that all agencies had to consult with SAGIA over new laws.[168] In addition to its (somewhat undefined) administrative functions, under the prince SAGIA has acted consistently, and to some degree publicly, "as a research body" to see how regulations could be improved.[169] SAGIA representatives voiced clear opinions about the merits of certain laws also outside of government offices.

Unlike the rest of the state, SAGIA occasionally tried to find allies outside the administration and did not hesitate to send drafts of laws to embassies or international companies.[170] It has been the government's staunchest proponent of more inclusive decision-making and consultation with interest groups. This was probably both an attempt to break the mold of secretive rule-making and to claim a stake for itself as a broker in future consultations. The SAGIA chief convinced the crown prince to issue a memo demanding consultation with all concerned parties before a law is issued, which SAGIA claimed as one of its bigger successes.[171]

SAGIA's "Obstacle Report"

Abdallah bin Faisal's most comprehensive effort at changing the administrative environment surrounding SAGIA probably was the 2003 "obstacle report." In April 2002, the Saudi press outlets cited plans for a "total review" of the investment regulations to remove obstacles to foreign investment, explaining that SAGIA intended to take into account views of commercial attachés in foreign embassies, foreign investors, businessmen, and CCIs.[172] The main point of departure for the eventual report was an August 2002 World Bank study about administrative barriers to investment in Saudi Arabia.

---

164. Sander interview; SAGIA 2003: 88.
165. World Bank 2002: 57.
166. Interview with SAGIA functionary I, May 2004.
167. *Arab News*, 10 April 2002.
168. Interview with SAGIA functionary I, May 2004.
169. *Arab News*, TOP100 supplement, 17 January 2004: 18.
170. Ibid.
171. Circular 7/B/15168, 19 October 2001; interview with SAGIA functionary II, April 2004; *Arab News*, TOP100 supplement, 17 January 2004: 18.
172. "Investment law set to undergo radical changes." *Arab News*, 10 April 2002.

The report started circulating in the second half of 2003 and engaged with numerous issues like delays in the issuance of licenses, various forms of red tape, arbitrary administrative decisions, and contradictory regulations. It pointed out that twenty-four laws were in conflict with the investment law, that real estate rules were not operational, that the bylaw's minimum investment thresholds contradicted the FIA itself, and that the MoC ignored the content of SAGIA licenses.[173] It repeatedly pointed out that both domestic and international capital was being lost to better-prepared neighboring jurisdictions.

Apart from the MoC, the report frequently mentioned the Ministries of Interior and Justice as the parties responsible for various difficulties. The report also contained broader observations on the country's social and cultural development, a thinly veiled attack on the religious interests in the state whose policies made Saudi Arabia inhospitable to foreigners.

The report's proposed remedies were wide-ranging and included a systematic publication of all agency decisions and decrees, closer coordination with SAGIA, and swift resolution of the above contradictions along the lines of FIA rules and principles.[174] The document was widely discussed in the bureaucracy, and some liberal voices saw its broad criticisms and relative frankness as contributing to the country's development.[175]

This was not the general perception in the public sector, however. The report created unease and apprehension in most of the agencies.[176] The MoC leadership reportedly attempted to have references to commercial registration removed from its list of "obstacles."[177] Already before the report, there had been complaints that SAGIA was trying to erode the powers of other ministries and that in its policies it "is seen to have little coordination with other Ministries."[178] A main effect of the report was to antagonize parts of the Saudi administration still further.

The crown prince followed the report with the creation of a committee to study investment constraints comprising MoCI, the Ministry of Economy and Planning, SAMA, a minister of state, and three members of the SEC advisory council. The eventual document emerging from this committee did take up quite a few items, expanding on general features of Saudi governance and various policy areas, although sometimes only in general terms. Among other things, it demanded limits on administrative discretion and increased transparency of procedures, the

---

173. SAGIA 2003: 11, 48, 54, 63, 73.
174. Ibid.: 18.
175. Interview with senior advisor to Saudi government, June 2004.
176. A number of obstacles in the SAGIA report supposedly "scared" other ministries. Interview with SAGIA functionary III, May 2004.
177. Interview with Saudi industrialist, March 2004
178. World Bank 2002: 159.

reformulation of other agencies' rules in line with those of SAGIA, a follow-up mechanism on all government decisions, the introduction of independent commercial courts and of an integrated competition policy, streamlining of unclear visa procedures, empowerment of the one-stop shops with time limits for all procedures, more transparency on available industrial land, the introduction of electronic communication between agencies, the continuation of tax holidays, and streamlining of customs procedures. SAGIA was tasked with coordinating these measures with the concerned agencies and reporting to the committee, which would in turn interact with the agencies and submit a report to the SEC.[179]

What was the practical effect of the obstacle report? In some cases, procedures indeed were simplified, although it was not always clear how much SAGIA had contributed to these outcomes. In early 2004, a new MoCI center for processing CRs was opened. Unlike previous procedures, applying for or renewing a CR no longer required a "Saudization" certificate from the MoL and a *Zakat*[180] certificate from the tax department, which considerably facilitated procedures. The new tax law has brought certain corporate tax procedures in line with international practice.[181] SAGIA was also authorized to issue VIP cards for certain businessmen to facilitate visa procedures. By 2004, SAGIA representatives claimed that investment was slowly becoming easier and that some obstacles had been removed.[182]

In many other regards, however, there was little movement, at least until WTO accession. SAGIA remained the only agency with clear licensing deadlines, MoC licenses continued to diverge from SAGIA ones, interagency coordination remained haphazard, little was heard about commercial courts, customs procedures remained complicated, and real estate remained difficult for foreigners to acquire. Not all of the successes SAGIA has claimed for itself have seen consistent follow-up, moreover. System-wide changes have proven especially hard to engineer: many agencies ignore the royal decree stating that SAGIA must receive other organization's legal and procedural documents.[183] Similarly, the SAGIA suggestion to all agencies to make their relevant laws available on their websites in several languages seems to have had little impact.[184] Certain measures, including the 2004 tax law, have openly contradicted recommendations of the committee report on investment obstacles. In specific areas, there was also (at least temporary)

---

179. Interministerial committee report on investment obstacles, Riyadh 2004 (unofficial English translation).
180. A religious tax levied, rather inconsistently, on the liquid capital of Saudi companies.
181. Interview with senior accountant, Riyadh, May 2004.
182. Interview with SAGIA functionary III, May 2004.
183. For example, SAGIA had to get the rather widely circulated draft bylaws to the 2003 capital market law "from the street." Interview with SAGIA functionary III, May 2004.
184. *Arab News*, TOP100 supplement, 17 January 2004: 18.

regression; for example, additional security rules complicated the application procedures for those business visas not covered by the VIP rule.[185]

Many interpreted Abdallah bin Faisal's departure from SAGIA in early 2004 as a tacit admission of defeat.[186] The FDI registered with SAGIA declined from an initial spike in 2001 of 31 billion SR of investments licensed to 6.5 billion SR of projects licensed in 2003.[187] Due to SAGIA's limited capacity to measure actual investment on the ground,[188] the figures are imprecise and probably much inflated but nonetheless indicate a general trend of disenchantment. Several stories circulated of international companies putting their operations on hold or cancelling projects after realizing that many of the reform promises remained unfulfilled.[189] Foreign investors quickly became more cautious as the word spread.[190] UNCTAD's *World Investment Report 2004* ranked Saudi Arabia 31st in terms of investment potential—but 138th in actual investment performance.[191]

SAGIA under Amr Dabbagh

Abdallah bin Faisal's successor, Amr Dabbagh, a young, prominent business leader from an old Hijazi merchant family, appeared on the scene suddenly.[192] Not a royal, he has been less vocal and aggressive vis-à-vis other government actors than his predecessor, but he devoted more attention to procedural details than the prince, who had been a promoter rather than a manager and had left SAGIA's internal structure in flux.[193] Under Dabbagh, SAGIA evolved from a patrimonial court to an administration with more clearly defined departments, tasks, and job titles.[194] Reported to enjoy good access to Crown Prince Abdallah through his friendship

---

185. In July 2004, a bureaucrat in the MoFA reportedly decided to cancel the issuance of visitors' visas altogether, which made the private sector go "crazy." E-mail communication with Western economic advisor, July 2004.

186. Interview with board member of Riyadh Chamber, April 2004; interview with Western investment banker, Riyadh, April 2004.

187. SAGIA monthly investment statistics (received summer 2004).

188. According to one estimate, the actual value of investments since March 2000 is less than 7 percent of the value of licenses granted. Discussion with Western economic advisor I in Riyadh, June 2004.

189. Several UK companies, for example, have tried to set up fully foreign-owned projects, which faltered due to visa and other administrative issues. Interview with Western diplomat II. A big poultry project reportedly fell apart and was relocated to the United Arab Emirates because an Iraqi businessman involved did not get a visa; cf. *Arab News,* 13 July 2003, quoting Okaz.

190. The impact of 9/11 is hard to gauge. If it was a significant factor, then this was in addition to the local perceptions of disenchantment with an unchanging administrative environment.

191. UNCTAD 2004, tables A.1.5 and A.1.7.

192. *Arab News,* 23 March 2004.

193. Discussions with SAGIA staff.

194. Discussions with Saudi and Western businesspeople, Riyadh, November/December 2005.

with one of his sons and gatekeepers, he also increased the hiring of senior personnel from the private sector and expanded the existing office infrastructure.[195]

One significant achievement SAGIA has claimed under Amr Dabbagh is a series of memoranda of understanding (MoUs) with other agencies signed in early 2005. Broadly supported by the crown prince, these aimed at streamlining administrative procedures and are, while much belated, impressive on paper—even if the fact that different government agencies would negotiate agreements among themselves struck some spectators as odd. The issues covered include shorter waiting periods for licenses, streamlined judicial procedures, easier availability of visas, facilitated import collection at Saudi ports, and incentives for investment in underdeveloped regions and for workforce Saudization.[196]

Since Dabbagh has taken office, Saudi Arabia has scored several much-publicized successes in attracting large project investment. In some projects, SAGIA has been directly involved. It has initiated the King Abdullah Economic City in the Western region, convincing UAE real estate giant Emaar to undertake its development, and it is promoting several other economic cities with local and foreign partners.[197] In 2005, Intel signed an agreement with SAGIA to set up a 100 million US$ high-tech investment company.[198] In the underdeveloped southern province of Jizan, SAGIA successfully promoted a global-scale shrimp farming project.

Actual FDI inflow into the Kingdom more than doubled to a reported 1.87 billion US$ in 2004 (the figure, for the first time, being based on actual survey work with the help of external advisors), and it skyrocketed to more than 20 billion US$ in 2007.[199]

In 2006, SAGIA announced the "10 by 10" initiative, which aimed at getting Saudi Arabia among the top 10 countries worldwide in the World Bank's "Doing Business" report by 2010. The report ranks the efficiency of national bureaucracies' interactions with business. Saudi Arabia has been crawling up the list recently, from 67 in 2004 to 16 in 2009.

SAGIA as Super-Wasta

What explains the stunning turnaround? Has SAGIA finally succeeded in restructuring the bureaucracy around it? Closer inspection shows that it rather has

---

195. Dabbagh reportedly is part of the "triple A" group around the king, the other two being the king's son Abdulaziz bin Abdallah and Abdallah Saleh Kamel, a son of one of the Western region's most prominent businessmen.
196. *Arab News*, 3 July 2005, 15 August 2005; discussions with SAGIA functionaries, Riyadh, December 2005.
197. Cf. the website for the project, available at http://www.kingabdullahcity.com/en/.
198. *Arab News*, 17 November 2005.
199. SAGIA 2005.

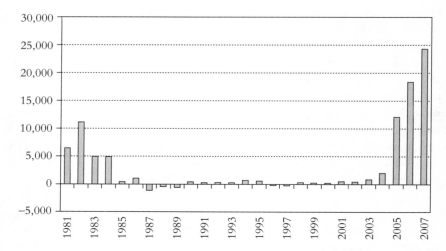

Fig. 5.2 FDI inflows until 2007 (million US$)
Source: UNCTAD.

learned to live with it. Rather than an immediate result of broad-based structural change in the administration, many of SAGIA's recent project successes have been a product of heightened case-by-case advocacy through high-level SAGIA functionaries: Since 2004, Amr Dabbagh and his deputies have taken to making discretionary phone calls to high-level administrators in other agencies to expedite procedures on an individual basis.[200] Due to the general backing of Crown Prince (now King) Abdallah, this individual intercession has been successful in several cases, and on some occasions, SAGIA has built individual alliances with high-level figures such as regional governors interested in local development.

This dynamism, however, is based on individual, personalized, case-by-case brokerage through SAGIA, with SAGIA itself acting as wasta or informal intermediary (arguably reducing the business opportunities of some private intermediaries). SAGIA even helps investors to figure out "ways around the rules."[201] Resources for high-level, personalized intercession are limited, however, and smaller to midsized foreign investors cannot always mobilize such high-level support.[202] Such actors tend to suffer from anonymous bureaucratic stonewalling just as before,

---

200. Barillka, Greenberg interviews. Dabbagh's deputy Awad Al-Awad has come to be known as "the fixer."
201. Barillka, Greenberg interviews.
202. Discussion with Saudi businessman, Italy, March 2006.

since the bureaucracy has seen little structural change. SAGIA has not imposed systemic change yet.[203]

SAGIA has not reported specific progress on the memoranda of understanding, which were never published. As of this writing, it remains unclear how well implemented they are and how many unintended follow-up problems they will generate. The general perception in early 2009 was that despite the introduction of some simplified procedures, bureaucratic behavior in general has changed little in most agencies.[204] In at least one case, a senior intervention seems to have thwarted the negotiation of a meaningful MoU, as Mohammad bin Naif, son and assistant of the minister of interior, has reportedly much watered down the agreement in order to preserve MoI discretion.[205]

On SAGIA prodding, some regulatory procedures have been formally streamlined, requiring fewer steps, lower fees, and less paperwork. With Abdallah's broad backing, Dabbagh and his senior executives were able to negotiate with Yamani's MoCI to simplify the commercial registration process, abolishing the authentication of company books by the CCI and the public notification in local newspapers and reducing official fees by 80 percent.[206]

Yet, core components of the process are still at the mercy of low-level bureaucratic discretion at the MoCI. Even with the reduced procedures, getting a CR can still take six months.[207] Individual-level discretion is hardly reduced: MoCI bureaucrats can still demand changes to the wording of a company's articles of association and make sudden demands for documents, such as papers stamped by the MoJ, that are not codified anywhere.[208] Changing a company's articles of association still tends to be a cumbersome process, and deputy ministers can still change administrative procedures through unpublished and sudden memos. Some interviewees believe that in the face of streamlined general rules, such ministerial intervention has increased.[209] Line agencies can still refuse to issue licenses

---

203. Barillka, Greenberg interviews.
204. Interview with senior expatriate lawyer, December 2005, Barillka, Greenberg interviews; SAGIA functionary I, December 2005.
205. Discussions in Riyadh, November/December 2005. The Ministry of Interior also tends to strong-arm the Foreign Affairs Ministry if it wants to prevent visas from being issued; interview with former British diplomat, London, June 2005.
206. World Bank 2008: 12.
207. One lawyer estimated two to three months to open a branch office and six to nine months for a limited liability company, that is, a real piece of FDI. Interview with senior Western lawyer, Dubai, February 2009.
208. Interview with Saudi lawyer, Riyadh, January 2009.
209. Ibid. Also interviews with senior Western lawyer.

to holders of SAGIA investment permits, and pieces of paper still often take weeks to move from one office to another.[210]

More transactions are now conducted in SAGIA's service centers, but this does not necessarily speed up procedures; in some cases, while ministers claim that procedural issues have been resolved, their representatives in the centers stall on decisions to avoid taking responsibility.[211] In the clientelist Saudi bureaucracy, even formal agreements on details of procedure can only go so far in changing actual practice.

And while the leaders of SAGIA and MoCI seem to have come to some agreement, the deeper fiefdoms of the Saudi state have proven harder to penetrate. Courts are still unaccountable, can ignore secular regulations, and struggle to enforce their own decisions.[212] Judges' commercial skills are limited, and arbitral awards are not implemented.[213] The offices of the notary public is still difficult to navigate; foreigners often must go "notary shopping" until they have found a cooperative one willing to accept the company's documents. Big Saudi businessmen, by contrast, can have "pet notaries" who might well come to their houses to do the paperwork.[214]

The atmosphere in the Ministry of Interior in 2009 seemed to be less obstructionist than five years before, possibly because Mohammad bin Naif, a younger and more urbane character, had taken over much of the day-to-day management from his father. Yet on basic issues it had not budged: procedures for real estate acquisition remained unclear, and it was still impossible to bring in a foreign managing director to establish a new company.[215] While a pared down negative list theoretically allowed foreign ownership of advertising, approval from the Ministry of Information—under strong MoI influence—was not forthcoming.[216]

How can we square this with the Kingdom's ranking as the world's sixteenth-best business environment in 2009? The answer reveals much about SAGIA's capacity as a well-managed agency—but not a capacity that could really be used to revamp bureaucratic structures. Saudi Arabia's first jump in ranks was a stroke of luck. It was largely a change in the World Bank's ranking methods from

---

210. Interview with Saudi lawyer, Riyadh, February 2007.
211. Ibid.
212. Niblock 2007: 167.
213. Interview with senior Western lawyer, Dubai, February 2009; the problem is already mentioned in SAGIA 2003: 77.
214. Interview with senior Western lawyer, Dubai, February 2009.
215. *Saudi Gazette,* 31 March 2007, 10 November 2006.
216. Interview with senior Western lawyer, Dubai, February 2009. Visa acquisition also remained problematic; cf. *Arab News,* 28 April 2008, 18 February 2007.

the 2005 to the 2006 report that explained the Kingdom's jump from 67th to 38th rank.[217]

The subsequent ascendancy was mostly SAGIA's doing. Under the "10 by 10" program sponsored by the king, SAGIA negotiated various formal changes in regulations with other agencies that corresponded exactly to the indicators of the World Bank survey. As the latter in many respects measured formal rules rather than their implementation, Saudi Arabia managed to climb rapidly.[218] Some of the changes did in fact simplify cumbersome procedures, but at the same time, the World Bank's indicators did not pick up on most of the core difficulties of doing business in the Kingdom.

SAGIA created a special department in charge of "national competitiveness" dedicated to measurement issues, and the bonuses of SAGIA executives depended on achieving specific rankings in the World Bank study.[219] Its success by 2009 reflects the organization's capacity in preparing information, interacting with World Bank delegations, and administering a complex package of formal regulatory change. SAGIA also ran a highly effective domestic and international PR campaign. Yet, its power over day-to-day practices in other agencies remains limited, and many of the formal changes—eliminating stamps, fees or advertisements, abolishing minimum capital levels—did not address fundamental bureaucratic obstacles. In its 2009 ranking, the World Bank estimated that it took twelve days to start a business, but for foreign-owned companies a more realistic estimate was six months. Saudi Arabia was ranked first in the world for registering property, but again, this did not take into account the specific travails of foreign investors.[220]

Saudi Arabia's ranking was the product of a concerted SAGIA campaign targeting a particular set of indicators. On other international rankings of bureaucracy and business environment, Saudi Arabia's status hardly changed. On the Heritage Foundation's index of economic freedom, it has fluctuated between 60 and 65 points since SAGIA's inception, with no apparent trend. The same is true about the World Bank's "governance indicators" on government effectiveness and regulatory quality.[221] In a 2006 survey of local and international business

---

217. According to the old indicators, it actually ranked slightly worse than the year before; the World Bank has now removed the old rankings from its website. Saudi Arabia also profits from the inclusion of tax and labor regulation as ranking criteria, because domestic businessmen pay low taxes and all businesses can hire and fire their foreign workers rather freely.

218. Oxford Analytica, Bank's indicators put to test, 26 June 2008.

219. World Bank 2008: 45.

220. The one area in which the Kingdom ranked badly, enforcing contracts, is the one that focused more clearly on the actual length of procedures. Saudi Arabia fell from rank 136 to 137 compared to 2008, with an estimate that trial and enforcement took an average total of 635 days.

221. Both are unrelated to the "Doing Business" ranking. The Kingdom ranked in the 51st and 52nd percentile respectively in 2007, below countries like Ghana and Romania, despite being in the top quarter of states in terms of GDP per capita.

leaders active in the Gulf, 52.2 percent of respondents ranked the Kingdom as the most difficult GCC state in which to do business (while almost 31 percent chose the answer "I don't know").[222] This was at a time when "Doing Business" was already ranking Saudi Arabia at the top of the GCC. While SAGIA implemented the "10 by 10" program, the Saudi private sector continued to complain about bureaucratic sluggishness and investment obstacles.[223]

Explaining FDI Successes

So where did the large FDI sums after 2004 come from? Some were related to SAGIA-promoted projects. Most of them, however, were in heavy industry or utilities, administered by separate agencies or state-owned enterprises. In 2006, the last year for which a detailed breakdown is available, SAGIA estimated total investment inflows at 14.3 billion US$. Of these, 4.4 billion US$ occurred in mining, oil, and gas, mostly related to the international gas consortia operating under Aramco's tutelage. 2.1 billion US$ were in refining and 3 billion US$ in chemical manufacturing—mostly through joint ventures administered by Aramco and SABIC, respectively. 1.7 billion US$ flowed into electricity and water, through large public-private power projects undertaken by the publicly owned Saudi Electricity Company and Marafiq, a public utility company tied to the RCJY.[224] 2.3 billion US$ went into finance, thanks to the opening of the insurance and investment banking sector under SAMA's auspices. 800 million US$ went into real estate and 500 million US$ into contracting. In all other sectors—light manufacturing, hotels, trade, media, services, transport, and communication—the investment flow was either negative or smaller than 11 million US$. The picture in 2005 and 2004 looked similar.[225]

In other words, although clearly a lot was going on and the post-2003 boom again attracted big international players, there occurred little diffuse, smaller-scale investment. Much of the investment looked not very different from that of the early 1980s, when Saudi Arabia saw the last wave of large, capital-intensive joint ventures with foreigners. Separate bureaucratic islands, many of them with strong track records since the 1970s or even earlier, administered the lion's share of the new projects. Players like Aramco, SABIC, and SAMA, however, have little to do

---

222. Leaders in Dubai 2006.
223. *Arab News,* 20 December 2006, 14 February 2007.
224. U.S. Energy Information Administration 2008.
225. There are only two variations. In 2005, real estate reached 1.4 billion US$, probably due to the kickoff of the King Abdullah Economic City. In 2004, transport and communication reached 840 million US$, thanks to an Etisalat-led consortium winning the bid for the second Saudi mobile phone license; cf. SAGIA 2006a, 2006b, 2007b: 16.

with the bureaucracy at large or with SAGIA. The foreign partners or licensees of these pockets of efficiency have traditionally operated through them rather than through the rest of the state.[226] Foreign investors diversified into new sectors such as electricity and finance, but along old patterns of enclave regulation and capital-intensive investment. Service, IT, and retail companies and light and high-tech manufacturers are still shunning the Kingdom. Judging from the detailed 2006 figures, there seems in fact to be less smaller-scale, diffuse manufacturing and service investment than in previous decades.

The number of SAGIA licenses averaged about 600 per year until 2006 and then suddenly rose to 1,389.[227] A significant share of this probably represents the legalization of cover-up businesses after the 2005 abolition of minimum investment levels. Most of the smaller investors are probably Arab expatriates already in the country.[228] Seventy-one percent of licensed finance is from Saudis, indicating that joint ventures still predominate.

Back to the Age of Projects?

Perhaps because reengineering the bureaucracy proved so daunting, SAGIA itself has become a promoter of large projects. However, its largest project to date, a cluster of new economic cities throughout the Kingdom, has in turn become entangled in interagency politics, once again underlining the difficult fate of late-comer institutions.

King Abdallah Economic City (KAEC), Amr Dabbagh's most ambitious venture, was the first city to be announced in 2005. The SAGIA board apparently was not consulted about the project, and one member subsequently resigned.[229] Stories spread about princely land grabs in the area intended for KAEC use. Yet, it bore the name of the king, so it became SAGIA's flagship project.

Although the UAE real estate company Emaar signed up to coordinate the expansion of the city, other private investors were expected to develop much of the basic infrastructure. These have been reticent so far, since the legal status of KAEC has been unclear. SAGIA hoped to give KAEC free zone status, i.e. a separate regulatory, licensing and fee regime, similar to the successful FDI enclaves in Dubai. Other agencies rejected the plan, however, and to date KAEC has been

---

226. Interview with Saudi lawyer, Riyadh, January 2007. Newly opened sectors such as finance, telecoms, or electricity are also administered through separate regulators.
227. SAGIA 2007a: 22, 33.
228. Interview with Saudi lawyer, Riyadh, January 2007; discussions with SAGIA functionaries, Jeddah, September 2008.
229. Interview with former deputy minister, Riyadh, February 2007.

subject to the same rules as all other businesses in the Kingdom, putting it at a disadvantage vis-à-vis the special zones in neighboring UAE.[230]

Other parts of the Saudi government clearly were unhappy with the project, and refused to be drafted into cooperating or providing services. A MoMRA deputy minister openly complained that SAGIA's new economic cities were not in line with the Kingdom's development plans and the national urban strategy.[231] In 2007, it was still unclear whether the Saudi ports authority would cooperate on KAEC's planned sea port.[232]

Similarly, it was not clear whether SAGIA's Jizan Economic City, announced in 2006, would have access to the industrial feedstock its planned petrochemical industries would need; as Aramco was not involved in planning the project, it reportedly refused to make the requisite gas available.[233] With many more players laying claim to sectoral regulation than in the 1970s, it now appears to be difficult to establish new enclaves along the lines of RCJY. SAGIA reportedly wants to apply "Aramco rules" for KAEC, allowing female driving and the mixing of genders. It is facing strong conservative resistance, however. The new King Abdullah University of Science and Technology announced in 2006, by contrast, is under the influence of powerful and well-established Aramco and hence will operate under liberal rules.[234]

In 2009, SAGIA was still trying to acquire free zone status for KAEC and had managed to win concessions on labor visas for the city, which are supposed to be administered by a separate, SAGIA-sponsored agency called Kader.[235] The jury on KAEC is still out—but even if it succeeds, it will signify a return to the 1970s model of enclave development.

Saudi Arabia's FDI figures since 2005 have been formidable. Yet, by and large this is not the result of a more coherent or streamlined national bureaucracy. Instead, the Kingdom has played on its historical strengths: autonomous pockets of bureaucratic efficiency outside of the regular state apparatus and large-scale project management—a paradigm toward which even SAGIA has been gravitating.

Wider-ranging reforms of the Saudi administration have been much more difficult. Despite considerable formal concessions, foreign investors outside of a limited number of enclave sectors are under de facto bureaucratic constraints similar to those in play a decade ago. In the absence of a powerful broker—of whatever kind—swift access to many Saudi markets remains elusive.

230. Interview with Saudi lawyer, Riyadh, January 2007.
231. *Saudi Gazette,* 5 April 2007.
232. Discussion with Saudi bureaucrats and bankers, Riyadh, January 2007.
233. Barilka, Greenberg interviews.
234. *Kuwait Times,* 21 August 2008.
235. Discussions with SAGIA functionaries, Jeddah, September 2008.

Conclusion

Cleavages within the Government

The foreign investment reform put the flexibility of the Saudi bureaucratic system to its most comprehensive test since boom times. Although the leadership and technocracy had seemed to agree on the reform project, contradictory views within the Saudi government soon became apparent: not so much over the reform in general than over its specific implications for various organizations. When more agencies got involved in spelling out details in the bylaw and the negative list, the reform started to look less radical. Numerous line agencies were able to cast at least a dilatory veto against the opening of their sectors. No broader reform coalition seemed to emerge in the cabinet. On the most senior level, the crown prince appeared as the lone driver of the initiative. With the exception of an occasional (and selective) push from above, commoner ministers would remain passive, stall initiatives on an individual basis, and defend their agency turf.

With his remarkable expedition of the lingering debate in early 2000, Abdallah demonstrated that as the most senior active leader, he had the capacity to propel policy-making if he exerted the necessary pressure. The negative list and subsequent developments indicate, however, either that agency and other interests persuaded him to desist from pushing the reform to its practical conclusion or that he was incapable of getting involved with technical details. His interventions were selective.

While some technocratic agencies started to cooperate with SAGIA selectively, other players with a deeper history of institutional autonomy remained beyond the organization's grasp. Although less directly concerned than the MoCI, the MoI has probably put up the stiffest resistance in political and administrative negotiations on FDI, looking at all things foreign from a security rather than an economic perspective. Due to its high rank in the informal pecking order, MoI representatives were hard to argue with.[236] While his son intervened at least at one juncture, it is unclear whether the minister was personally involved on specific issues. What is more important is that MoI personnel derive their power largely from his seniority; his mere presence has a structural effect. Similarly, the historical status of the judicial system as a separate ideological sphere gave it much leeway to ignore FIA reform efforts. SAGIA had close to zero leverage over the largely autonomous MoI and MoJ. Conversely, the biggest chunks of FDI have been channeled past SAGIA through older, better established islands of efficiency like SABIC and Aramco.

---

236. Interview with SAGIA functionary I, May 2004.

The Fate of Latecomers: SAGIA

SAGIA itself has been an administrative island and an institutional client of Abdallah that is very different from its surroundings. It represented a typical attempt to solve an emerging problem by creating a new organization. But SAGIA needed to trespass on the territory of existing players willy-nilly, as providing a legal framework for FDI proved an essentially cross-cutting task—an exceedingly difficult one in a setting where bureaucratic territoriality reigned supreme and SAGIA's authority was, insofar as it was effective, mediated through the superior level.

SAGIA was created at a moment when administrative turfs had already been claimed. The royal status of its first leader may have helped in providing a space for free debate, but it does not seem to have helped much in the face of resilience of established agencies. SAGIA's focus on policy advocacy, some of it public, probably was also a reaction to the otherwise crowded and impenetrable institutional landscape. Despite expectations that it would be responsible for all licensing and would control all investment-related procedures, it never acquired such lead agency functions.[237] The SAGIA license that now precedes the commercial license signifies, most of all, an additional layer of administration.

At the same time, on a policy level, its board—like most interministerial councils—reproduced the very segmentation of state agencies it tried to contain. On the administrative level, the "one-stop shop" did the same: The local representatives who did little more than collect papers for their respective agencies were conduits in a segmented system in which communication mostly happens vertically, within the segments. Although it is not easy to choreograph interagency cooperation on a working level anywhere, other countries like Portugal, Hungary or Malaysia have managed to create one-stop shops with real regulatory authority.[238]

By most accounts, SAGIA activity has been well-intentioned, and despite its initial lack of internal structure, allegations of manipulation or secrecy have been rare. It has been a reasonably well-"insulated" agency with independent recruitment mechanisms and a clear reform agenda. Attempts at institutional insulation are meaningless, however, if old veto players in a fragmented system cannot be overruled.[239] SAGIA did not enjoy exclusive access to the top level; it was one of many supplicants. At times, gatekeepers and advisors from important families around the crown prince reportedly hampered access and shielded SAGIA input from him.[240]

---

237. Bfai Unternehmerbriefe, 15 March 2000; Montagu 2001: 38.
238. World Bank 2002: 65.
239. Cf. Chibber 2002.
240. Discussion with SAGIA staff, summer 2004.

Defending the Turf on Lower Levels

Splits over policy on the cabinet level do not explain the full range of problems that SAGIA encountered. "Low-level" resistance and incapacity was the main issue it had to deal with on a day-to-day basis.[241] Much of it was a result of the bureaucratic inertia of the Saudi distributive state, in which recruitment and often promotion are nonmeritocratic, incentives for initiative lacking, decisions centralized, and sanctions for underperformance hardly applied—not in one but many parallel bureaucratic bodies. Unlike the project- and agency-based development of the 1970s, FDI reform involved immovable mid-level bureaucrats in many organizations.

In many cases, individual administrators, just like the leaders of their institutions, seemed to develop a proprietary sense of their posts and an active interest in defending existing mechanisms of discretion and control, continuing their application unless directly ordered otherwise from above.[242] Many of the instruments of "predatory" business regulation in terms of licenses and documentation requirements have not disappeared. Even after SAGIA, "the government still wants to run the show," as has been its habit since the oil boom.[243]

In the strictly hierarchical Saudi system, complications and delays are more easily engineered than a straightforward decision for which responsibility has to be taken. This creates opportunities. Beyond direct bribery and nepotism, slow-moving, predatory, or contradictory regulation sustains the broad class of middlemen between administration and business that lives off unequal access to the opaque, segmented, and informally penetrable state—often themselves drawing on senior-level patronage.

Formal restrictions on access to the system, as in the case of minimum investment levels, tend to provide additional space for middlemen. The limited capacity of most Saudi government agencies to keep tabs on what businesses are actually doing further increases the scope for games of pretense and manipulation. It is on this micro-level that societal networks—largely incapable of influencing the state on the meso-level—end up having great influence on policy implementation and state behavior.

Insofar as SAGIA has fulfilled its mission of attracting FDI, it has often done so by itself working around established rules and bureaucratic habits rather than

---

241. Interview with SAGIA functionary II, April 2004.
242. Montagu (2001:7) cites a socioeconomic explanation by a long-term expatriate: "The little guy in the middle or bottom [of the bureaucracy] has had no pay rise for 15 years.... The guys in the middle are trying to make a truly Byzantine muddle to slow up the pace of reforms, because they fear they will be further marginalised and impoverished."
243. Interview with board member of Riyadh Chamber, Riyadh, May 2004.

through bureaucratic reform. This reflects a certain informal suppleness of the Saudi system on a senior level but is a far cry from systemic change—certainly on the mundane bureaucratic level.

The Private Sector: Individual Development and Collective Lethargy

Saudi private sector actors can be part of the intermediary networks but can also suffer from them. Quite frequently, both can be true about the same actor, depending on the context. That business actors have different positions in such games probably explains some of the diffuseness of private sector opinion. There is, however, a clear trend within Saudi business to demand smoother and more predictable administrative procedures. This results from the increasing complexity of business operations and the growing autonomy of the private sector from public material patronage. Much of Saudi business has evolved far beyond mere middleman status and sees bureaucracy as more of an obstacle than a source of rents. It is also a function of the increasing size and complexity of the Saudi system in which gaining informal access has become costlier for all but the most powerful players, continuing a trend visible since at least the 1960s.

Such individual maturity has for the most part not translated into coherent collective policy demands on foreign investment issues, however. The FIA saw an unprecedented degree of open consultation on invitation of the regime. Yet, like on that other great issue of international opening, the WTO membership, no clear overall position emerged from the business community.

There were clear demands and voices on individual questions. The industrialists' position in favor of lower taxes for their international partners was a sign of relative sectoral sophistication, and the private sector is less blatant in asking for subsidies and protection than in previous decades.[244] In the end, however, Saudi business still exhibited a diffuse and individualized tendency for conservatism and a fear of losing established stakes. As a meso-level player, Saudi business remained a policy-taker toward the initiatives of the patron state, playing out its engrained role, and state actors reserved the prerogative to decide on last-minute policy changes.

---

244. On protectionist demands in the 1980s, see *Saudi Arabia Monitor,* May 1987: 24.

Chapter 6

Eluding the "Saudization" of Labor Markets

The foreign investment reform discussed in the last chapter was not able to break or remold existing structures of authority and clientelism in the Saudi state; instead, SAGIA settled on working through and around them. It was, however, a specific initiative tied to a specific new agency. One might argue that the policy simply got off to a bad start, being a victim of SAGIA's outsider position, and perhaps was not the leadership's highest priority. This chapter will analyze the policy of "Saudiizing" the national labor market, a policy implemented by long-established ministries, which the leadership clearly regards as crucial for the Kingdom's economic and political stability. Nonetheless, as a complex and in many regards cross-cutting regulatory issue, the problems it encountered remarkably resemble those of the FIA.

Much of the time, Saudization has been less about reforming the bureaucracy than about developing regulatory control over labor markets. This means that the focus of this chapter is somewhat more on state attempts to influence business behavior and to gather information and less about administrative streamlining. Despite this different thrust of the policy, however, we will see a tableau unfolding that in many respects resembles that of the FIA story: administrative fragmentation; parallel policies and lack of cabinet-level coordination; sudden, top-down royal interventions; unaccountable and seemingly immovable mid-level actors stalling decisions and pursuing their own agendas; and businesses developing sophisticated techniques to work around interventionist regulations and a bureaucracy they do not trust.

Moreover, although the Saudization issue is played out mostly between *domestic* businesses and the bureaucracy, we will encounter an inordinate number of brokers with privileged administrative access helping less well-connected businessmen avoid state-imposed labor market rigidities. Saudization made low levels of regulatory penetration and state-business cooperation even more salient than the FIA case. The shifting and unenforceable labor regulations recall the muddle of procurement and national privilege rules of the 1970s—but this time the regulatory clutter threatened national business rather than creating material opportunities for it. Rule-avoidance through brokers has been defensive rather than rent-seeking.

Unlike the FIA case, the government dropped a plan to create a dedicated institution for Saudization issues in 2004. Instead, the administration underwent a possibly unprecedented step of meso-level consolidation, as the crown prince brought labor market responsibilities together in a newly empowered Ministry of Labor (MoL) under a client of his. This step helped to unify labor visa policy, the most important macro-instrument on the Saudi labor market. It created a focal point of policy-making and responsibility, giving the MoL more control over labor market issues than SAGIA will probably ever have over foreign investment matters. At the same time, however, the politically strengthened MoL faltered in its ambitious attempts to micro-manage labor market structures, as its administrators were neither able nor motivated to regulate individual businesses consistently and to gather information from them. Cooperation between bureaucracy and business has remained low. Over a decade of Saudization, state and business never managed to find a credible accommodation on how to implement the policy, and the interest of business in—again—anything but dilatory lobbying is arguably undermined by the weak regulatory capacity of the state.

Background: The Saudi Labor Market

Like the other Gulf rentiers, Saudi Arabia has a long history of employing vast numbers of expatriates, which had largely crowded nationals out of the private labor market in the 1970s boom.[1] Due to increasing underemployment among nationals, reducing the share of foreign labor in overall employment has been the declared aim of the Saudi government ever since. In the third five-year plan (1980–84), the government targeted a reduction to 1 million expatriates. Their number actually increased from 1.3 million in 1979 to 2.7 million in 1984 and

---

1. The percentage of Saudi workers in Riyadh collapsed from 22.1 to 11 percent from 1976 to 1978 alone; cf. *Bahth 'an asbab 'uzuf ash-shabab as-sa'udi 'an al-'amal fil-qita' al-khas* (1399, Riyadh Chamber of Commerce and Industry): 9 (A study of the reasons for the Saudi youth's reluctance to work in the private sector), IPA, Folder 189.

continued to grow to 4.5 million by 1994.[2] In the second half of the 1990s, Saudization became more urgent. While the population continued to grow at more than 2 percent a year, budgets were squeezed, rendering untenable previous public employment guarantees for the increasing pool of Saudi graduates.[3]

Segmentation of the Labor Market

The government has been largely Saudiized since the oil boom. By the mid-1990s, 94 percent of the public workforce reportedly was Saudi. However, less than 7 percent of private sector employees were nationals.[4] Although the government reportedly began to hire again in 1999 after an imperfectly enforced three-year freeze, even freehanded Defense Minister Prince Sultan has admitted that the bureaucratic overstaffing was a problem.[5] The modest increases in the public payroll were not capable of accommodating the increasing numbers of young Saudis entering the job market.[6]

In 2001, the Central Department of Statistics (CDS) estimated that 3 million Saudis and as many foreigners were employed in the Kingdom.[7] According to Ministry of Civil Service records, 630,000 of the Saudis were government employees, leaving some 2.4 million Saudis in private employment. As late as 2003, however, the Ministry of Labor mentioned only about half a million private Saudi employees,[8] and unofficial estimates of state employment have been considerably higher, attributing almost half a million employees to the Ministry of Interior alone.[9] The discrepancy results from the uncoordinated statistical activities of the Ministry of Labor, CDS, and other agencies and shows that basic information is not harmonized even on pivotal political issues like national employment.[10] All of the estimates, however, agree that foreigners still outnumbered locals on the private labor market—even the high-end estimate of nationals only corresponds to about a third of the population of employment age.

2. Krimly 1999: 260–61.
3. As befits a distributive state, the provision of job opportunities is enshrined in article 28 of the 1992 basic law.
4. *MEED*, 6 April 1996: 55.
5. *Arab News*, 8 August 2004.
6. Increases have been hovering between ten and twenty thousand posts per year; cf. Statistical Yearbook, various issues.
7. According to the Saudi Central Department of Statistics' labor force survey, available at www.planning.gov.sa.
8. SAMA Annual Report 2003: 49. The same report also presents the higher figure in a different chapter, 303.
9. Saudi-British Bank, Giving a Boost, *SABB Notes*, 7 February 2008: 5.
10. Saudi economists have publicly expressed doubts on official labor statistics; cf. *Al-Watan*, 19 August 2007.

There is little dispute that the private sector will have to be the main source of future national employment. Although demographic growth slowed in the 1990s, there still are up to 200,000 graduates entering the Saudi job market each year, and estimates of newly available jobs in the early 2000s were as low as 20,000.[11]

The main economic obstacle to higher private Saudi employment has been a segmented labor market, a direct outcome of rentier state building. The public sector has been offering higher wages and more security for less effort to nationals, while in the private sector they have to compete with low-wage expatriates.[12] According to a 1996 survey, the average Saudi on the private labor market earned 5,700 SR per month, while the average expatriate had to make do with 1,900 SR.[13] Still, wages for Saudis remained higher in the civil service for all levels of education compared to private employment.[14] So Saudis, uncompetitive on the private labor market, had strong incentives to wait for a public job. Even remaining idle under such circumstances was seen as preferable to a low-paying private job.[15] Many family networks still are tight enough to sustain unemployed younger relations.

Even when young Saudis look for jobs in business, employers are often reluctant to hire them, as the religiously dominated education system scarcely equips them with the foreign language and numerical skills required for anything but basic tasks.[16] Expectations of public employment also lead to the selection of university majors irrelevant to the private sector, mostly in the humanities and social sciences. With the exception of well-paying islands of efficiency such as SABIC or Aramco, much of the state apparatus hires university graduates indiscriminately. The cavalier attitude of young Saudis to education and employment more generally also is a sign that the labor market is often not that much of a market: most young Saudi employees are placed through wasta, the connections of senior relatives and friends who know prospective employers.[17] Jobs for younger protégés are one of the main informal bargaining tokens in the informal brokerage networks in Saudi business and government.[18] Due to the personalized nature of this type of brokerage, it is limited to "strong" closely knit social networks and is hence a rigid and exclusive mechanism.

11. SAMBA 2002: 3.
12. Diwan and Girgis 2002.
13. Abdalwahid al-Humaid, *Tanmiyya wa tatwir al-quwa al-'amila fil-mamlaka al'arabiyya as-sa'udiyya* (Riyadh, 1422): 23 (Development of the work force in Saudi Arabia), IPA, folder 767 (354, 538).
14. Diwan and Girgis 2002.
15. *Riyadh Daily*, 18 June 2003.
16. Daghistani 2002: 60.
17. Montagu 2001: 22.
18. Interview with former Saudi deputy minister, May 2004.

All these factors make it difficult and costly to increase the Saudi share in specific types of private employment, as various Saudization decrees have required since the mid-1990s. Against the background of a fragmented and soft administration, this has led Saudi business to explore numerous ways around the rules.

Institutions and Interests in Labor Market Policy

Before looking at specific Saudization decrees and initiatives and their digestion in Saudi state and society, the chapter will give an overview of the actors and interests involved in labor market policies. Not least because job creation was a popular issue, numerous interests have attempted to claim the mantle of Saudization.[19] Policy instruments such as national employment quotas, sectoral recruitment bans, and charges on expatriate labor have been used by a wide variety of institutions. Saudization suffered from a "fragmented policies" syndrome similar to the one the health and education sectors have witnessed since boom times.

The Ministry of Labor

The main government player for executing Saudization formally is the Ministry of Labor (MoL, Ministry of Labor and Social Affairs until early 2004). It consists of a central administration in Riyadh and some "labor offices," regional outlets that report to a deputy minister in Riyadh. The MoL registers work contracts, is supposed to check the implementation of Saudization rules, and issues certificates to ascertain their fulfillment. Only Saudis have access to the MoL to do official business there.

MoL officials are responsible for approving applications for the importation of laborers. They can grant partial approvals and attach various strings to them. Similarly, when a new business is set up, MoL officials decide the percentage of Saudis to be hired for each professional category. Until 2004, applications were forwarded to the Ministry of Interior for further approval. Ministry officials check company payrolls and can inspect workplaces to ascertain compliance with Saudization.[20]

By the summer of 2001, the MoL's database covered 2.1 million employees. Many businesses, however, respond slowly to the ministry's information requests,

---

19. Saudization and job creation have been among the few issues on which the Saudi media outlets have attempted to create some pressure; cf. *Arab News,* 8 January 2004; *Arab News,* 11 May 2004.
20. *MEED,* 10 March 1995: 40.

a good share of them incorrectly, and some not at all.[21] At least until 2004, the database did not keep track of employees in private households, who were not under the purview of the MoL.

The Ministry of Interior

For many years it was not clear which institution actually was the lead agency on Saudization policy. In addition to the MoL, there was a "Manpower Council" (MPC), chaired by Minister of Interior Prince Naif, as a supreme labor policy body. In addition to some bureaucrats, the MPC included three Chambers of Commerce heads, and private sector representatives sat on in its preparatory body.[22]

Through his chairmanship, Naif styled himself as the champion of Saudization and pushed the issue aggressively within government. Other ministers, especially those without a Western economic education, have followed suit.[23] Once again, broadly interpreted security concerns—in this case the control of foreign residents—have led the autonomous MoI empire to get deeply involved in matters of economic regulation.

Until 2004, the Recruitment Affairs Department of the Ministry of Interior checked lists of visa applications handed over from the MoL against its list of allowable national quotas. If the department gave its go-ahead, it charged 2,000 SR per permit through a designated bank.[24] Human resources companies through which Saudis could recruit domestic labor received their licenses from the MoI.[25]

With its policing resources, the MoI has also been involved in controlling labor markets. In addition to rounding up *hajj* and *umrah* overstayers and illegal immigrants, it has taken part in workplace inspections together with MoL representatives.[26]

Parallel Actors

Whereas any business planning to import labor through official channels inevitably had to deal with the MoL and MoI, there was a whole range of further agencies involved in specific labor markets, often in an ad hoc fashion. These included the Saudi Arabian Monetary Agency (SAMA), the Ministry of Commerce, the

---

21. *Arab News,* 16 July 2001; *Saudi Gazette,* 8 October 2005.
22. *Saudi Gazette,* 24 February 2004; Humaid, *Tanmiyya wa tatwir:* 30.
23. Interview with chief economist of Saudi bank I, March 2004.
24. World Bank 2002: 32.
25. Interview with Western diplomat I, April 2004.
26. Cf. *Saudi Gazette,* 23 February 2004; interview with member of Majlis Ash-Shura I, Riyadh, April 2004.

Ministry of Defense and Aviation and its sub-agencies, the Ministry of Health, and Saudi Aramco.[27] Each organization had historically assumed regulatory authority over "its" sectors.[28] The more powerful among the regional governorates have also issued Saudization decrees, implemented by regional Saudization committees.[29] The specifics of such individual policies were seldom consistent with the broader Saudization mechanisms and targets.

It is not only the Saudization rules that were governed by many overlapping bodies. Until 2004, as many as fourteen different government agencies were allowed to issue visas.[30] In 2003, the number of work visas given was to about 800,000, of which the MoL issued only an estimated 300,000.[31] This meant that the MoL had little control over the aggregate size and the composition of the expatriate labor force. The Ministry of Interior in particular issued a vast number of visas.

Private Sector: Interests and Capacities

Even more than the other economic reform moves since 2000, Saudization is a state-driven project. Although there is a diffuse and widespread societal interest in increasing national employment, labor interests have not been allowed to organize since the 1950s. This leaves only business as interest group. The private sector has been careful not to openly oppose the popular target of increasing national labor as such in the light of the broad if unorganized backing in the press and among the populace for the cause of nationalization. But business players have protested almost every individual administrative measure of Saudization. Says one: "Everybody is against the way it has been done."[32]

The individual economic interests of most Saudi businesses militate against Saudization. Saudi employment quotas and limits on expatriate hiring increase costs, decrease flexibility, and can impair productivity. That said, the sensitivity of sectors to Saudization differs. Contracting and construction, for example, have traditionally been highly labor-intensive sectors relying on low wages, much

---

27. Looney 2004; *Saudi Gazette,* 8 April 2007.
28. In addition to decrees from the Council of Ministers, the Manpower Council, and the Ministries of Interior and Labor, Daghistani's study on Saudization (2002) includes edicts from the Ministries of Municipal and Rural Affairs, Information, Civil Service, Communications, Education, and various governorates.
29. Cf. Daghistani 2002: 279.
30. Interview with member of Majlis Ash-Shura I. Another interviewee gave a figure of sixteen agencies. Interview with board member of Riyadh Chamber.
31. Interview with member of Majlis Ash-Shura I.
32. Interview with senior Saudi banker.

like smaller retail and service businesses.[33] Industry usually is more capital-intensive, and industrialists have claimed to be ready for Saudization.[34] However, employing Saudis in larger manufacturing plants has posed considerable problems in practice too.[35]

Whatever inter-sectoral differences there are, they have led to clear sectoral coalitions or policy proposals even less than in the FIA case. Instead, in interviews, in the chambers, and on conferences and symposia, business representatives have asked in the most general terms for more flexibility in the implementation of Saudization, better training of Saudis, a more open labor law, and clearer policy tools. The "means and manner" of Saudization are perceived as unclear; time and again, private sector representatives have complained over the lack of information on labor policies and on the labor market in general.[36] The business community has not been forthcoming with coherent counterproposals, however.

The clearest demand is that the government should not push for Saudization too quickly, as Saudis are poorly equipped for many jobs.[37] On a symposium on human resources development in Riyadh in 2004, one Saudi industrialist demanded privileged domestic market access as a reward for employing Saudis.[38] Some have asked for more cheap labor and have complained that expenses for residency requirements and the like are too high.[39]

Saudization, and How to Avoid It

What follows is a brief account of the Saudization decrees and their implementation, with emphasis on the negotiations between the administration and the private sector. Institutional fragmentation of the government, low monitoring capacity, and a haphazard policy-making process have made the implementation of Saudization impossible for the private sector to predict.

---

33. Margins of many of the smaller businesses are rather low, and competition is over price rather than quality. Interview with board member of Riyadh Chamber.
34. Interview with member of Majlis Ash-Shura II, Riyadh, May 2004; interview with former Saudi minister, March 2004.
35. Lehner interview.
36. For statements from representatives of leading business families Kanoo, Zamil, Qahtani, and Humaidan, see *Arab News*, 13 July 2003; *Arab News*, 21 April 2004, citing Okaz.
37. Interview with Saudi industrialist, March 2004.
38. Industrialist Abdalrahman Al-Zamil on the Human Resources Development Forum in Riyadh, 24 May 2004 (author's notes). Speaking on this MoL-sponsored event, Zamil showed statistics with sharply increasing Saudi staff in his enterprises, which however appeared to be still under the then official quota of 35 percent.
39. *Saudi Gazette*, 28 December 2003.

With exclusive upward accountability of government agencies, little interagency coordination, ambiguous rules, and occasional jolts of unpredictable top-down interference, Saudization policies were deeply fragmented on a meso-level. On a micro-level, the combination of bureaucratic inactivity and large de facto discretion has further undermined the consistency of labor market regulation. In this setting, businesses often preferred individual, informal, and even illegal strategies to cope with bureaucratic demands, often resorting to brokers of various kinds. Governmental ambitions to micro-manage specific job types and sectors have led to opportunities for predatory regulation and corruption on a considerably larger scale than in the FIA case. The government's Saudization demands were often incoherent but nevertheless grew continuously, particularly after 2001. When pressure increased to the degree that loopholes became harder to find, business protests became louder and more systematic. However, private sector lobbying consisted of little more than ad hoc pleas for postponement and flexibility.

In 1995 the cabinet issued the first comprehensive, target-based Saudization decree on the recommendation of the Manpower Council. It was called "Decree 50," and it became the cornerstone of subsequent labor nationalization policies. It mandated an annual increase of five percentage points in the Saudi labor share for all businesses with more than twenty employees and reserved certain administrative occupations for Saudis.[40] The MoL under Ali Al-Namlah, known as a stubborn Saudiizer, was one of the decree's main proponents.[41] The Ministry of Interior was to start enforcing the quotas in December 1995, and the government formally stopped the issuance of visas for certain job categories. Violations of Saudization rules were to be punished through non-invitation to government tenders, denial of new work permit requests, refusal of loans, and the refusal to process administrative formalities like amendments to the commercial registration.

Government agencies issued further decrees on restricted job categories and work permit limits in 1996, 1997, and 1998, often merely repeating and sometimes contradicting what had come before. Some of the pronouncements came from the MoL, others from the MoI, and still others from the Council of Ministers.[42] The 5 percent minimum quota and the attached sanctions were reissued in 1998, apparently restarting Decree 50, and Prince Naif explained that the MoI would issue new lists of exclusively Saudi jobs.[43] As the MoL and MoI

---

40. Decree 50, 21/4/1415; Humaid, *Tanmiyya wa tatwir:* 29–30.
41. Discussion with senior Saudi bureaucrat, June 2004; interview with former Saudi minister, December 2003.
42. Reuters, 27 October 1996, *Gulf Times,* 22 July 1997.
43. AFP, 20 August 1998; Manpower Council, decree 78/Q, 17/1/1419. Among the jobs reserved for nationals was the one of mu'aqqib, or small-time bureaucratic paper broker (see chapter 4); Manpower Council, decree M/1, 11/11/1415.

appeared to operate in parallel, it was not always clear which agency was setting the rules.

In the 2000s, the pace of Saudization decrees increased—even if they were increasingly out of step with formal Decree 50 requirements, making it an even more symbolic piece of legislation than the FIA.[44] In July 2001, the Council of Ministers decided on the gradual Saudization of cleaning and maintenance jobs, according to which every business that wanted to qualify for government contracts—uniquely important for maintenance firms—was to have at least 5 percent Saudi employees.[45] The rate would rise to 50 percent within nine years, similar to, but out of sync with, Decree 50 rules.[46] The MoL was tasked with enforcing the decree and reporting to Prince Naif at the Manpower Council.

In a little-publicized decision, the Ministry of Labor in 2003 commenced a separate Saudization quota for contractors at 10 percent, far below Decree 50 aims but far above the actual levels of Saudi labor in the sector—apparently a result of dilatory lobbying by Saudi contractors.[47] Minister Al-Namlah repeatedly announced the full nationalization of twenty-five additional job categories in retail by February 2004.[48] The target was far from achieved by February 2004, so the policy was simply announced again.[49] Explanations by high-level MoL bureaucrats about reserved jobs and deadlines contradicted prior declarations.[50] As we will see, contradictory, constantly reworked, and patchily enforced rules would be an important inducement for systematic evasion.

In addition to regulating by quota, the government tried to modify the price of expatriate labor by repeatedly increasing fees for work and residency permits (*iqama*) after 1995.[51] Involving simple administrative transactions, this policy was easier to implement than the various sectoral Saudization decrees, but by increasing costs it also created incentives for seeking shortcuts and brokerage to avoid fee payments.

---

44. SAGIA, predictably, opposed the unenforceable and cumbersome decree; cf. SAGIA 2003: 21.
45. Similar decrees had already been issued for specific contracting agencies; cf. Manpower Council, decree Q/8, 25/6/1416.
46. *Arab News,* 10 July 2001.
47. Interview with Saudi industrialist, Riyadh, March 2004; Hajjar interview; Ministry of Labor and Social Affairs, *Al-kitab al-ihsa'i as-sanawi 1423/24:* 18 (Statistical yearbook 2003).
48. *Saudi Gazette,* 16 October 2003. MoLSA reports mention twenty-two categories; *Al-kitab al-ihsa'i:* 18.
49. *Saudi Gazette,* 29 February 2004.
50. Ibid., 19 January 2004; *Al-kitab al-ihsa'i:* 18.
51. Royal decree M/8, 5/5/1421; *MEED,* 6 April 1996: 30; *Arab News,* 28 August 2000. Often employers illegally deducted the fees from the wages of their expatriate employees; cf. *Arab News,* 13 April 2002.

To Every Ministry, Its Policy

In addition to the MoL, the MoI, and the Council of Ministers, many individual administrative players have gotten involved in Saudization in specific sectors and geographical areas, spelling out the details of a broadly defined policy in often contradictory ways, similar to the government procurement and national privilege rules of the 1970s. Banks have repeatedly been put under pressure by the Saudi Arabian Monetary Agency (SAMA) to hire more Saudis.[52] Accountants have been under pressure from the Ministry of Commerce (MoC) and the Saudi accountants' society, which is under MoC supervision.[53] The Presidency of Civil Aviation (PCA), which is attached to the Ministry of Defense under Prince Sultan, has asked foreign airlines to hire at least 50 percent Saudis for their operations within the Kingdom.[54] The Ministry of Health (MoH) has stipulated Saudization conditions for the licensing of private hospitals, even as MoL dictated nationalization rules for jobs in the MoH-controlled public hospitals.[55] Aramco has set Saudization quotas for its contractors.[56] In addition to sectoral agencies and public companies, governors of regions preside over their own Saudization committees.[57] They would issue sectoral decrees that applied only in their own jurisdiction, but not in other governorates.[58] The criteria for most of these agency-level or regional interventions have been unclear.

Diffuse Saudization pressure from the top leadership, increasingly concerned about underemployment, created a need to be seen to be doing something on lower levels but did little to integrate policies. That commoner ministers reported to different principals further undermined coherent policy-making.[59] While different agencies tried to regulate what they considered their own spheres, coordination through the cabinet and cross-cutting implementation strategy were absent, resulting in policy drift, interspersed by occasional, uncoordinated responses.

---

52. Interview with human resources manager of Saudi bank, Riyadh, May 2004; Smith 2003.
53. Interview with senior accountant.
54. *Gulf News,* 19 May 2004.
55. *Arab News,* 29 May 2004.
56. The quotas were at 25 percent in early 2004, with an annual increment officially set at 5 percent; cf. *Saudi Gazette,* 23 February 2003. Aramco has formed a special body with its contractors to oversee Saudization; cf. *Arab News,* 26 January 2004.
57. *Saudi Gazette,* 29 February 2004. Prince Salman in particular is the supposed "Saudization champion" for Riyadh, negotiating Saudization measures with local businesses. Interview with human resources manager of Saudi bank, May 2004.
58. Cf. *Saudi Gazette,* 30 December 2003.
59. Civil Aviation, for example, is under Prince Sultan; the minister of health has been a client of the governor of Riyadh, Prince Salman; Minister of Education Mohammad Al-Rasheed has been close to the Tuwaijri family, who have acted as gatekeepers to Crown Prince Abdallah; and Hisham Yamani in the MoCI again has enjoyed Sultan's patronage.

Since the inception of obligatory Saudization in 1995, decisions have frequently been restated, quotas reset, deadlines postponed, and implementation broken up into phases. Princes repeatedly called upon the private sector to contribute their share to national employment but did not attend to policy details.[60]

Ambiguous Rules and the Ways around Them

Private sector protest against Saudization measures was subdued for several years after Decree 50 was issued. Much of this is due to the patchy implementation of the decree. In recent years, however, Saudization rules have inflicted "great pain" on many Saudi businesses, even though the legislation was not taken seriously at first.[61]

The exact meaning of Decree 50 has been shrouded in confusion. Some senior technocrats took the decree as merely indicative, reflecting an ideal target. That has not been everyone's interpretation, however.[62] Former bureaucrats complain both that Decree 50 is difficult to implement across the board and that the private sector has not taken the government's intentions seriously.[63] At the same time, even technocrats skeptical of rigid quota regulations carp that "the government is not serious in implementation."[64] Mutual expectations of compliance in state and business were low.

Implementation indeed has been highly inconsistent. By far the largest share of Saudi companies could not live up to Decree 50 requirements. De facto administrative discretion therefore reigned supreme when it came to evaluating Saudization performance and targets for individual businesses—with few formal recourse mechanisms for concerned businesses.[65]

Discretion on various levels only increased when government agencies continued to produce inconsistent rules and figures. At a meeting with foreign businesses in 2001, for example, the deputy minister of labor gave an obligatory Saudization quota that contradicted the figures his own MoL bureaucrats were officially demanding of businesses at the time. In fact, however, MoL representatives often settled for much lower rates than any of the figures discussed in public.[66] Official

---

60. AFP, 26 April 1998.
61. Interview with human resources manager of Saudi bank, May 2004.
62. Interview with former Saudi minister, Riyadh, March 2004.
63. Interview with member of Majlis Ash-Shura I; interview with former Saudi minister, Riyadh, March 2004.
64. Interview with member of Majlis Ash-Shura I.
65. Government agencies were supposed to give precedence to Saudiizers in project tenders, but the criteria remained unclear. *Al-kitab al-ihsa'i:* 18.
66. Lehner interview. The interviewee likened the Saudi government to a "blind man in a tunnel," moving forwards with little orientation and correcting its direction each time it hits a wall.

global Saudization quotas continued to be set on an ad hoc basis from year to year but remained far above what was feasible for most businesses.[67] Various post hoc sectoral exceptions, often reacting to business complaints, only complicated the picture.[68]

In the absence of clear or realistic figures, administrators have been able to strike individual deals.[69] Unaccountable mid-level bureaucrats had to use their discretion almost by default, as senior civil servants, despite their formal involvement in most decisions, could not effectively control the large-scale administration of labor imports on a day-to-day basis.[70] Labor offices have used fuzzy criteria, like the fact that a company places local job ads, to establish its willingness to Saudiize (it is said that many ads include deliberately wrong fax numbers).[71] The imposition of remedies has also been desultory: a labor office representative, for example, may decide that a certain number of Saudis have to be trained before more labor can be imported.[72] Individual quotas are often imposed on an ad hoc basis.

Labor offices also have to decide whether employees are qualified for specific jobs, which means that large international companies have to submit résumés and diplomas to labor office bureaucrats. In order to minimize risks and comply with pressures from their superiors, bureaucrats have erred on the side of rigidity on Saudization and have often been slow to make decisions.[73]

Although many MoL bureaucrats may have been well-intentioned, they lacked both the qualification and the data to make a realistic assessment of labor needs and the feasibility of Saudi employment.[74] They had little to no private sector experience, were on an intermediate level of seniority, and had no incentive to engage with business productively.[75] They are described as unresponsive by virtually everyone in the private sector. In the absence of specific personal links, mutual distrust has prevailed.

For several years, none of this was an existential problem for most businesses. As overall political pressure for Saudization was still moderate during the 1990s,

---

67. *Arab News* supplement, 17 January 2004: 30.
68. *Al-kitab al-ihsa'i as-sanawi:* 18.
69. As early as 1996, it was reported that the MoI would not grant block visas if a company were not perceived to make a sufficient Saudization "effort," a policy necessitating bureaucratic interpretation; cf. *MEED,* 6 April 1996: 56.
70. The labor offices hand essentially every work visa request on to high-level committees in Riyadh. Interview with board member of Jeddah Chamber of Commerce and Industry, Jeddah, September 2008.
71. *Arab News,* 24 April 2004.
72. Interview with human resources manager of Saudi bank, May 2004.
73. Interviews with various human resources managers.
74. *Arab News,* 8 August 2004; "Although some are well-intentioned, they are out of touch in the MoL." Interview with senior expatriate lawyer, May 2004.
75. Interview with human resources manager of Saudi bank, June 2004.

most businesses managed to slip around regulations and assessments relatively easily. In 1999, Saudization efforts were still "lightly policed."[76]

Insofar as personal persuasion was not enough to handle Saudization demands, businesses had several strategies at their disposal to handle Saudization formalities. Many business owners could leverage direct and indirect connections in the MoL, the MoI, or other agencies to obtain Saudization certificates, work permits, or visas.[77] Although there were complaints about unequal implementation—for example, similar stores being accorded different quotas—this did not lead to any organized protest.[78]

Beyond immediate use of networks and patronage, there were a few more technical or indirect solutions to elude Saudization and labor import regulations. These usually relied on the low supervisory power of the MoL, which as late as 2005 reportedly had only twenty inspectors responsible for 300,000 businesses.[79] They would often draw on informal connections in an indirect fashion.

Trade in Labor...

If employees with low qualifications were needed, employers—especially smaller ones—could resort to hiring expatriates on the "free visa" market, that is, from brokers who employ them on paper only. Since they are hired informally, they do not show on the official payroll.

The workings of the free visa market are rooted in the highly unequal access to labor import allowances in Saudi Arabia, which goes back to the 1970s boom, if not earlier. While acquiring visas can be cumbersome without good networks, some players enjoy privileged access.[80] The MoI in particular seems to have functioned as a large-scale patronage machine distributing block visa grants to favored clients. Many royals have been involved in this trade,[81] and labor block visas have been granted to unlikely figures such as tribal shuyukh in villages.[82] With significant economic interests at stake, the provision of labor import licenses has become a business "on an industrial scale."[83]

---

76. *MEED*, 17 September 1999: 8.
77. One interviewee, for example, went to see the deputy minister to get some urgent visas for his company, and personal calls, gifts and favors are used regularly. Interview with human resources manager of Saudi bank, May 2004.
78. Hajjar interview, April 2004.
79. *Arab News*, 20 December 2005.
80. World Bank 2002: xi.
81. One anecdote tells of the governor of the Eastern Province giving three thousand visas to a befriended businessman just before giving a speech on a Saudization conference.
82. Interview with chief economist of Saudi bank I, March 2004.
83. Interview with Western diplomat I, April 2004.

By law, the importing sponsor is supposed to be the employer of any expatriate worker, which is why primary importers often set up letterbox companies.[84] A large portion of unskilled labor is resold once it has been imported. The actual employer then pays fees to the official sponsor, who acts as a broker making the regulatory resources of an otherwise unaccountable labor bureaucracy available. The fees have been estimated at around five to six thousand Saudi riyals per year in the case of housemaids, who are frequently "bought" on the free visa market by richer expatriate families who stand little chance of importing labor through official channels.[85] The fees are often higher for other workers. Although the serious "block visa" trade is limited to a few well-connected individuals, many ordinary Saudis also register small companies on paper in order to import a few workers for resale.[86]

A second variant of the free visa trade, important especially for nondomestic labor employed by Saudi nationals, is the official transfer of sponsorship briefly after an expatriate has entered the country, in this case for a one-time lump sum. A leading Saudi industrialist estimated in 2004 that 70 percent of the six hundred thousand iqama transfers in one year happen within two months of arrival.[87] In both cases brokerage tends to be fully monetized. Changing sponsorship can be costly—up to 10,000 SR—and can also involve intermediaries in or around the labor bureaucracy.[88] In this case, the worker at least enjoys a formal work relationship.

...and Other Avoidance Mechanisms

The free visa market—complemented by the market of illegal visa "overstayers"—has its limits when it comes to more specific labor needs.[89] Full-scale Saudization

---

84. An October 2001 decree abolished the official designation of "sponsor," replacing it with "employer," but changed little about the actual relationship of bondage; Council of Ministers decree 166, 12/7/1421.

85. Interview with chief economist of Saudi bank II, March 2004. At least in the case of non-Saudi families, "no maid has her original sponsor as employer." Interview with Western diplomat I, April 2004.

86. Interview with board member of Jeddah Chamber.

87. Abdalrahman Al-Zamil reportedly confronted the new Minister of Labor Al-Gosaibi with these figures in a chamber discussion in early 2004. Interview with board member of Riyadh Chamber.

88. Interview with senior GCC functionary, Riyadh, January 2007.

89. "Overstayers" come into the Kingdom on hajj or umrah visas and then go underground, crowding the lower tiers of the Saudi job market even more. They are more likely to work in odd, menial jobs (washing cars, plumbing, peddling drinks on the street) than have a permanent employer. They are regularly rounded up by the MoI and deported, but the problem persists. Corrupt policemen reportedly afford them protection; cf. *Saudi Gazette*, 31 December 2005. The number of overstayers was estimated at one million in 2001; Montagu 2001: 25.

of specific job categories such as accountants or secretaries has led to great bottlenecks, as qualified Saudis willing to work in these low-prestige jobs are almost impossible to find.[90] Recourse to brokers is not always available in such cases. Ambiguous rules and the labor administration's low regulatory capacity can, however, create opportunities to work around state demands. One such way has been to hire expatriate accountants and secretaries under false job descriptions. Even senior Western managers have entered Saudi Arabia on more easily obtained shepherd's visas.

The MoI has declared that it knows about the practice, has announced fines, and has even threatened those involved with jail sentences.[91] However, no systematic investigation of such misrepresentations has taken place, and they are difficult to monitor. A two-month grace period for businesses and employees to correct such occupational mismatches passed with little fanfare in 2000.[92] In late 2005, many businesses were still resorting to the practice.[93] Insofar as job categories are not completely closed to expatriates, job title descriptions can also be changed post hoc through using connections in the responsible agencies.[94]

An even safer, if somewhat costlier, course of action is to hire Saudi nationals just so they can appear in the payroll records—usually with a low salary and no actual work to do. Such fake employees are sometimes friends or relatives of the business owner, and they often do not show up at the workplace at all, brokering and commoditizing nothing beyond their nationality.[95] In fact, some are explicitly told not to come to the workplace.[96] Sometimes employers also create "straw men" by showing the IDs of former job applicants to government inspectors.[97]

In family conglomerates, the pro forma shifting of employees between company units has proven a useful tool for obscuring the actual employment record. Employers also use the splitting of companies to dodge Saudization demands, especially in the case of rules that apply only above a minimum workforce size. In other cases of creative labor force accounting, employers use companies that have in fact ceased to exist but still have an official commercial registration.[98]

---

90. Interview with senior expatriate lawyer, May 2004.
91. *Arab News*, 16 July 2000.
92. Ibid., 23 August 2000.
93. Discussions with businessmen and bureaucrats in Riyadh, November/December 2005.
94. Interview with human resources manager of Saudi bank, May 2004.
95. World Bank 2002: 45; *Arab News*, 20 December 2005.
96. *Arab News*, 10 August 2007.
97. Ibid., 24 April 2004. In one case known to me, a Saudi *mutawwa* (religious policeman, not usually known for financial expertise) was formally hired as accountant, while a Sudanese expatriate employed under a different job description did the actual work.
98. Lehner interview.

Like in the free visa case, in the field of "virtual Saudization" an informal market has emerged in which some players commoditize their access and administrative patronage within the opaque and fragmented bureaucracy. One method is to have workers formally employed by another firm that functions only as intermediary and receives a fee from the actual employer. In other cases, tasks are in actual fact outsourced to other firms—for example, menial office work—although these firms do not necessarily have a higher Saudization rate.[99] Under many of these outsourcing constructions, the contractor enjoys administrative patronage, usually through connections with high-level regime figures, which prevent him from being sanctioned for low Saudization.[100]

The techniques outlined here mix creativity and wasta in varying proportions. Personal networks and the general venality of officialdom have allowed the skirting of unclear and inconsistent regulations. The Saudi press decries many such practices but without pointing the finger at high-level patrons. The opacity of actual Saudization practices has put most foreign businesses in the Kingdom at a disadvantage.[101] Similarly, new Saudi entrants with less capital and fewer networks to leverage struggle with labor market rigidities.[102]

Formal Blockage and Informal Markets

With un- and underemployment becoming more and more salient in the Kingdom, the government increased its Saudization pressure in the years after 2000. Although loopholes did not disappear, they became more difficult to find and often cost more to exploit. Saudization became a more acute issue for a greater number of players.

Although visa numbers remained high in 2001/2002, for many private sector actors it took even longer to acquire them than before, hampering especially labor-intensive businesses like construction.[103] Increasing Saudization pressure flung the door for corruption wide open, because the vast majority of companies in most sectors were not living up to the formal Saudization requirements.[104]

---

99. *Arab News*, 23 March 2004.

100. Interview with chief economist of Saudi bank I, Riyadh, December 2003. Good links to the MoI have played an especially prominent role here. Interview with senior Saudi banker.

101. Barillka interview. Some diplomats think Saudization is deliberately deployed against foreigners. Interview with Western diplomat III, Riyadh, April 2003. Labor offices supposedly have tried to "kill" specific companies, not necessarily foreign ones. Interview with human resources manager of Saudi bank, May 2004.

102. Unable to acquire visas (that is, unable or unwilling to use wasta), some young entrepreneurs have been forced to give up their business plans. Discussion with Saudi industrialist II, Riyadh, April 2004.

103. *MEED*, 22 February 2002: 36.

104. Interview with Saudi industrialist, March 2004.

Higher fees for expatriate labor import and maintenance increased the opportunities for payoffs.[105]

Transactions are reported to be quite straightforward, by Saudi standards. Factory managers are said to carry cash to labor offices, whose representatives visit factories to collect cuts up to twice a year. This includes inspectors on lower levels of seniority, who then ensure a benevolent Saudization assessment.[106] Larger businesses reportedly submit visa applications in bulk, which tends to reduce payments and make them more predictable.[107] Payments are sometimes made through brokers; as corruption becomes more common, these are not always required.[108] The MoI has also been criticized for venality. Even the Saudi press has carried stories of an iqama fraud in which names mysteriously disappeared from passport department computers in Riyadh.[109]

Labor office "Saudization certificates," a crucial component of the enforcement drive that started in 2001 and 2002, quickly became hotly sought after on informal markets, as they were required for various administrative transactions, such as renewal of commercial registrations.[110] Buying such certificates, either directly or through an intermediary, was often preferred to full-scale Saudization efforts according to challenging and elusive criteria.

Crackdowns and Grace Periods: Private Sector Responses

With increased Saudization pressure, costs and bottlenecks increased across the board for most businesses. At this point, with individual evasion becoming harder, collective protests from the private sector became more vocal. Contractors in particular protested through the Chambers to the MoI.[111] In various labor policy meetings, businessmen vented their anger about ill-prepared and opaque policies, lack of trained Saudis, loss of comparative advantage, and absence of consultation.[112] In the face of unimplementable policies, the government has in several cases withdrawn from specific decrees following private sector protests.[113] However, protests and ad hoc consultations have not led to a consensus on a formula for Saudization policy.

---

105. Interview with chief economist of Saudi bank I, March 2004.
106. Ibid.
107. Discussion with Western economic advisor I, Riyadh, June 2004.
108. Interview with Saudi industrialist III, Riyadh, March 2004.
109. *Arab News,* 29 August 2004.
110. *MEED,* 18 May 2001: 19; German-Saudi Liaison Office for Economic Affairs, *Gulf Economy,* 4/2002: 25; Lehner interview.
111. Interview with member of Majlis Ash-Shura II, Riyadh, June 2004.
112. *Arab News,* 19 October 2003.
113. Interview with board member of Riyadh Chamber.

The complexity and economic sensitivity of the Saudization issue would appear to call for close alignment of public and private action. But government and private sector have coordinated little. Calls for better communication by the private sector have led to repeated meetings with MoL representatives and princes in chambers and other fora, but neither this nor the private sector representation via the Manpower Council has led to continuous and/or comprehensive policy negotiations. Exchanges were even more ad hoc than in the FIA case.[114] Although Saudization is discussed everywhere, the discussion is not politically structured in terms of negotiating a strategy: "There are just people on TV putting out theories on Saudization; there is no real dialogue."[115]

The government had to repeatedly scale back its aspirations on Saudization. Even Prince Naif had "heard enough from the private sector to reduce his expectations."[116] This has not led to any systematic evaluation of private sector input or comprehensive reassessment of policies, however. Rather, business has continued to muddle through and engage in dilatory lobbying—both with some degree of success. There seemed no space for a credible policy agreement to which both sides would be willing, and able, to adhere.

By the summer of 2003, actual Saudization rates were reported at 18 percent in transport and storage, 11 percent in wholesale and retail trade, and 15 percent in other industries.[117] According to another source, rates were generally higher in establishments with more than twenty employees (16 percent as opposed to an overall average of 13 percent).[118] All of these rates were far below the goals of the raft of repeatedly revised decrees. The government had succeeded in increasing the costs associated with Saudization but had fallen far short of generating the number of Saudi jobs envisaged in its plans.

Sledgehammer Saudization: The Case of the Travel Agents

Saudi labor market policy saw further revisions, this time institutional, in 2004, which produced a significant (if partial) success by bundling authority in the MoL and temporarily reducing the number of new expatriates. Before analyzing how Abdallah tackled the meso-level problems of labor administration, however, we will

---

114. *Saudi Gazette,* 22 January 2004; *Arab News,* 4 May 2000, 30 March 2004.
115. Interview with human resources manager of Saudi bank, May 2004. He told me that his attempt to organize a coherent dialogue failed due to the complete unresponsiveness of the labor offices.
116. Interview with former Saudi deputy minister, May 2004.
117. *Arab News,* 6 August 2003.
118. Ibid., 31 July 2003. The MoL presents an even lower rate of 10.1 percent Saudization for 2004.

look at a specific case study of state-business bargaining on a sectoral level, one that illustrates fluidity and fragmentation in even very delimited policy areas.

The best-reported case of a specific Saudization initiative probably is that of the travel agents in Riyadh governorate, subject to great and sudden pressures of Saudization in the spring of 2004. The case lucidly evinces the lack of strategy in both business and government, characterized by repeated knee-jerk reactions and inability to sustain continuous cooperation based on credible regulation and mutual trust.

Sales and ticketing jobs in the travel industry seemed to be part of the aforementioned series of retail Saudization decrees, which involved a successive increase of quotas over three years. At least this was what business expected. On 21 February 2004, however, MoL and MoI inspectors, together with representatives of the Riyadh governorate, suddenly cracked down on travel agencies in the capital, expelling and even arresting expatriate front-office staff members.[119] The following days saw further arrests, and front-office employees of the prominent travel company Kanoo were shaved against their will and prepared for deportation.[120] The decision for the crackdown seems to have come from the royal level.[121] In the weeks following the roundups, expatriate travel agents, afraid of further arrests and possible deportation, served customers from outside their offices, about two hundred of which were closed.[122]

The aim of the unannounced crackdown was apparently an instantaneous Saudization rate of 100 percent.[123] According to the retail decrees, a figure around 25 percent was what many Saudi travel agency owners had expected. There had also been an MoL memo instructing travel agents to achieve a rate of at least 40 percent, with another decree by the Riyadh governorate setting a February deadline.[124] In an interview in late February, the deputy minister of labor himself mentioned a 100 percent rate, but without specifying a deadline.[125] What exactly was demanded remained unclear.

In early March, travel agency owners formed a negotiation committee and met Prince Sattam, deputy governor of the Riyadh region.[126] As the sector included

---

119. IPS, 26 March 2004; *Saudi Gazette,* 6 May 2004. The squads included *"mujahedeen"* from the MoI, a particularly rough police force. Interview with Nasser Al-Tayyar, chairman and chief executive officer, Al-Tayyar travel group, Riyadh, December 2005.
120. They had to pay for the cost of shaving; cf. *Arab News,* 4 March 2004.
121. Interview with member of Majlis Ash-Shura II, May 2004.
122. *Saudi Gazette,* 24 February 2004, 25 February 2004.
123. Ibid., 2 March 2004, 3 March 2004, 10 March 2004.
124. *Arab News,* 25 February 2004.
125. *Saudi Gazette,* 29 February 2004
126. As the travel sector did not have a standing committee in the CCIs, this was ad hoc, demonstrating the suppleness of informal links. Tayyar interview.

several big business names such as Kanoo and Tayyar as well as some princes, getting access was relatively easy.[127] The primary aim was to negotiate postponement of Saudization.[128] Business representatives complained that there were no suitable Saudis for the jobs and blamed the government for lack of training efforts.[129] The agents also started discussions with the Riyadh Chamber, the Supreme Commission for Tourism (SCT), and national technical education body GOTEVOT over future education plans.[130] Immediately after the meeting with Sattam, a travel business training scheme was announced.[131] Officials admitted that GOTEVOT had been ineffective in producing the right graduates for travel businesses.

By this time, the detained employees of Kanoo and the other large agencies had been released again. The quota issue had not been clarified yet, however, and travel agency managers complained that some government agencies wanted 35 percent Saudization, whereas others demanded 100 percent.[132] Without bureaucratic coordination mechanisms, diffuse top-down demands for Saudization results had led to fragmented policies.

In addition to the discussions with Sattam, leading travel agents appealed to Defense Minister Prince Sultan and Prince Naif in early March, using the good services of Prince Sultan bin Salman, son of the Riyadh governor and secretary general of the Supreme Commission for Tourism (of which Defense Minister Sultan is formally chairman).[133] Several of the bigger actors also voiced their complaints in the press. Interviewed by the *Saudi Gazette*, Nasser Al-Tayyar claimed to have been totally unaware of any deadline for complete front-office Saudization. He complained of severe "miscommunication" and recounted that officials from the governorate and the MoL had failed to show up in a coordination meeting in the Chamber in December. In Jeddah and Dammam, he asserted, authorities supposedly had gotten in touch with business owners, avoiding clashes as in Riyadh.[134] The general manager of Kanoo claimed that all of his employees had correct papers. The case, he said, would be taken not only to Naif but also to the king himself.[135]

Travel businesses also sent a petition to Prince Salman, governor of Riyadh, pointing out that the proper Saudization rate was 35 percent, not 100 percent.

---

127. Discussion with Saudi businessman, Italy, March 2006.
128. *Saudi Gazette,* 2 March 2004.
129. Tayyar interview. One major representative of the business claimed that he had spent 6 million SR on training six hundred Saudis, without retaining a single one; cf. *Arab News,* 11 March 2004.
130. *Saudi Gazette,* 24 February 2004.
131. Ibid., 3 March 2004.
132. Ibid., 4 March 2004, 8 March 2004.
133. *Arab News,* 10 March 2004; Tayyar interview.
134. *Saudi Gazette,* 4 March 2004.
135. Ibid., 8 March 2004.

Full Saudization, it was argued, was for receptionists, and there were no receptionists in travel, only ticket agents and other specialists.[136] The government's ambition to micro-manage very specific types of jobs had invariably led to ambiguities.

On 5 March it was reported that officials had verbally informed businesses about a temporary halt of the Saudization campaign. Moreover, the SCT was to take over control of the sector from the Presidency of Civil Aviation (under the Ministry of Defense).[137] However, further arrests happened on 8 March, including in the offices of the Tayyar and Al-Fursan travel chains, where supposedly a significant number of Saudis were working. "The whole travel industry at least in Riyadh [was] in turmoil and confusion over the raids." There was more talk of petitions to Defense Minister Prince Sultan and the Riyadh governor, "seeking their immediate intervention" and patently trying to bypass the unaccountable bureaucracy.[138]

In an interview published on 10 March, Nasser Al-Tayyar told the *Saudi Gazette* that Sultan bin Salman of the SCT had made his own recommendations to Defense Minister Prince Sultan based on a petition he had received from the travel agents. A letter from the defense minister reportedly went to Naif after the former had approved the proposal to solve the ongoing crisis in the travel industry.[139] Tayyar related that royal broker Sultan bin Salman and his deputy (a figure from the important Sudairy family close to the Al Fahd) had done a great deal to help them and to bring the situation to the attention of the "higher authorities." The proposal from the travel agencies suggested gradual Saudization under SCT supervision, whose Saudization committee up to this point had been ignored.[140] Details were not available.[141]

One day later, Tayyar again appeared in the press, protesting that some expatriate employees were still being detained. The travel agents, he said, were seeking an urgent meeting with the crown prince. Apparently there had been no concrete orders yet to stop the raids. The Council of Saudi Chambers was to raise the issue with high officials.[142] Although rumors of a grace period circulated, some employees remained in detention, and many businesses remained closed.[143]

In mid-March, the picture remained unclear. The issue was referred to the SCT, but at the same time, Deputy Minister of Labor Mansour announced that

---

136. Ibid., 6 March 2004.
137. *Arab News*, 5 March 2004.
138. Ibid., 9 March 2004.
139. *Saudi Gazette*, 10 March 2004.
140. Tayyar interview.
141. *Arab News, Saudi Gazette*, 10 March 2004.
142. *Arab News*, 11 March 2004.
143. Ibid., 13 March 2004; *Saudi Gazette*, 15 March 2004.

the Saudization drive would continue, that the government had not changed its stand, and that there would be no three-month grace period.[144] New rumors spread that sales jobs in other sectors soon would be tackled, and computer shop owners received letters telling them to Saudiize at a 50 percent rate within six months.[145]

On March 20 finally, Sultan bin Salman announced that travel agents would receive a two-year grace period until full front-office Saudization. Vocational training programs in the sector would be coordinated by the SCT. The MoI was said to investigate the detentions.[146] A few days later, travel agents met with SCT representatives to plan the Saudization of the sector. Tayyar again thanked Sultan bin Salman for his role in resolving the crisis.[147]

Further details emerging in the following days included a target of fully Saudizing semi-skilled jobs within 2004 and an aim of 81 percent overall Saudization by the end of the grace period, with at least 70 percent of ticketing staff being Saudiized.[148] Sultan bin Salman promised more comprehensive plans for Saudization within three months, involving GOTEVOT and other institutions.[149] Finally, it seemed, a coordinated solution was under way, and formal channels to establish mutual commitments had been established, based on relatively equal negotiations.

In subsequent weeks, however, the response of travel businesses to the SCT's information requests was lackluster.[150] After the first deadline to submit data on employees on 5 April had passed, a new one was set for 18 April, and noncomplying offices were threatened with closure. The response again was lukewarm, and the SCT and the PCA announced penalties to be imposed from 11 May on.[151] The PCA, still responsible for the administration of certain ticketing aspects, did indeed pull the plug on the online operations of a few travel agents.[152] After the spring 2004 brouhaha, however, little more was heard of travel business Saudization, apart from an August 2004 announcement that the transition period would last for three years, with an eventual aim of 88 percent.[153] In September 2005 it was reported that despite joint training efforts, the travel sector was still lagging

---

144. *Saudi Gazette,* 15 March 2004.
145. *Arab News,* 16 March 2004.
146. *Saudi Gazette,* 21 March 2004.
147. Ibid., 24 March 2004.
148. Ibid., 25 March 2004.
149. *Arab News,* 29 March 2004.
150. Ibid., 14 April 2004.
151. Ibid., 6 May 2004.
152. Ibid., 12 May 2004.
153. *Saudi Gazette,* 25 August 2004.

far behind its (already revised) goals.[154] In December 2006, the 81 percent Saudization aim resurfaced again, this time with 2009 as the target date.[155]

The travel agents' tale offers a condensed account of the poor coordination, low regulatory penetration, use of political patronage, and post hoc policy adjustments that have been characteristic of sectoral Saudization. One might be tempted to blame it on the idiosyncrasies of the Riyadh governorate and the rushed way in which the policy was implemented. However, many similar problems have been occurring over a much longer period of time in other sectors targeted by specific Saudization initiatives, most notably in the cases of the gold and vegetable suqs and in the taxi sector.

All of these cases feature repeated sudden crackdowns, unclear Saudization targets, and repeated postponement of deadlines.[156] The main difference to the travel agents' case probably is that players in other sectors usually had less collective bargaining capacity, being small and often cover-up businesses. This might be why they have resorted more frequently to employment of Saudi nationals as "stooges," hide-and-seek games between expatriate workers and inspectors, and individual exemptions based on low-level wasta.[157] Whatever the differences, in no case has there been successful transfer of day-to-day operations to Saudi hands across the board. Businesses usually waited out the issue and, in case of repeated crackdowns, often either closed down or preferred to shift to a different sector instead of tackling Saudization strategically. Police officers reportedly are sluggish and prefer to chase away illegal employees rather than take them in, unless there are immediate "orders from above" to crack down.[158]

How Weak Regulation Creates Ad Hoc Veto Coalitions

The travel agent case also demonstrates the state's lack of administrative credibility and, closely related, the weak cooperation between government agencies and the private sector. One reason for that appears to be the disaggregation of Saudization policies across different agencies—each with its own informal networks and different modes of power, depending on the nature of royal backing. Although private sector representatives asked Prince Naif to assign a clear responsibility for

---

154. Ibid., 15 September 2005.
155. *Khaleej Times*, 22 December 2006.
156. One MoI decision explicitly mentions protests by gold merchants which led it to reschedule sectoral Saudization; MoI cable 1B/5317, 2-3/6/1422.
157. *MEED*, 8 November 1996: 29, 15 September 2000: 37; *Arab News*, 4 June 2003, 12 September 2003, 7 February 2004, 7 March 2004; *Saudi Gazette*, 4 May 2003, 15 November 2003, 23 February 2004, 17 August 2004.
158. *Arab News*, 29 May 2005.

Saudization policies, until the reorganization of jurisdictions in spring 2004 this had not happened.[159]

Expectations in the private sector about future Saudization policies have been highly volatile. When gold suqs and travel agents were suddenly addressed in February 2004, the Saudi press openly wondered what had happened to the other twenty-five retail sectors which were officially slated for Saudization, as no apparent effort was made to check quotas there. A spokesman for the Jeddah Chamber of Commerce claimed that quotas were different across retail sectors. Shopkeepers had no idea what the quota would be for the upcoming year. Expatriate employees tended to be relaxed about Saudization pressures, citing the previous attempts to Saudiize vegetable and gold suqs in predicting that nothing serious would happen.[160]

Heightened political pressure from senior royal figures usually was the main cause of the sudden crackdowns, which only served to highlight the capacity deficits of government agencies, however. Although such crackdowns were able to shut down a large number of businesses temporarily, calls for 100 percent Saudization proved unimplementable; cover-up businesses continued to exist, and after a shock period, many businesses reverted to individual tricks and loopholes to avoid Saudization. A serious policy debate, taking into regard salary levels, working conditions, standards of training, and the interests of young Saudis, did not take place in any sector.

The low standard of compliance with the law among many Saudi businesses—unless immediately threatened—illustrates the low formal regulatory penetration the Saudi state has achieved in Saudi society. Due to patchy implementation, favoritism and unpredictable (meso-level) institutional policies and (micro-level) administrative behavior, many business owners tend to feel that acting according to the rules would make them "suckers" in an unfair game.[161] Despite much talk and occasional negotiations, motivation for compliance has been low. All of this privileges individual solutions and undermines incentives to organize and engage in sustained collective bargaining.

As the Riyadh travel agents included several important business players, they could organize themselves as an ad hoc veto coalition after the fact. They concentrated most of their efforts on lobbying with senior princes, even those with no formal responsibility for labor issues, thereby bypassing bureaucratic channels, Majlis Ash-Shura, and other, more institutionalized actors perceived as rigid and unresponsive. It was evident that the problem would be resolved (or postponed)

---

159. Interview with human resources manager of Saudi bank, May 2004.
160. *Arab News,* 9 March 2004.
161. Levi 1988: 68.

much more swiftly through structures of senior princely patronage than through other institutional means. Sultan bin Salman acted as policy broker between senior royal patrons and business clients. Even Tayyar's "modern" use of the press was probably a means to make the issue salient for princes eager to appear as arbitrators. When push comes to shove, politics is done through intercession with the court, the core locus of political authority and patronage. After the intercession succeeded and the immediate pressure decreased, the travel business coalition seems to have quickly disintegrated.

This example of princely arbitration is all the more ironic, as in all probability it was a princely decision, probably from Salman, that caused the turmoil in the first place.[162] Here as in other cases, people around the royal decision-makers possibly were too afraid to say out loud that the travel agency measure would not be implementable.[163]

The short attention span on sectoral Saudization issues in part results from senior princes' limited capacity for detailed follow-up. Here as in other policy areas, royals are prone to make the occasional strong statement and then completely forget about the issue. Saudi bureaucracy more generally is usually incapable of sustaining permanent pressure in implementing difficult regulations, as its instruments of information-gathering are limited, its sanctioning mechanisms blunt, and the incentives for bureaucrats to follow up on specific cases limited.

A New Attempt: Institutional Consolidation in 2004

Despite the onslaught of quota decrees and hiring bans, little had changed about the broader institutional framework for Saudi labor market policy by 2004. The creation of an (underperforming) Human Resources Development Fund in 2000 was seen as an instance of tinkering rather than a full-blown institutional reform.[164]

Only in early 2004, after almost a decade of fragmented policies, did a more fundamental institutional reform take place. In January, the head of the HRDF admitted that the incremental 5 percent rule was "not...a practical tool" and suggested that an independent body might be required to take care of

---

162. There are many more examples of appeals to senior players bypassing other institutions. *Umrah* companies taking care of religious pilgrims, for example, have gone through the crown prince personally to settle a dispute over fines imposed by the passport department (at the MoI under Naif); cf. *Arab News*, 25 March 2004.

163. Interview with member of Majlis Ash-Shura II, May 2004.

164. The fund uses some of the income from higher labor fees to subsidize training and first years on the job but has struggled to find takers. Discussions with Saudi bureaucrats, May 2004.

Saudization.¹⁶⁵ It seemed as if the government was about to create a SAGIA equivalent for labor market issues.

What happened instead was more fundamental and, since the oil boom, unprecedented: a substantive reordering of responsibilities between the MoI and MoL. The recruitment administration, which had been dishing out hundreds of thousands of visas, was moved from the MoI to MoL and combined with the labor offices.¹⁶⁶ As the MoI was preoccupied with the tense internal security situation in the Kingdom, Crown Prince Abdallah seemed to seize the opportunity to clip its wings, playing on the political urgency of Saudization matters, which had become the "talk of the town" and a great concern for the leadership.¹⁶⁷

In principle, the reordering substantially enhanced the MoL's role as lead agency on labor and foreign worker issues. There was a customary delay in the actual transfer of the administrative assets, and the passports department stayed with the MoI.¹⁶⁸ However, the stage was basically set for a bundling of visa policies in one institution, which would enable the reduction of foreign workers. This was crucial: over the years it had become clear that the low regulatory capacity of the Saudi state would never allow meaningful Saudization as long as there was a surplus of foreign workers that could reenter the labor market through various informal back doors. Without limiting the aggregate labor supply, all the micromanagement of quotas and job categories was likely to create evasion and corruption, at best shifting surpluses from one sector to another.

Not everyone in the private sector was happy about the transfer, as it gave greater power to MoL bureaucrats; some businessmen commented that, unlike MoI staff, MoL administrators were "still hungry," alluding to MoL corruption.¹⁶⁹ Established links to the recruitment structures in the MoI broke down ("they [MoI] knew my needs," in the words of one businessman).¹⁷⁰ Abrogating established and "sticky" informal links seemed to make the shift costly for established players.

Splitting the Ministry of Labor and Social Affairs

In March 2004, what had thitherto been the Ministry of Labor and Social Affairs was divided into the Ministry of Labor and the Ministry of Social Affairs,

---

165. *Saudi Gazette*, 24 January 2004.
166. Ibid., 27 July 2004.
167. Interview with senior advisor to Saudi government, April 2004.
168. Interviews with members of Majlis Ash-Shura I and II, Riyadh, April/May 2004.
169. Interview with human resources manager of Saudi bank, May 2004; discussion with Saudi businessmen at family gathering in Riyadh, June 2004.
170. Interview with human resources manager of Saudi bank, May 2004.

signifying a renewed priority for labor market issues. The Manpower Council ended its independent existence, being absorbed into the new MoL.[171] Its secretary general stayed on as labor undersecretary.[172]

In April 2004, Ghazi Al-Gosaibi, previously minister of water and electricity, was appointed minister of labor.[173] Gosaibi had a track record as reformer and was said to have, at least to some degree, "cleaned up" opaque institutions like the Ministry of Health before. Of the commoner ministers, he is seen as probably the most strong-willed. He was close to then–Crown Prince Abdallah, and the general reading was that Abdallah had brought him in to sort out touchy visa matters.[174] No other commoner, it seemed, would have been capable of confronting the MoI.[175]

Gosaibi soon announced tough measures, clarifying that the recruitment of expatriates had to be limited. He promised full cooperation with the private sector but threatened Saudization "violators" with sanctions.[176] Gosaibi openly denounced the trade in visas.[177] In a Council of Chambers meeting in late April 2004, he asserted that 70 percent of visas are sold on the black market. Following previous trends, he explained, 750,000 visas would be issued in 2004, a number that was too high for him to accept.[178] Gosaibi warned businesses not to "blackmail" him with capital flight and announced that he would endorse all visa allowances personally, strongly implying an end to all abuse.[179] A lot of his tough talk, some thought, was directed at the MoI.[180] To gain control over his own bureaucracy, he even forbade assistant deputy ministers to meet anyone in other government agencies without his explicit permission.[181]

The MoL set up a blacklist of companies that did not fulfill their employment and training obligations; they would no longer receive MoL services such as visas or residence permits.[182] In May, Gosaibi sent letters of warning to two hundred particularly egregious companies in which he announced a six-month grace period.[183] It was also reported that one company had 48,000 employees, only

---

171. *Arab News,* 23 March 2004.
172. *Saudi Gazette,* 6 May 2004.
173. Royal decree A/88, Um Al-Qura, 12 April 2003: 1.
174. Discussion with Saudi industrialist II, April 2004.
175. Discussions in Riyadh, summer 2004.
176. *Arab News,* 28 April 2004; *Saudi Gazette,* 20 April 2004.
177. *Arab News,* 14 April 2004.
178. Ibid., 29 April 2004.
179. Ibid., 1 May 2004; interview with board member of Riyadh Chamber.
180. Interview with member of Majlis Ash-Shura II, May 2004.
181. Discussion with senior Saudi bureaucrat, Riyadh, April 2004.
182. *Saudi Gazette,* 30 May 2004.
183. Ibid., 31 May 2004.

10 percent of whom were Saudis. Presumably this was one of the large contractors that so far had not encountered sufficient direct pressure.

Measures of the New Minister

Gosaibi was willing to confront vested interests. However, it soon became apparent that the style of his policies was not all that different from those of his predecessors. Although he had some success in reducing the raw number of visas, predictability and consistency of Saudization policies remained low.

Gosaibi attempted to stop the visa trade with several measures. In one case it was an immediate carryover from his predecessor; in February 2004, the MoL had officially ended the transfer of sponsorship except for big businesses and expatriates who were "technically and professionally qualified." The aim was "combating tampering with sponsorship law," which would put an end to "visa trade" and "underhand dealings." Foreigners with low qualifications would no longer be imported in bulk and resold.[184]

Small enterprises predictably complained over the blunt discrimination but were incapable of organizing their protest. Business had not demanded a proper sponsorship transfer mechanism before, and apart from requests to postpone the measure, no coherent policy proposals emerged from the private sector.[185]

In late March 2004, local papers reported that foreigners on free visas were still pouring into the Kingdom, simply changing their job descriptions for 1,000 SR after arrival (presumably to switch to an officially transferable profession).[186] Business sources confirmed that the recruitment of unskilled manpower continued due to wasta.[187]

Gosaibi re-decreed the sponsorship transfer stop for unskilled workers in late April.[188] The implementation was patchy, however. While small businesses apparently managed to get visas transferred in the Khobar labor office, the Dammam labor office in the same province (Eastern Province) rejected them.[189] In July, the MoL again banned transfer of sponsorship apart from some qualified professions after taking over the responsibility for transfers from the MoI's passport offices.[190]

---

184. Ibid., 21 February 2004.
185. Interview with senior advisor to Saudi government, April 2004.
186. The numbers remained relatively limited, however; in no year were more than thirty thousand job descriptions changed. Ministry of Labor and Social Affairs, *Al-kitab al-ihsa'i as-sanawi* (Statistical yearbook), various issues.
187. *Arab News,* 23 March 2004.
188. *Saudi Gazette,* 30 April 2004.
189. Ibid., 8 June 2004.
190. *Saudi Gazette,* 23 August 2004.

More or less the same measure kept on being re-announced. Gosaibi also officially banned the sale of work permits and prohibited all forms of human trade the same month.[191] But with connections and money, it was reported, free visa movements were still possible.[192]

While Gosaibi was struggling to establish control, the press openly reported past abuses of the system—an implicit indictment of the MoI, among other players. In August 2004 the MoL admitted that some persons owned fifty companies only "on paper," and a source admitted that abuse had occurred in passports offices.[193] Just when the MoL appeared to take over full control from the MoI, however, it delayed the sponsorship transfer ban by six months, "to give citizens enough time to correct their situation in agreement with the law."[194] The feedback was modest, and hundreds of thousands of expatriate workers remained trapped between their de jure and their de facto employers.[195]

Another harsh MoL measure to restrict labor imports and free visa trade was the ban of labor visas for small enterprises. Despite previous denials by his deputy, Gosaibi announced in late April a recruitment stop for establishments with fewer than ten employees.[196] Labor office sources made clear that large numbers of small businesses were fake, and Gosaibi openly linked his ban to the matter of letterbox companies.[197] Hit hard by the ban, some small business owners asked Gosaibi to review his decision, but there was no concerted protest.[198] Businessmen with urgent labor needs probably claimed to have ten or more laborers so as to receive recruitment permits.[199]

The new rule was blunt and stifling for anyone trying to start a new enterprise. In late May, the MoL granted a three-month grace period for recruitment by small enterprises (reconfirmed in early June), while the implementation from labor office to labor office was somewhat erratic.[200] At the same time, and potentially undermining Gosaibi's rigor, the Majlis Ash-Shura endorsed legislation for short-term, seasonal work visas.[201]

When the grace period ended, nothing was heard of a total recruitment ban for smaller enterprises. However, the MoL imposed a rule by which employers

191. Ibid., 18 July 2004.
192. *Arab News*, 25 July 2004; interviews with businessmen in summer 2004 and late 2005.
193. *Arab News*, 25 August 2004; *Saudi Gazette*, 23 August 2004.
194. *Saudi Gazette*, 26 August 2004.
195. Ibid., 14 September 2004.
196. *Arab News*, 9 March 2004.
197. Ibid., 4 June 2004.
198. *Saudi Gazette*, 20 May 2004; *Arab News*, 23 May 2004.
199. Interview with SAGIA functionary I, May 2004.
200. *Arab News*, 25 May, 7 June, 8 August 2004; various interviews with businessmen.
201. *Arab News, Saudi Gazette*, 6 July 2004.

could submit only one application for one individual business for a period of two months, which led to loud protests among Saudi businessmen, who often head diversified conglomerates.[202] The MoL again stated that it was trying to suppress the import of surplus labor likely to be traded subsequently.[203] It also decreed that no new staff could be imported by trucking companies with less than twenty-five trucks.[204] The measures again were blunt and did considerable collateral damage to Saudi business. The brouhaha over visa and labor restrictions died down somewhat in the remaining months of 2004, although generally the acquisition of visas had become more difficult.

One of the effects of the more rigid labor market was that individual expatriates were even more exposed to the whims of their employers, as they were now unable to leave them.[205] Although Gosaibi has also created a unit within the MoL to deal with complaints from expatriate employees, such malpractices as withholding of wages, deduction of administrative expenses from salaries, physical abuse, and restricted freedom of movement were still widespread. Labor courts remained overburdened. Gosaibi's overcentralization of authority led to great rigidities in visa administration. His low trust in his own subordinates was probably well justified, but as he was in fact not capable of checking all applications himself, much de facto discretion remained on lower levels.[206]

Micro-management of the labor market did not appear to yield the intended Saudization results. Many of the administrative measures had unintended, deleterious economic consequences and had to be abandoned or postponed. The blunt, deeply interventionist instruments hit all businesses indiscriminately, seriously impeding the operations of many companies. Unable to monitor effectively, the MoL was unable to prevent free visa trade without affecting legitimate actors. As more targeted controls remained elusive, Gosaibi managed to antagonize substantial parts of the Saudi business community, and despite his declared will to coordinate, little emerged in terms of a concerted public-private labor market strategy.[207] All in all, private sector representatives think, Gosaibi had grossly underestimated the problems he faced.[208]

---

202. *Arab News*, 23 August 2004.
203. *Saudi Gazette*, 26 August 2004.
204. Ibid.
205. *Saudi Gazette*, 19 July 2004.
206. The United Arab Emirates have seen an even more open attempt of a new minister to bring the labor administration under control, with several employees suing the government for wrongful dismissal. For more comparative details on strikingly similar regulatory policies and problems in other GCC states, see Hertog 2006a.
207. In early 2006, he was dubbed the "most hated minister" in Saudi Arabia. Discussion with Saudi businessman, Italy, March 2006.
208. Interview with Saudi industrialist, Riyadh, May 2004.

Although the MoL did not have the capacity to implement its own micro-level targets without great collateral damage, the 2004 restructuring of ministerial responsibilities allowed it to assert broad control over foreign labor. Officially, only the MoL issued visas. At least in the first few months, the policy was not too strictly implemented.[209] And although the assent of the MoL has become much more important, the passports department of the MoI still has a final say on visa decisions.

This is a veto power rather than a positive power, however, and the MoI seems to have lost much of its role. Total work permits for expatriates in 2004 totaled only 423,172, compared to 565,124 the year before (see table 6.1).[210] The 25 percent reduction indicates that the bundling of responsibilities in the MoL had a strong aggregate effect: 2004 was an economic boom year in which recruitment usually would have increased. Enjoying Abdallah's immediate patronage, Gosaibi has been able to refer even senior princely requests for visas directly upwards to Abdallah instead of directly yielding to pressure—and Abdallah himself has tended to refuse such requests.[211] Although some senior figures in the MoI reportedly were still involved in the visa trade[212] and the MoI still issues occasional labor-related decrees, its day-to-day role has been strongly curtailed.[213]

The reduction in labor imports continued into 2005, when the total number of work permits for foreigners was 352,924, below the levels even of the stagnant 1980s and 1990s. Simultaneously, in Hijri year 1425 (2004/2005), the number of registered Saudi employees in the private sector reportedly grew by more than 100,000, to 801,780, an unprecedented increase.[214] Saudization and visa policies are not the only factors expanding national employment, but scarcity of expatriate labor under boom conditions probably has contributed to the sudden jump. Salaries for some expatriates were reported in mid-2005 to have gone up by 50 percent since the MoL had started to tighten the reigns in 2004.[215] The macrobalance of the Saudi labor market seemed to tilt, making Saudi employees more attractive.

---

209. Interview with member of Majlis Ash-Shura I, June 2004.

210. The MoI reportedly issued 684,201 recruitment visas, down from 832,244 the year before; cf. *Arab News,* 26 February 2005. The discrepancy might be due to the fact that not all of the visas were used. In any case, the trend is the same.

211. Discussion with Saudi businessman, Italy, March 2006.

212. In late 2005, free visa trade was still rather common. Interview, GCC functionary; discussion with Saudi political researcher, Oxford, November 2005.

213. The "Prince Naif Price for Saudization" has survived the reorganization however; cf. *Arab News,* 28 August 2005.

214. *Saudi Gazette,* 17 October 2005. By the official MoL/SAMA count, the increase from 2004 to 2005 was from 486,000 to 623,000, which is even more impressive; cf. SAMA, various yearbooks.

215. *Saudi Gazette,* 18 June 2005.

TABLE 6.1.
Number of labor permits issued per year

1997	531,434
1998	420,928
1999	415,619
2000	516,826
2001	538,892
2002	597,272
2003	565,124
2004	423,172
2005	352,924
2006	716,347
2007	1,141,601
2008	970,805

Sources: Annual reports and statistical yearbooks of the Ministry of Labor (Arabic, various years).

The policies might have continued were it not for the contracting boom that set in toward the end of 2005 involving involved large-scale infrastructure and housing projects. Although employable in mid-level service and manufacturing jobs, most Saudis are still unwilling to work on construction sites, where salaries can be as low as 500 SR per month. As the 2006 budget included large capital outlays and the private sector was investing heavily in real estate, large-scale labor imports were inevitable.[216]

Contractors put considerable pressure on the government, as the Ministry of Finance, concerned about escalating project costs, reportedly did.[217] Labor bottlenecks slowed down projects considerably.[218] Abdallah, king since August 2005, gave in. Work permit issuance doubled in 2006, only to increases by another 50 percent in 2007 after the cabinet decided to ease visa procedures for companies working on government projects.[219] The backlog in projects was overcome, but due to the MoL's weak supervision, many imported workers also slipped into other sectors.[220]

Due to the exigencies of the boom, the attempt to restrict the inflow of foreign labor proved temporary. With its macro-instruments compromised, the MoL

---

216. SAMBA 2005.
217. Discussion with senior Saudi bureaucrat, Riyadh, January 2009.
218. Discussion with Saudi bureaucrat and real estate investor, January 2007.
219. *Khaleej Times,* 1 December 2006.
220. Half of the workers imported in 2007 were in the "basic engineering" category, corresponding to construction, compared to only a third in 2005; cf. Ministry of Labor, *Al-kitab al-ihsa'i as-sanawi* (Annual statistical yearbook), various issues.

continued to pursue Saudization through less aggressive tinkering with quotas and job categories. It repeatedly had to change its aims along the way. The Saudization quota for contractors was further reduced to 5 percent, and the government backed off a 50% Saudization quota for gold suqs.[221] Rules for sponsorship transfer underwent further changes.[222] Pronouncements from the minister of labor and his deputies often contradicted each other.[223] The MoL imposed sanctions on selected violators, but not by any discernible logic.[224] As practically no one fulfilled the Saudization requirements, bureaucratic discretion continued. Saudization successes were modest, as Saudi private employment increased only gradually after 2005.[225] The cabinet issued a new labor law in 2005 that changed little about basic labor market mechanisms but reaffirmed an unrealistic 75 percent Saudization requirement that had been on the books since the 1950s.[226]

Unlike labor agencies in some of the smaller GCC states, the MoL did not touch the main cause of uneven access to labor resources, that is, the sponsorship system that prevents internal mobility of foreign labor. Free visa trade therefore continued. Reports of bribery and manipulation in the labor bureaucracy continued to circulate, and visas received for big government contracts were informally resold.[227] Although the MoL had reduced direct block visa grants to regime clients, its bureaucrats had little capacity or incentive to crack down on letterbox companies and overimporting by legitimate companies.[228] Some companies that used to run real operations shifted entirely to the visa trade.[229] In May 2008, the MoL stumbled across a free visa scandal in which six companies had resold fourteen thousand visas.[230] Despite the minister's tough reputation, he has not cleaned up the ranks of the MoL.

By early 2009, Saudization remained a distant goal. Yet, the situation was more hopeful than in the early 2000s. There now was only one responsible agency, whose rules, although shifting, were clearer than the pre-2004 muddle that included half the government in labor policy decisions. Other agencies could no longer undermine official policy through large-scale block visas.

---

221. *Arab News,* 25 January 2007, 5 December 2006.
222. *Saudi Gazette,* 31 January 2007.
223. Interview with Saudi lawyer, Riyadh, January 2007.
224. *Arab News,* 18 July 2005; *Saudi Gazette,* 23 December 2006.
225. The official figure increased to 766,000 nationals by 2007; cf. SAMA annual report 2008: 233.
226. Confidential report (suspension of Akhbar Al-Dhahran), 27 February 1956, Mulligan Papers, folder 50, box 2.
227. *Arab News,* 2 February, 11 April 2007.
228. In 2006, the MoL still had only some three hundred inspectors; cf. *Arab News,* 3 August 2007.
229. *Arab News,* 11 April 2007.
230. *Saudi Gazette,* 21 May 2008.

As the exclusive issuer of work permits, the MoL would in principle be able to reimpose control over the labor market after the boom, gradually increasing the price of foreign labor through reduced imports. As a supply-oriented macro-instrument, this would not require the inspection and supervision capacities that quotas and prohibitions impose, capacities the MoL lacks. It is too early to tell whether this will happen. From 2008 on, the government-licensed National Society for Human Rights and the MoL also started a debate about the abolition of the sponsorship system, which could eliminate the free visa phenomenon by allowing expatriates to change employers freely, similar to the Bahraini labor reforms since 2005.[231] With labor policy competence centralized in the MoL, it was proposed, a government-controlled labor import mechanism could replace individual sponsorship. As of this writing, no decision had been made, however. Abandoning the futile micro-interventions in the labor market seemed to go against the control instincts of both the minister and his bureaucrats.

Conclusion

Given the imbalances of the Saudi labor market, Saudization would have been a challenge for any government. Nonetheless, it is remarkable how badly coordinated and unrealistic many of the Saudi policy measures were. State elites, although agreeing in principle, never moved beyond the most general consensus that Saudization is necessary and were unable to develop a viable and coherent cross-cutting strategy.

Like the foreign investment reform, Saudization faltered over a fragmented state and parallel spheres of administrative authority. Unlike the FIA story, however, there were few organizational interests that were clearly poised against specific aspects of the new policy.[232] The messy character of Saudization, rather, was a result of different agencies simply "doing their own thing" in the Saudization field, in which somehow everyone, including several senior royals, wanted to be seen as engaged. At least when it came to Saudization quotas and the regulation of specific job categories, it was lack of coordination rather than determinedly opposed interests that led to disjointed policies on the cabinet level. Only in the area of visa policy were there clear organizational interests, particularly within

---

231. Discussion with senior MoL bureaucrat, Riyadh, January 2009.
232. One exception has been SAGIA, because Abdallah bin Faisal has opposed quota-based nationalization of the job market, regarding it as an illiberal investment obstacle. Discussion with Arab economic advisor in Riyadh, April 2003.

the MoI, which reportedly wanted to retain its autonomous visa authority as a patronage mechanism, as it had been doing for decades.

Parallel policy-making in the Saudi government contributed significantly to the fog surrounding Saudization. The involvement of senior princes with short attention spans made the policy process even jerkier. Measures were often rushed and not thought through. Senior businessmen could intercede with various senior players to second-guess policies, further undermining the predictability of state behavior and its cross-cutting coordination. The involvement of regional governorates, whose competences and relations to other agencies have historically been vague, further complicated matters.

In an interventionist tradition going back at least to the 1970s oil boom, the institutions involved have used blunt and ill-defined regulatory instruments. The paternal regulatory ambition of the government, however, did not square with its limited capacities, as it proved incapable of making Saudi business follow its rules or share basic information. Unclear rules and limited regulatory powers made policies even less predictable and gave the inert, unaccountable, and unmotivated mid-level bureaucrats in the MoL considerable de facto discretion. Administrative clientelism and networks of brokers thrived on this, boosted rather than undermined by the increasing scarcity of freely available expatriate labor.

Ambitious attempts to micro-manage specific markets through quotas and prohibitions have brought about two unintended consequences: choking sectors and—rather like the FIA bylaw's minimum investment levels—increasing informal dealings. The latter have usually been the most efficient way of coping with the otherwise unaccountable labor bureaucracy and its predatory regulation. Low-level state-business networks further scuppered the coherence of national labor policies. More strongly than in the FIA case, informal interests on the micro-scale have penetrated the state, and corruption has played a more salient part in the associated brokerage. In the 1970s, brokerage and rule avoidance still played an important role in spreading the state's wealth. In the case of Saudization, businesses have for the most part resorted to such means simply to avoid the state's regulatory pressure. Ambiguous and unimplementable rules have become a threat rather than an opportunity.

Business is at the core of Saudization. It is the target of regulation, and through its contribution to national employment, it is supposed to play a bigger developmental role than at any point before oil transformed the Saudi system in the 1950s. Saudi business has been comparatively well organized; unlike other social groups, such as Saudi labor, the government in principle had a negotiation partner.

Low levels of formal trust, reciprocity, and predictability have characterized state-business relations on labor issues, however. A lasting accommodation has not come to pass. This is true on a micro-level as far as compliance with formal rules is concerned; the only networks based on higher levels of trust are informal ones created to maneuver around formal policies. It is also true on the meso-level, where the to and fro of state policies has caused insecurity and higher business risk, preventing meaningful and binding dialogue. At least until 2004, the absence of an agreed forum of negotiations exacerbated the situation. Despite numerous consultative meetings with bureaucrats and princes, business would simply not know to whom to appeal, as responsibilities and hence institutional credibility were unclear and fragmented.

With credible commitments hard to come by, the safer path for business was to procrastinate or to lobby for grace periods and postponements—or failing this, to pursue individual avoidance strategies. The state's low regulatory capacity further undermined interest in a lasting, collective policy agreement, as even credibly negotiated meso-level deals were unlikely to be consistently implemented by the bureaucracy.[233] The collective interest of business in a coherent policy package seemed even weaker than in the FIA case. Beyond ad hoc veto coalitions through senior brokers, business gave almost no constructive policy input, and policy-making retained its paternal character despite the notionally central role of business. As in the FIA context, the private sector has repeatedly asked for clearer and smoother administrative procedures but not forwarded any coherent proposals.

Unlike the FIA case, Saudization has seen a solid consolidation of administrative authority. The MoL reform under Gosaibi has demonstrated that in specific sectors, senior leaders can overcome policy fragmentation if they invest sufficient political capital. It also shows that reshaping existing organizations probably is a more effective, if more painful, means for integrating policies on a meso-level than the creation of new agencies. So far, however, the MoL reorganization seems to be the exception rather than the rule. Many observers of Saudi Arabia were surprised by Abdallah's capacity to trespass on MoI turf.

The reduction in the inflow of foreign labor from 2003 to 2005 was a remarkable feat that would have been inconceivable without the 2004 consolidation of jurisdictions. It proved temporary, however, and it remains to be seen whether the MoL can reassert control over aggregate visa issuance after the contracting boom. Even then, however, the MoL would have to cope with its limited regulatory capacity on the micro-level. Its dogged attempts to micro-manage labor markets have been compromised by its enduring deficits in data-gathering, monitoring, and enforcement. Gosaibi's typical attempts to centralize decision-making might reduce corruption as the risk of detection increases, but it will not contribute much to the MoL's regulatory capacity or to smooth state-business cooperation.

233. For further evidence on how individual rule-avoidance can undermine collective policy negotiations, see Hertog 2006a.

The Saudization case demonstrates both the uses and limits of meso-level bureaucratic reform. As long as the MoL cannot resist the typical Saudi bureaucratic temptation to get too "involved," it is likely to do more harm than good. As opposed to the project developments of the 1970s, Saudization in its current shape, whether concentrated in one agency or not, requires a degree of regulatory power that the MoL has never had. The only viable approach to Saudization, however—focusing on immigration policy as such and liberalizing the internal labor market for expatriates—seems to cut against the interventionist instincts of Saudi technocrats reared during the boom decades.

Chapter 7

The Fragmented Domestic Negotiations over WTO Adaptation

> *"The WTO is great for organizing, like a mother. There's a lot of 'I'm tidying your room and keep it that way.' In Saudi Arabia we grew too quickly and we are not organized…We need someone to organize us"*[1]
> —Hijazi merchant

Rich rentier states like Saudi Arabia enjoy the privilege of conducting most economic policy negotiations domestically: Since the 1990/91 Gulf War, the Saudi government has not contracted any international debt, and World Bank, the International Monetary Fund, and other actors are present as paid consultants, not as creditors that can impose political conditions on the Kingdom's policy-makers. Accordingly, labor market issues have been negotiated within Saudi Arabia, and issues of foreign investment regulation until 2003 also were decided domestically—against a background of international competitive pressures, to be sure, but without the involvement of external political actors. The same is true of other economic reforms not taken up in detail in this book, such as Saudi privatizations and tax reforms.

There has been one big exception to the domestic nature of Saudi economic policy-making: the Saudi negotiations over WTO accession and the policy changes induced by it. During WTO negotiations between 1995 and 2005, foreign agents in the form of leading WTO member states played a pivotal role in negotiating specific conditions of regulatory change in the Kingdom. This allows a test of how the segmented Saudi political economy interacts with coherently

---

1. Cited in Montagu 2001: 48.

articulated external policy demands, one that this chapter takes up. The case study concentrates more on the runup to WTO accession because this was when external conditionality affected the Saudi system. The subsequent domestic digestion of WTO-induced changes, which follows familiar patterns, is addressed more briefly.

Despite its somewhat different focus, the case will in important respects look similar to the other two chapters. WTO negotiations have involved a complex bundle of cross-cutting regulatory reform issues, concerning—in principle—a large number of government agencies. The Ministry of Commerce (MoC) was tasked with handling the WTO portfolio and appeared as a lone fighter in a fragmented and hierarchical bureaucracy often indifferent to its demands. For a considerable time, the WTO issue had little impact on fiefdoms such as the customs administration or the judiciary. Rather unlike young and flexible SAGIA but similar to the Ministry of Labor, the MoC moreover had to make do with weak technocratic resources, further hampering its management of the WTO's complex technicalities.

No clear coalition within the fragmented state seemed to emerge, either in favor of or against WTO-related reforms, leading to policy drift and stalemate. Similarly, the private sector did not find a coherent position on the matter, despite being directly concerned—quite unlike the numerous precedents of trade-related business mobilization in other countries.[2] The only time that certain segments of the business community acted vigorously on a WTO-related policy issue, it was to thwart a symbolic liberalization measure. In a generally fragmented setting, no broader coalitions emerged.

After several years, however, it became clear that the United States would not budge on its harsh accession conditions. Once the Saudis had realized that becoming a WTO member would only become more complicated with time, the force of external conditionality kicked in. The leadership finally gave a clear top-down signal that U.S. demands would have to be fulfilled. As the policy bundle was non-negotiable, in all important respects assembled externally, and much more clearly articulated than any domestically negotiated reform program, government agencies saw no alternative to quickly falling into line and formally adopting numerous regulatory changes. Fragmentation of the system suddenly had a reverse impact on the policy process, as it left little scope for organized resistance within bureaucracy and business. Technocrats seized the chance to push through several reform projects that had been scattered around the government's numerous drawers for many years, surreptitiously including some non-WTO issues for good

---

2. One might think of the recent debates about the Qualified Industrial Zones in Egypt and Jordan or lobbying over free trade agreements in Morocco.

measure. Implementation so far has been imperfect. External pressure decreased after accession, and while some of the regulatory changes required only a royal decree, others demanded that scores of bureaucrats on the micro-level alter their daily routines—a type of change that is much harder to bring about. Nonetheless, the scope of WTO-induced change was unprecedented.

A Note on Saudi Arabia's Trade Interests

A capital-rich importer of goods and services for many years, Saudi Arabia has, at least formally, been a relatively open economy for a long time. Capital flows are free, and its currency is fully convertible. There are a few formal restrictions on international trade, with protective tariffs for certain industries—up to 100 percent in some cases, but usually not above 20 percent.[3] Certain imports require official licenses, and as in other Gulf Cooperation Council states, distribution of international goods and services traditionally is done through local "agents." Until 2005, foreign investment in commerce and some other important services like banking and insurance was strongly limited. All told, however, in comparison with Middle Eastern states with a longer history of import substitution, formal restrictions on trade and services have been modest.

Attempting to diversify its industrial base, the Kingdom has had several good reasons for joining the WTO. Among them would be increased legal security for those of its industries that enjoy a comparative advantage internationally. Most prominent among them are energy- and feedstock-intensive sectors like petrochemicals and its associated downstream industries.[4] SABIC, which had emerged as one of the world's leading petrochemical players by the early 2000s, needs free access to international markets and security from anti-dumping measures. Though oil exports still dwarf nonoil exports by a factor of about nine, nonoil exports have increased greatly since the early 1980s, eclipsing those of any other Arab country.

With its reinforced drive for economic diversification and foreign investment, Saudi Arabia's interest in being seen as a serious, credible, and accountable international economic player has increased. Since most major economic powers and all other GCC states had joined the WTO by 2000, membership had become a sine qua non of such credibility.[5]

---

3. U.S. Foreign Commercial Service 2004.
4. Ramady and Mansour 2006: 194–95; Wilson 2004: 76ff.
5. *MEED,* 8 November 1996: 53; interviews in Riyadh, 2003–05; Niblock, 2006: 138.

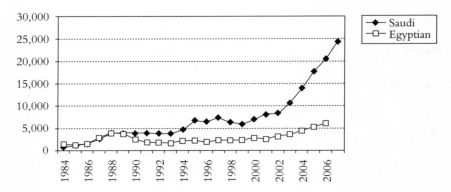

Fig. 7.1 Saudi and Egyptian non-oil exports compared, 1984–2006 (million $)
Source: SAMA (2007) and World Bank Development Indicators.

The Kingdom would quite probably have found it easy to join the General Agreement on Tariffs and Trade (GATT) in the 1980s, when its economy was less open, but the United States was much more relaxed about membership and Saudi Arabia was perceived as full ally in the Cold War.[6] As chapter 4 has shown, however, economic policy-making had reached a stalemate with the fiscal crisis of the 1980s. The Kingdom submitted its GATT application as late as 1993, and serious negotiations did not start before 1996, in the wake of another fiscal retrenchment, which had also precipitated the other reforms of the late 1990s.

Domestic Saudi Interests: The Segmented State

At the time, the leadership was moderately optimistic about joining the organization soon. However, the negotiations and the associated reforms ran into a problems that, as we shall see, reflected primarily the domestic political economy of Saudi Arabia—more than the Kingdom's "objective" interests or, at the time, the international negotiations themselves.

Most of all, the way WTO issues were digested domestically reflected the segmentation of Saudi institutions of government. Only specific players, most of all the Ministry of Commerce, pursued WTO-related reform matters. Although there is a sizeable bridgehead of Western-educated technocrats with a liberal, pro-trade orientation in the upper echelons of Saudi bureaucracy, this group

---

6. Interview with Saudi industrialist, March 2004; Hoekman and Kostecki: 67.

is scattered, tied up in institutional hierarchies, and hence not organized as a political force.

As with any important policy decision, the one to go ahead on WTO issues was made by senior princes. The attitudes toward economic openness vary among the leading royals: Whereas Abdallah has generally been seen as a driver of reforms, Minister of Interior Prince Naif has been more guarded, always wary of foreign intrusions. He has publicly clarified that WTO would not force Saudi Arabia to give up its faith and culture, trying to cultivate a stern and conservative image as a protector of Saudi authenticity and national interests.[7] Once again, even if Naif has not been much involved in the minutiae of daily regulation, his senior position has made the Ministry of Interior (MoI) largely impenetrable to other Saudi agencies. Although it may not have worked actively against WTO accession, its strong and autonomous role in economic regulation has made wholesale implementation of more open rules more difficult.

Officially, the cabinet charged the Ministry of Commerce (MoC) under Usama Faqih with the WTO portfolio. The MoC has never been a particularly powerful player among Saudi institutions, however. When some other agencies voiced specific reservations on economic opening—usually trying to defend their respective sectors—the MoC found it hard to force its way.[8] The defense of agency interests has often trumped the common liberal outlook of a sizeable number of senior bureaucrats. Although hundreds of them were academically socialized in the West and in principle have a greater understanding of issues of economic liberalization than technocrats in most other Middle Eastern states, they had no channels or incentives to coalesce, despite the government's nominally pro-WTO policy from 1993 on. Agencies kept their documents secret and discouraged communication with other institutions.

Many of the reforms discussed in the WTO context had been discussed domestically for many years, including capital market reform, intellectual property enforcement, regulatory reform of various sectors, competition regulation, bureaucratic capacity-building, FDI reform, and service liberalization. Until 2003, however, the demands of the WTO often had little impact on these circular debates.

The segmentation of different institutions and their interests undermined the functioning of coordinating bodies. As U.S. and EU demands in the negotiations were broad, many agencies were represented on the Saudi negotiation delegations to Geneva. An interministerial committee held numerous meetings to

---

7. *Arab News*, 25 April 2000.
8. Interview with senior advisor to the Saudi government, June 2004.

discuss the strategy of negotiation and the reports of specialized committees.[9] In practice, however, the MoC was often left alone with WTO issues. Insofar as the more powerful Ministry of Finance (MoF) and the MoF-related Saudi Arabian Monetary Agency took responsibility for commodities and service negotiations, this again does not seem to have ushered in broader coordination.[10] The MoF has a reputation for being secretive, and its relationship to other agencies is often tense.[11]

Other agencies were often unaware of WTO requirements in their regulatory practice, and Faqih complained that he did not have the full authority for negotiations.[12] When issues like foreign investment regulation, taxation, subsidies, or product standards entered the general reform debates in the late 1990s and early 2000s, WTO criteria were often not taken into consideration. WTO issues were often perceived as a "Ministry of Commerce problem" within the administration.[13] Horizontal communication was deficient, as agencies reported mostly upwards to their royal superiors instead of coordinating with their peer agencies.

The Majlis Ash-Shura, which could in principle have functioned as a clearing board, started addressing WTO issues only in late 2003, seven years after the negotiations had started.[14] Even then, it reacted only to government reports, as apparently the cabinet wanted to present results rather than open issues for debate.[15] Over the years, several pieces of Majlis legislation failed to take WTO requirements into account.[16]

Bureaucratic Capacity Limits

A typical piece of Saudi rentier bureaucracy, the MoC has suffered from its own capacity constraints. Adjustment capacity has been low, small-scale resistance to change strong. Lacking reliable statistics, the MoC apparently had no capacity to gauge the impact of WTO-induced changes on specific markets.[17]

---

9. SAMA Annual Report 2003: 47.
10. Ibid.: 47–48.
11. The closing accounts of the Saudi Arabian government, for example, have not been available since 1989, not even within the government.
12. Interview with senior advisor to the Saudi government, Riyadh, June 2004; interview with SAGIA functionary I, May 2004.
13. Interview with SAGIA functionary II, April 2004.
14. Interview with senior Saudi banker.
15. Interview with former Saudi minister, March 2004.
16. Interview with SAGIA functionary III, May 2004.
17. Interview with Western diplomat I, April 2004.

Former bureaucrats as well as Saudi businessmen feel that the staff needed to handle the highly technical WTO issues has been lacking in the Saudi administration.[18] Even though the government is overstaffed and has accumulated general economic expertise since the 1970s, qualified specialists below the senior level are few and far between, in the MoC as well as other parts of the state, most notably the judiciary. As tasks grew more complex, the Saudi pool of good generalists became insufficient. An internal government memo by a senior advisor in the late 1990s actually recommended that the negotiations be put on hold in order to build up technical capacity.[19]

As on many other complex policy issues such as the foreign investment act, the Saudi government has made use of international advisors on WTO matters, including a group of UNDP specialists and international law firms, building up almost a "shadow government" on WTO matters along lines familiar from the 1970s contracting boom.[20] Such technical missions have usually changed little about the lower levels of bureaucracy, where imparting specialized skills on staffers has been an uphill battle.[21]

Several agencies remained impenetrable to the WTO-related requirements of technical adaptation in trade and service regulation. These included the largely self-contained Ministry of Interior (with its intrusive goods and facilities inspection regimes), technical units within the Ministry of Commerce itself, and the autonomous Saudi courts system which, even when WTO-compatible legislation was enacted, lacked expertise on (and, often, interest in) complex issues such as intellectual property rights and has been highly reluctant to enforce international settlements. Despite repeated announcements over the years, there were no commercial courts until 2005.

The MoF-affiliated customs administration was in some ways a microcosm of WTO-related problems. Procedures were often arbitrary, appeals had to be made orally, decrees remained unpublished, rumors of corruption were widespread, and clusters of bureaucratic brokers peddled their services around the customs offices. Customs, moreover, had no authority over security inspections and quality testing, for which the MoI and the Saudi Arabian Standards Organization (SASO) were responsible, respectively, further complicating procedures. International technical cooperation missions had at best a partial impact on customs practices.[22]

---

18. Interview with former Saudi minister, March 2004.
19. Interview with former Saudi bureaucrat, Riyadh, November 2005.
20. UNDP 2004; interview with Western Diplomat IV, January 2007.
21. Interview with Western diplomat II, Riyadh, May 2004; SAGIA 2003: 34.
22. U.S. Department of State 2001; ESCWA 2001: 70ff.; World Bank 2002: 154; SAGIA 2003: 44, 61. SASO itself has three uniformed board members, which demonstrates the MoI's power.

Domestic Saudi Interests: A Scattered Private Sector

Despite its relatively liberal trade policy tradition, the Saudi government appeared structurally unprepared for WTO adaptation. Looking at the Saudi private sector, at first glance the picture appeared much more encouraging; many of the large Saudi business groups are among the most impressive in the Middle East, not only in terms of size but also relative managerial sophistication. Large industrial groups have become increasingly export-oriented, serving regional and Asian markets.[23]

One might have expected Saudi business to play a proactive policy role on WTO issues, whether pushing for rapid accession or voicing specific criticisms of the undertaking. However, as we shall see, in some ways the private sector mirrored the fragmentation of the administration, which has been divided and, more important, under-organized and under-informed on WTO matters. Although apprehension over WTO was widespread, it hardly congealed into meaningful programmatic statements, with policy positions remaining even more diffuse than in the case of foreign investment reform.

Attitudes to WTO among Saudi businessmen were distributed over the full spectrum of opinion, from those open-heartedly embracing international integration to those calling for rekindled industrial protection.[24] Some sectors have been enjoying various forms of protection, and the prediction is that some businesses will suffer under the opening.[25] However, in the Saudi debate it never became clear which specific sectors were seen as most endangered or, conversely, which ones could expect to profit most from WTO. Far from what trade theory would lead us to expect, clear sectoral interests have not evolved.[26]

Investors in export-oriented heavy industry could in principle have profited from trade liberalization thanks to Saudi Arabia's comparative advantage in cheap feedstock. They did not act as a lobbying group at any point, however.[27] Among owners of lighter manufacturing industries which have been protected with tariffs of either 12 or 20 percent, there were worries about competition, but again, a clear sectoral position does not seem to have emerged.[28] When leading industrialist Abdalrahman Al-Zamil—himself prominently involved in heavy industry—openly asked for infant industry protection and for privileged domestic market

---

23. Luciani 2005.
24. Discussion with Saudi diplomat in Oxford, February 2005; *Arab News* supplement, 17 January 2004: 17.
25. Ramady and Mansour 2006: 193–94; SAMBA 2006.
26. Frieden and Rogowski 1996.
27. Discussion with Western economic advisor, December 2005.
28. Interviews with SAGIA functionaries III and IV, Western diplomat III, June 2004.

access as a reward for employing Saudi nationals, this did not appear to reflect a concerted policy position of Saudi industrialists or a broader lobbying campaign, but rather an ad hoc, personal initiative.[29]

With its trade agency privileges, Saudi commerce is one sector that could be infringed upon by WTO, and agents are considered an interest group.[30] Accounts of how anxious commercialists are toward WTO vary widely, however, and with an exception, to be discussed below, there were no concerted demands.[31] The same is true of agriculture and contracting.

Although desultory lobbying attempts on specific tariff issues seem to have happened, and there were many "little letters" and smaller meetings of protest, broader political campaigns have not occurred.[32] The dislike of WTO has not led to the formulation of clear counterproposals, whether from the private sector in general or from its subsectors.[33] Instead, a generalized fear spread across parts of the private sector that large multinational enterprises might steamroll over local business.[34] Some businessmen like Saleh Al-Kamel dressed their concern that economic globalization is a potential danger in "moral" and Islamic language, recalling the occasional nationalist sniping against foreign investment reform.[35]

The Chambers of Commerce and Industry, quite vocal on other policy questions, have been cautious in their statements on the WTO.[36] Business voices clearly in favor of membership were, again, individual ones, like that of Khalid Juffali, senior figure in a large family conglomerate with trade and industry interests.[37] On balance, it seems that skepticism was more pronounced than pro-trade sentiment, but business players did not act as lobbying group either way.[38] Neither did subgroups of business engage in systematic lobbying, indicating that the private sector's fragmentation approximated atomization and therefore was deeper than that of the state, where at least agency-based interests existed.

---

29. Abdalrahman Al-Zamil on the Human Resources Development Forum, 23–25 May 2004, Riyadh (author's notes).
30. Interview with SAGIA functionary III, May 2004; *Arab News* supplement, 17 January 2004: 14.
31. Interview with member of Riyadh Chamber board, May 2004; interview with former Saudi minister, March 2004; interview with Majlis Ash-Shura member II, May 2004.
32. *Arab News,* 18 May 2000, 13 April 2004; *Saudi Gazette,* 16 March 2004; discussion with senior Saudi bureaucrat, December 2005.
33. Interview with senior economist of Saudi bank I, December 2003.
34. *Arab News* supplement, 17 January 2004: 28; Malik 1999: 259ff.; Ramady 2005: 316.
35. *Gulf Business,* May 1998.
36. A Riyadh Chamber study showing that Saudi Arabia will lose from WTO accession seems to have had no further impact;. For many years, few people turned up at Chamber meetings on WTO issues; cf. Malik 1999: 34, 259ff.
37. *Washington Post,* 10 March 2001.
38. Interview with senior economist of Saudi bank II, December 2003.

The Business Debate: Little Information, Little Interest

The most striking feature of the on-and-off debate about WTO membership in Saudi Arabia probably was the low level of information and tangible policy issues; indeed, "in many cases there [was] sheer confusion."[39] The diffuse sense of peril among many businesses was based on little knowledge about the actual meaning of trade negotiations.[40] Frequently, Saudi business saw technical reforms as something best left for the government to deal with.[41]

Many complained that the government had not provided enough concrete information on the WTO.[42] However, the private sector itself did little to research the implications of WTO or to augment Saudi businesspeople's generic knowledge about the technicalities of adjustment. As late as 2004, the only publication on the matter was apparently an old pamphlet by Adel Faqeeh's Savola Group, entitled "Saudi Arabia and WTO membership: The Promise and the Peril," which took an anti-WTO line.[43] It indicated that membership could lead to forced imports of alcohol and prohibitions on local ownership of broadcast media.

Such statements were part of the "spectacular amount of disinformation about the WTO" described by *Arab News*.[44] Rumors spread that the WTO would force the total opening of Saudi borders, free access to the holy sites for nonbelievers, and engender the replacement of religious *Zakat* by income tax. Widely circulating scare stories also featured foreign-run striptease joints, pork imports, and forced mixing of genders—all strictly prohibited in the Kingdom and anathema to Saudi culture.[45] The American assurance that such cultural exceptions to free trade would be tolerated did not stop the rumors.[46]

The scare stories appear all the odder considering that the actual danger to the Saudi private sector from the WTO, although hard to gauge in detail, is moderate at most.[47] Saudi business by and large is relatively less protected and more sophisticated in managerial terms than its peers in other Middle Eastern countries. Moreover, domestic capital resources are formidable and would help to shield the domestic economy from large-scale foreign takeovers. In other Middle

---

39. *Arab News* supplement, 17 January 2004: 17.
40. *Gulf Business,* May 1998.
41. Interview with Western diplomat II.
42. *Arab News* supplement, 17 January 2004: 28; *Saudi Gazette,* 4 September 2003; *Arab News,* 7 September 2003.
43. *Arab News* supplement, 17 January 2004: 17.
44. Ibid.: 13.
45. Interview with SAGIA functionary I, Riyadh May 2004.
46. *Arab News,* 1 August 2003.
47. SAMBA 2006: 4.

Eastern countries, WTO membership, although usually acquired more easily, has not changed business practices much.[48]

Once again, individual sophistication in the private sector seemed far ahead of its collective maturity as interest group. As so often, this left the bureaucracy to deal with policy specifics. And in fact, it was the more efficient parts of the state that seemed to defend sectoral interests the most clearly: SAMA defending privileges of domestic banks vis-à-vis other agencies and SABIC pushing for WTO accession in order to gain secure access to petrochemicals markets while defending its access to cheap gas feedstock.[49] SABIC had developed a sophisticated international marketing strategy for which it needed diplomatic backing. Traditional islands of technocratic efficiency, the two institutions had the clearest policy positions and the best understanding of their objective interests.

Disjointed State-Business Dealings

Fragmentation was also the dominant theme in the interaction of government and business on WTO issues. It was not an absence of talk about WTO issues as such that prevented the communication of policy interests—indeed, the topic was debated repeatedly in the chambers, and MoC figures addressed various business audiences.[50] "Lots of paper" was produced.[51] Deputy Minister Fawaz Alamy, the Kingdom's chief negotiator, claims that he had more than two hundred meetings with business representatives.[52] Early on, a group was formed within the Chambers to consult with the MoC.[53]

Its activities petered out, however, and the symposia did not significantly clarify interests on WTO.[54] Ultimately, the attempts at public-private consultation were desultory. Their outreach effects within the business community were limited, not least due to the low coherence of the private sector as a policy actor.

Just as important, and more than in the other cases, the government seemed unwilling to engage in meaningful dialogue. Discussions were noncommittal, ad hoc, and limited to small audiences, and there was no institutional framework that could have given them any binding character. With unclear decision-making structures and responsibilities in and around the government, it was not obvious

---

48. This is particularly the case with other GCC countries.
49. Discussion with Western economic advisor, Riyadh, December 2005.
50. Interview with senior advisor to the Saudi government, June 2004; interview with former Saudi minister, March 2004.
51. *Arab News* supplement, 17 January 2004: 17.
52. *Arab News*, 20 January 2004.
53. SAMA Annual Report 2003: 48.
54. Interview with senior advisor to the Saudi government, Riyadh, June 2004.

how useful business consultations with the MoC would be. Further undermining trust, government agencies kept most of the substantive content of the membership negotiations secret, fearing a public backlash fomented by media and business interests.[55] The actual advantages of WTO membership were little advertised.[56]

Whereas the private sector was part of a somewhat dysfunctional debate, it appears that the rest of the Saudi public had no debate at all (give or take the occasional speculation about the impending onslaught of pork and alcohol). The lack of public education by the government was criticized in the press, but the press itself did little to flesh out tangible issues of concern to Saudi citizens.[57] The Majlis Ash-Shura, which is supposed to represent public interests, was little more than a passive recipient of the government agencies' WTO policies.

Whereas the absence of public debate on specific business-related issues such as the foreign investment reform case is understandable, the public should in principle have had an interest in WTO demands that would broadly affect prices and standards in consumer products and services. However, as the history chapter has shown, potentially concerned interests—consumers, workers, and so on—have simply not been organized in Saudi Arabia.

The WTO Negotiations: Crucial Junctures and External Conditionality

The account so far has been largely static, reflecting the repetitive and stagnant domestic negotiation process up to 2003. As we will see, however, after this date Saudi Arabia saw rapid and significant changes on the legislative level—most of which were externally driven. To understand these, a closer look at the chronology of negotiations is necessary, with a special focus on the shifting impact of U.S. and EU demands.

Negotiations until 2003 under Minister Faqih

In early 1997, less than a year after negotiations had started, the hope was to gain admission before the end of the year.[58] Negotiations turned out to be more complex than the Saudi side had initially thought, however, not least due to the

---

55. Interview with SAGIA functionaries III and IV, May 2004; interview with chief economist of Saudi bank I, December 2003.
56. Interview with SAGIA functionary II, April 2004.
57. Muhammad Omar Al-Amoudi, Saudi agenda and WTO, Arabview, 5 January 2001, available at www.arabview.com/articles.asp?article=75.
58. *MEED*, 17 January 1997: 4.

breadth of the issues broached, including complex issues of service liberalization and intellectual property rights.

Saudi market access offers drew criticism from the United States and other countries. The proposed binding tariff rates were seen as too high, often far above actually applied levels. Important service sectors were still missing from the offer, and Saudi Arabia appeared reluctant to sign a government procurement agreement.[59] It seemed that the MoC had put together the Saudi offer somewhat naively, not anticipating how many policy areas WTO negotiations actually affected. Many of the concessions demanded would have gone beyond the administrative turf of the MoC; as outlined above, other agencies at the time cared little about WTO questions, which had the effect of compromising the MoC's negotiating authority. The Ministry of Agriculture clung to its subsidy regime, SAMA defended banking and insurance, while the Ministry of PTT wanted to keep full control over the telecoms market.[60]

The talks lingered on for several years, one sticking point being the Saudi demand to be treated as a developing country, which would give it generous transition phases.[61] The MoC appeared out of touch, issuing repeated announcements of impending accession that were denied by everyone else.[62] In the summer of 1999 there was much speculation that if entry did not happen by early 2000, the issue would be downgraded.[63]

The WTO issue indeed seems to have been put on hold on the Saudi side in 2000, as the MoC had apparently exhausted itself, with little results, and a decision was made to build technical capacity. After the disappointing November 2000 working party meeting, Saudi Arabia "did not answer the phone for two years" when American negotiators wanted to debate WTO issues.[64] No dates for meetings were set, although a large number of open issues would have been up for debate, including insurance, customs regulations, tariffs, and rules for settling commercial disputes.[65] In November 2001, Deputy Minister Alamy predicted at least two more years of negotiations.[66]

---

59. *Journal of Commerce*, 20 November 1998; Wilson 2004: 91.
60. Interview with senior Western lawyer, Dubai, February 2009; interview with former deputy minister, Riyadh, January 2009.
61. Smith and Siddiqi 2000.
62. William M. Daley, Usama Faqih, press conference, 16 October 1999 in Riyadh, available at www.usembassy-israel.org.il/publish/peace/archives/1999/october/me1018a.html; *MEED*, 17 September 1999: 9.
63. Ibid., 2 July 1999: 3.
64. Montagu 2001: 49; interview with Western diplomat I, March 2004.
65. *MEED*, 16 March 2001: 20.
66. *Arab News*, 5 November 2001.

The Fight over Commercial Agencies

Minister of Commerce Faqih appears to have made at least one attempt to break vested interests in the segmented domestic setting. In late 2000, the trade press reported that new sponsorship and commercial agency laws were in preparation.[67] In 1962, the Saudi government had made commercial representation of foreign exporters though local Saudi "agents" obligatory, and many of the large business families still hold agencies of major international manufacturers.

A change in these regulations was unlikely to revolutionize Saudi commerce. Foreign producers usually do not get involved in local distribution and retail even if they are allowed.[68] Moreover, in many cases the relationship of foreign exporter and local distributor is well established, and many of the large Saudi agents run quite professional businesses.[69] Nonetheless, agencies have high symbolic value for many Saudi businessmen, who made their first millions as distributors. Many royal family members have made fortunes with agencies as well.[70] Although some Saudis look at the agency privilege with a critical eye, there is a strong feeling among businessmen that agencies should be Saudi.[71]

WTO negotiation partners made it clear, however, that existing regulations would have to be loosened.[72] A draft bill emerged from the MoC that tied the abolition of monopolistic commercial agencies in their present form to new competition regulations. The law went through the Chambers—where it apparently was approved by a functionary not understanding its technical implications—and reached the Majlis Ash-Shura.[73]

At that point, when it became clear that monopolistic agencies were about to be "busted," at least on paper, several large distributors awoke to the challenge and started organizing a counter-campaign. The main players included Abdulrahman Al-Jeraisy, head of the Saudi Council of Chambers of Commerce and owner of several hundred agencies, and Abdullatif Jameel, a Jeddah businessman who among many other things controls the important Toyota distributorship in the Kingdom. Al-Jeraisy "went wild" and, with Jameel, organized Chamber delegations that complained to ministers, princes, the crown prince, and the royal court.[74] From his offices, Jameel organized a "grassroots campaign" to mobilize

---

67. *MEED,* 1 December 2000: 38.
68. Interview with senior advisor to the Saudi government, June 2004.
69. According to a senior businessman, no one will lose his agency; cf. Montagu 2001: 60.
70. Interview with Western diplomat I, June 2004.
71. Interview with former Saudi minister, March 2004; Niblock 2006: 134.
72. Interview with chief economist of Saudi bank II, December 2003.
73. Interview with Western diplomat I, June 2004.
74. Interview with former deputy minister, May 2004.

concerned businessmen, drawing on advice from international law firms.[75] The law was duly brought down in the Majlis Ash-Shura. Apparently Faqih had lacked the senior-level backing to resist the onslaught.[76]

At the time, royal interest in specific WTO-related reforms and technicalities seemed to be muted. Before 2002, there was only one (characteristically abrupt) top-down initiative related to trade liberalization. In May 2001, most Saudi tariffs were unexpectedly slashed to 5 percent, based on a decision in the Supreme Economic Council chaired by Crown Prince Abdallah.[77] Although the move seems to have been more immediately related to the envisaged GCC customs union, it was expected that the move would help with the WTO portfolio.[78]

Negotiations under Minister Yamani

Otherwise, little happened. However, in a limited cabinet reshuffle in April 2003, Hashim Yamani, previously minister of electricity and industry, replaced Faqih as minister of commerce. Yamani, who holds a Harvard Ph.D. in physics, enjoyed a better reputation among businessmen as a pragmatic technocrat willing to communicate policies. By the time he was replaced, Faqih was held in low esteem by many in the business community; some accused him of "disinformation" on the WTO issue.[79]

With Yamani's appointment, many Saudi and foreign businessmen perceived a significant acceleration of progress on WTO matters.[80] In part, this seemed a product of his better management skills. At least as important, however, the leadership—specifically the crown prince—had apparently decided to force WTO issues.[81] Yamani's appointment may have resulted from this decision.[82] The Kingdom finally seemed to be coming to terms with the external pressure exerted

---

75. Interview with Western diplomat I, June 2004; discussion with advisor to Abdullatif Jameel, Riyadh, January 2009.

76. The law was also said to have been badly constructed in technical terms. Interview with former deputy minister, March 2004; *Arab News*, 4 January 2004.

77. *MEED*, 29 June 2001: 26.

78. Interview with senior advisor to the Saudi government, June 2004; *Arab News*, 28 May 2001; Wilson 2004: 32.

79. Interview with senior Saudi banker. Robert Jordan, U.S. ambassador in the Kingdom from 2001 to 2003, commented that certain senior bureaucrats were less committed to the WTO than Crown Prince Abdallah, probably a reference to Faqih; remarks, Middle East Policy Council conference, January 2006, available at www.mepc.org/public_asp/forums_chcs/42.asp.

80. *Arab News* supplement, 17 January 2004: 29; interview with former minister, March 2004.

81. Interview with Western diplomat II; interview with senior Western lawyer, Dubai, February 2009.

82. According to an EU diplomat, "It was a political decision on the Saudi side to move ahead. Dr. Yamani is firmly in favor of membership. That is a big change." *Arab News* supplement, 17 January 2004: 29.

mainly by the United States and with the fact that, fair or not, membership would inevitably come with significant strings attached.

The royal leadership had, in principle, been for WTO membership all along but had, as is customary, left considerable scope for the commoner cabinet and bureaucracy to flesh out the details. Now, by way of royal fiat, international demands were directly translated into change on the ground, leaving much less scope for interagency politics. The sudden adaptation proceeded in two main steps. First, the Kingdom signed its bilateral WTO accession agreement with the EU in July 2003, making substantial concessions on FDI liberalization in services, which had emerged as the biggest issue in the negotiations due to the great potential of finance and telecommunications for foreign investors.[83] The agreement allowed for foreign majority shares in both, encroaching on the turf of line agencies such as SAMA and the Communications and Information Technology Commission (CITC), the new telecoms regulator.

In all likelihood, such steps were only possible thanks to high-level royal backing, in turn induced by the persistence of international demands. The changes cut more deeply than previous domestically negotiated reforms. In the same month, Yamani traveled to Washington to sign a bilateral trade and investment agreement, which served as prelude to the second step, the bilateral agreement with the United States.[84] This would turn out to be even tougher and wider-ranging than anticipated, deepening the Saudi feeling that the international community was moving the goalposts.[85] U.S. demands were wide-ranging, including issues of general legal and bureaucratic transparency, intellectual property rights, and further service liberalization.[86]

Regulatory Change since 2003

This delayed the final agreement, but less than one might have expected. By September 2005, the United States and Saudi Arabia had struck a comprehensive final deal. By that time, external pressure combined with the early 2003 royal decision to cave in had contributed to several significant attempts of legal reform, which now consistently followed international demands as opposed to intra-government politics. Adaptation measures that should have happened years before—and had

---

83. *Saudi Gazette*, 30 August 2003; *Arab News* supplement, 17 January 2004: 15. Banking has been one of the main areas the Saudi state wanted to protect; cf. Wilson 2004: 91.
84. AMEINFO, 6 July 2004, available at www.ameinfo.com/news/Detailed/42278.html; *Arab News*, 9 August 2003.
85. Oxford Analytica, 2 November 2005; Wilson 2004: 91–92.
86. *Arab News* supplement, 17 January 2004: 29; *Financial Times*, 26 February 2004.

often been debated, independent of the WTO, for many years—were tackled under external induced royal pressure and Yamani's leadership.

Few broader measures of legal reform had been taken before 2003. However, significant laws on intellectual property, trademarks, and copyrights were issued in 2003 and 2004.[87] In the summer of 2003, the government issued a new capital markets law that provided for the creation of the formally independent Capital Market Authority, which came into being in 2004, and prepared the ground for the entrance of foreign investment banks. In the summer of 2003, the cabinet approved a long-expected insurance law, and in June 2004 it approved a competition law that provided for a new competition council, whose members were appointed in October 2005.

Moreover, as mentioned in the foreign investment case study, SAMA has licensed several foreign banks since 2003, and in 2004, a rumor circulated that the retail sector would be opened up to a maximum foreign ownership of 75 percent, again encroaching on the established turf of national business.[88] Although Saudi officials are reluctant to admit that the WTO forced such measures, they generally concede that these steps helped greatly in WTO negotiations.[89] Tellingly, the government tackled most of the "touchy" areas like finance and commerce only once they had come up in the negotiations.[90]

The final accession deal announced in November 2005 involved substantial concessions in several areas, including 60 percent foreign ownership in banking, 60 percent foreign ownership of insurance, and 75 percent foreign ownership of distribution within three years.[91] Insurance liberalization encroached on the turf of the religious establishment, as the supposedly Islamic precepts of "cooperative insurance" underwent repeated revisions to adapt to U.S. demands.[92] Through an "importer of record" clause, direct sales to Saudi buyers are now possible before customs, potentially circumventing local agents.[93] Although the 30 percent rule on Saudi participation in local tenders remained intact, bidders are no longer required to employ Saudi agents for government contracts.[94] Saudi Arabia also made numerous technical concessions on trade procedures and standards.

---

87. *Arab News,* 6 July 2004.
88. Interview with SAGIA functionary III, May 2004.
89. Interview with senior advisor to the Saudi government, June 2004. The more candid interview subjects admitted the great influence of the WTO.
90. Moreover, the press as well as officials engaged in the negotiations consistently mentioned most of the measures in a WTO context.
91. WTO 2005a, 2005b.
92. Greenberg interview.
93. SAMBA 2006: 9.
94. Ibid.: 24.

The legal changes involved many agencies. Although several of them have run into customary implementation problems, that the package of external demands was so broad gave the resulting reform drive unprecedented breadth.[95] As heightened foreign demands translated directly into policy thanks to the cabinet's acceptance of the WTO deal, interagency politics now mattered much less. Indeed, the fragmentation of the bureaucracy meant that there was little scope for broader resistance during and after the late phase of negotiations. Bureaucratic veto players who may have procrastinated on individual technicalities had received a clear signal from the top that the externally imposed—and externally judged—package would have to be accepted and implemented in its totality.[96]

Similarly, there appears to have been little organized business protest against any of the measures. As it was clear that a royal decision to move ahead had been made and WTO membership was seen as inevitable, business remained largely silent on most of its ramifications. Some cautiously positive post hoc comments emerged from the private sector after accession had happened—as tends to occur in the Kingdom once a decision has been made—but no coherent policy statements or analyses of the situation.[97] As the Saudi press complained in early 2004, a moment when the leadership's capitulation to U.S. demands had become obvious, with few exceptions it has been "an almost impossible task" to get those who were opposed to the WTO to talk—unlike in earlier years, when the debate was disorganized, but some anti-WTO voices were audible: "It soon became obvious they [anti-WTO businessmen] did not want to raise their heads above the parapet, knowing that the authorities are in favor."[98]

While We're at It: The WTO as Midwife for Unrelated Reforms

No matter how upset Saudi technocrats are about the squeezing their old American allies gave them, many freely concede that the WTO process helped accelerate several reforms within the Kingdom.[99] Deputy Minister Alamy himself explained that WTO membership will speed up privatization (not a WTO issue per se) and

---

95. Interview with Western diplomat I, December 2005.
96. A seasoned Western observer involved in the negotiations stressed the crucial role of an external "judge" for keeping the reform package coherent and preventing it from fizzling out among the different government agencies. Interview with Western diplomat, Riyadh, December 2005.
97. *Arab News,* 10 November 2005, 12 November 2005. There was merely one Jeddah Chamber of Commerce representative predicting a flood of cheap goods; cf. *Gulf Times,* 14 November 2005.
98. *Arab News* supplement, 17 January 2004: 17. This is very much my own experience, as I encountered great reluctance in the private sector to talk about interests that might be perceived as protectionist or regressive; Saudi businessmen often want to be seen as leading the nation's development.
99. Interview with senior advisor to the Saudi government, Riyadh, June 2004.

lock in reforms.[100] Leading businessmen from the Jomaih and Alireza families have stated that the WTO has actually helped the government to push through some awkward measures.[101] Indeed, many of the steps taken now had been held up before through interagency cleavages, endless deliberation, and lack of business support, notably the opening up of finance and commerce, which had remained on the "negative list" for foreign investors after the 2000 FDI reform.

Liberal Saudi technocrats by themselves have not been able to engineer a pact for cross-cutting reforms, as their interests remained tied up within their local institutional and patronage structures, and they lacked negotiation and interest aggregation mechanisms. With the WTO-induced pressure, translated domestically through a royal decision to move ahead, liberal Saudi technocrats were able to pursue policies that had been languishing in their drawers—and debated in interministerial committees—for many years. Suddenly, fragmentation of the institutional landscape was working in favor of reforms. The opening of sensitive sectors, reform of intellectual property rights, competition legislation, and creation of sectoral regulators became possible almost overnight. The enthusiasm of many Saudi administrators about recent measures, however imperfectly implemented, is remarkable. That the push for most measures came from outside is often underplayed. Accession was of course a royal decision at the end of day. But the core ingredients of WTO reforms were defined externally and hence not left to the mercy of the Saudi technocracy on which the royals leaders otherwise rely and whose fragmented interests are played out when policy details are negotiated.

Remarkably, the government seems to have pushed through some domestic reform measures under the WTO banner that were not strictly linked to the WTO's technical requirements, including reform of the Saudi court system.[102] The creation of a circuit of commercial courts was finally announced in the spring of 2005. They had been debated for decades, but only now did the leadership make an attempt to undo the thorough fragmentation of the quasi-judicial administrative bodies that had been created to bypass the sharia courts.

The Fine Print: Implementing WTO

After 2003 there was a more forceful drive to adapt to WTO requirements, as the leadership had recognized the inevitability of certain core changes. Royal will to adapt, on its own, is not enough to reengineer bureaucratic realities, however. Major rule changes that could be dealt with on the cabinet level or through

---

100. *Arab News* supplement, 17 January 2004: 16.
101. Ibid.: 17.
102. Interview with Western diplomat I, December 2005.

efficient agencies such as SAMA have happened swiftly, be it tariff reductions, the promulgation of new sectoral laws, or the issuance of banking and insurance licenses. The reforms of banking and the capital markets under SAMA and its spinoff, the Capital Market Authority, proceeded successfully and smoothly, although they required detailed regulations.[103] Significant foreign capital has flowed into investment banking since 2004.

In other parts of the state, however, implementation of more diffuse and complex demands was often wanting on the lower echelons. Administrative transparency was one of the main U.S. gripes in the negotiations.[104] Clear procurement standards were part of this, but it has been well-nigh impossible to force behemoths like the Ministries of Interior or Defense to change their ingrained practices, the new procurement law notwithstanding. Similarly, the U.S. demand that all trade-relevant legislation be translated into good English[105] has thus far foundered over the widespread practice of keeping important ministerial decrees unpublished.[106] While ministers and their deputies are easily called to account or frightened into submission, the leadership lacks the means to discipline the wider reaches of its sprawling bureaucracy in a sustained fashion. Vested interests in opacity and fragmentation exist in many agencies. But even in the absence of such interests, passivity reigns and incentives to get involved with new technical matters are often lacking. A leading Saudi industrialist has complained over "virtual laws," issued quickly in the course of WTO adjustment but impossible to implement in the existing structures.[107]

The slow struggle to improve customs procedures and the strong, cross-cutting role of the wayward Ministry of Interior have already been mentioned.[108] Intellectual property enforcement has been similarly difficult for existing institutions. The field involves a variety of different organizations with often limited capacity, including the Ministry of Commerce, the Ministry of Information (under MoI influence), and King Abdulaziz City for Science and Technology (directly under

---

103. Interviews with investment bankers, 2005–09.
104. *Arab News* supplement, 17 January 2004: 17; Montagu 2001: 55.
105. Discussion with Western economic advisor, Riyadh, May 2004.
106. Although Yamani has welcomed the American request for clarification of certain laws and regulations, as it would benefit the Kingdom, a 2004 op-ed claims that the very request shows the failure of the ministry, considering that the WTO has been knocking on the door for such a long time. The author goes on to claim that Saudi officials are deliberately acting in obstructionist ways to twist the arm of the WTO; cf. *Arab News* 5 January 2004, citing Al-Riyadh. The Working Party Report mentions a Saudi commitment to create a website on which all trade-related regulations will be published. This has not yet happened; it would be surprising if information on ministerial-level decrees were complete.
107. Interview, March 2004.
108. A business survey in late 2007 still ranked Saudi non-tariff trade barriers as much worse than those of all other GCC members; cf. *Gulf News*, 8 April 2008.

the king). U.S. software firms have been worried about rampant piracy in the Kingdom.[109] Although authorities have occasionally cracked down on illegal street markets, the bloated Saudi state does not possess the regulatory reach in society to prevent software smuggling. Dodging of official regulations is rampant among many small businesses, formal and informal, immersed in a culture of avoidance. Although there is a new patent law, the King Abdulaziz City for Science and Technology responsible for assessing patents has until recently lacked technical capacity.[110] Though external pressure has given the WTO adaptation process cohesion on a broad policy level, the process has been top-down and has met capacity deficits and administrative fragmentation on the level of day-to-day administration.

Only in February 2007 did the MoCI set up a technical agency to deal with WTO issues.[111] Even so, the government continued to lack a high-powered sounding board for WTO questions similar to the Office of the U.S. Trade Representative to follow up on individual questions.[112] Some of the reforms seemed to fizzle out. Details were left hanging, such as the regulations of the agencies, which continued to reserve commercial representation to nationals—without such representation, however, foreign ownership in distribution companies made little sense.[113]

The reform of the court system, one of the Kingdom's deepest fiefdoms, proceeded slowly and haltingly. Almost no information was available until October 2007, and the details of the new system of specialized courts announced at that time were still unclear as of this writing. King Abdallah reshuffled senior figures in the religious and judicial establishment in February 2009, perhaps reacting to the resistance his reform efforts had met. If Abdallah should manage to impose secular regulations on Saudi courts, this would open a whole new chapter in the history of the Saudi state. It has not happened yet, however. The conservative establishment has come under much pressure after 9/11 and the domestic terror campaign of 2003 through 2005. It is too early to tell whether this will really allow the leadership to reshape their institutional bargain with the religious forces.

Conclusion

The WTO case addresses in particularly clear-cut fashion the puzzle of why everyone in Saudi Arabia seems to agree on reforms but so little seems to happen.

---

109. U.S.-Saudi Business Council, Business Brief 2/2004; interviews with Western businessmen.
110. U.S. Department of State 2001. In 2008, Saudi Arabia remained on the U.S. 301 watch list of countries with weak intellectual property protection.
111. *Arab News,* 20 February 2007.
112. Interview with Western diplomat V, Riyadh, February 2007.
113. Interviews with various lawyers, February 2009.

Policy outcomes are not a result of individual attitudes but result from the interplay of institutions, namely the hierarchies and the meso-level fragmentation of the Saudi state. Efforts to adapt to WTO rules for many years were locally circumscribed and got stuck between different government organizations or on the lower levels of the bureaucracy. The lack of horizontal linkages between administrative fiefdoms meant that even reform-minded bureaucrats were subject to institutional constraints. Against this background, the Ministry of Commerce, itself an agency with strong capacity limits, was incapable of developing a realistic agenda, forging a reform coalition or acting as a credible negotiator. There are better-functioning organizations in the Saudi state than the MoC, but these, if engaged at all, were interested in defending their specific, limited interests as veto players, be it pro-liberalization in the case of SABIC or defense of their sectors, as in the case of SAMA. Up until 2003, the fragmentation of the Saudi bureaucracy characterized WTO adaptation attempts. The royal leadership did not devote continuous attention to trade matters, its activism limited to a sudden tariff reduction in 2001.

The politics of Saudi business in many ways reflected the fragmentation of the administration, which suffered from misinformation and lack of policy-making capacity. Expecting the state to lead, it witnessed little systematic and integrated collective debate and never articulated its apprehension over WTO into coherent lobbying on a broader front. The only time collective action happened, it involved veto lobbying on a symbolic issue. In the WTO context in particular, Saudi Arabia seemed to lack a public forum to settle policy debates and build coalitions, whether within the state or between state and business—although the absence of other organized interests in society could arguably have made an accommodation easier. Private sector expectations were still cast in a paternal mold: business acted as client and policy-taker. It might be far ahead of the state in terms of individual managerial capacities, but on the level of organized politics, the state still dominates.

Fragmentation had led to policy drift as long as external conditions had not been fully accepted by the leadership. However, as soon as Abdallah took a top-down decision to go along with an integrated pack of *external* demands, the fragmented bureaucracy rapidly fell in line and acquiesced to measures that had been discussed for many years but had never come to fruition domestically.[114] Similarly, after government pressure increased in 2003, the disorganization of Saudi business meant that there was little organized resistance to the top-down reforms. Instead, international demands were directly translated into the domestic arena by royal decision.

---

114. By the admission of a senior bureaucrat, the WTO helped the government think about and implement many things; discussion, Riyadh, December 2005.

The fragmentation of state and business, and of the relations between them, can hence mean policy stalemate, but it can also make for rapid policy adjustment when there is focused external pressure. Of course, most reforms are reactions to some kind of pressure. However, in most cases, this pressure has been so diffuse as to leave wide scope for domestic actors to negotiate policy details, allowing the policy process to fragment and, in many cases, fizzle out. This is true both of the diffuse (international) competitive pressure that instigated the FIA reform and of the (domestic) labor market pressures that led to incoherent and top-down royal demands to Saudiize. The WTO case was different in that it involved an externally defined, comprehensive, and detailed policy package—in effect allowing external actors to act as a surrogate cabinet that could integrate adjustment policies on the meso-level.

The WTO case study strongly affirms institutional explanations of economic policy. An analysis of "objective" sectoral interests or interests related to the relative prices of production factors contributes little to our understanding of how business and government behave.[115] It was not rational economic interests or individual bureaucratic attitudes that determined policy outcomes but rather organizational structures of authority. Policy perceptions, lobbying, and policy decisions were shaped, or prevented from taking shape, by the fragmented political economy of the Kingdom.[116] The opposite outcomes of the pre- and post-2003 periods highlight that it is institutions that determine how interests are translated into decisions, in a way that "objective" sector- or factor-based explanations fail to explain.

The externally steered adaptation process has mostly affected formal regulations and has circumvented the meso-level fragmentation of the bureaucracy. The top-down adjustment has met capacity deficits on the micro-level of day-to-day administration, however. It remains to be seen how much WTO accession can change about the fabric of the Saudi state in the foreseeable future. Deeper fiefs, such as the court system, will pose a particular challenge, but lower-level clientelist bureaucracy will be hard to budge also in other parts of the state. In its fragmentation, passivity, limited regulatory capacity, and assorted networks of brokerage it is probably more change-resistant than any other feature of the Saudi system. In the coming years, WTO compliance demands and efforts to build enhanced regulatory powers will put this resilience to an important test. Now that membership is secured, one fears, fragmentation of policies and institutions may reign supreme all over again.

---

115. Frieden and Rogowski 1996.
116. As Kathleen Thelen (1999: 374–77) has argued, institutions can determine rather than just channel interests.

Chapter 8

Comparing the Case Studies, Comparing Saudi Arabia

When the Saudi state embarked on its first comprehensive economic reform drive in two decades in the late 1990s, it did not look so bad, at first glance. It was still relatively well-funded, corruption—despite the widespread myth—was less endemic on the administrative working level than in many other Arab states, bureaucratic education levels were reasonable, and the higher echelons of the technocracy contained a host of highly qualified figures with a liberal outlook. From a technocratic perspective, policy-making did not seem unduly encumbered by organized social interests that could thwart reforms, and despite an oligarchic structure of princes on top of the state, the system was rather neatly centralized. So why is it that the record of reforms between 1999 and 2009 looks so decidedly patchy? Why did this authoritarian and centralized state struggle so much with imposing regulatory change on bureaucracy and business?

This puzzle is not easily captured with existing theories of economic policy-making in rentier states or developing countries in general, which generalize too much to explain complex outcomes. This book has argued that it is the clientelist and fragmented nature of the Saudi system, both formal and informal, and the accompanying dominance of vertical links that explain specific coordination failures and bureaucratic capacity problems. Although general, this statement has more specific implications on the meso- and micro-levels of bureaucracy and state-society relations than previous explanations. The three in-depth case studies have spelled out the specifics of this system, its resilience, and the occasional ability of bureaucrats and businessmen to overcome or work around its limitations.

The Case Studies and the History of the State

Bureaucratic clientelism, steep hierarchies, stop-and-go policy-making, autonomous administrative spheres, networks of brokers, and the absence of credible interest groups in society all can be traced back for many decades. Their emergence and solidification has been described in detail in the history section of this book. Although it grew tremendously over a few decades, the Saudi state in many respects has changed much less than the literature to this point has had it. The main change has been the bureaucratization of the system, accompanied by a great increase in complexity through new layers of administration and the extension of the state's distributional structures all over society.

This seems to have greatly reduced the suppleness of the Saudi system. How much it has been reduced was arguably a matter of speculation until the late 1990s. Despite a fiscal crisis starting in the mid-1980s, the princes did not dare tackle any comprehensive policy reform once the state apparatus had stopped its rapid growth. It took fifteen years until the leadership, Crown Prince Abdallah in particular, set out to reengineer Saudi economic policies. The reforms since the late 1990s were the first comprehensive test of the Saudi system since it grew to full maturity in the oil boom. Three of these reforms were addressed in the second half of this book.

Somewhat simplified, the main challenge was to shift from efficiency in distribution to efficiency in regulation. Unlike the project developments of the 1970s, the task was now to provide an environment for independent investment, job creation, and international economic integration. This turned out to be much harder than expected, as numerous unintended consequences of early institution-building became salient. Bureaucratic clientelism, ambiguity of rules, and networks of brokers had been useful in the context of public service expansion and distribution in the 1970s and 1980s, when the state aimed at giving large numbers of individuals a stake in the system. These inherited features often became intractable hindrances, however, when it came to reshaping the regulatory environment, when (in the worst cases) brokerage would transmogrify from a tool of distribution to a tool of predation. Similarly, meso-level fragmentation of the state turned out to be much more problematic once growth was supposed to be driven not by the state, with its sectorally limited development projects, but by a private sector that required consistent regulation.

Underlying Policy Patterns

The three case studies vary considerably on both the explanatory and context variables. The protagonists of reform varied: SAGIA was a new, insulated agency that could afford to hire good staff outside of rigid civil service rules, whereas the Ministries of Labor and Commerce were older players with much more baggage of formal bureaucratic clientelism.

The thrust of reform varied as well. Although all three cases involved regulatory change of some kind, SAGIA and the Ministry of Commerce were more clearly poised against other agencies than was the Ministry of Labor. Saudization involved fewer open conflicts of interests between agencies but was rather pursued in parallel by various organizations. Moreover, whereas foreign investment reform and WTO adjustment were aimed first of all at bureaucratic reform, Saudization primarily aimed at regulating business behavior.

Finally, the fragmentation of actors involved in the reform process varied. All cases started with a large number of organizations, but the Saudization case saw a consolidation of jurisdictions, and the institutional dynamics of WTO reforms were changed through the overriding imposition of an external reform package.

The similarity of outcomes up to 2003 is striking. It was only when institutional fragmentation decreased in the Saudization and WTO cases that their results—predictably—deviated from those seen in the FIA case. All cases saw incompatible and often autonomous behavior of vertically divided state organizations, unclear and overlapping rules, repeated delays, dilution of policies in the implementation stage, low regulatory capacity, informal brokerage of state resources, occasional jolts of royal policy activism, and paternal and state-driven patterns of policy-making despite great expectations toward the role of the private sector.

Although the context of policy had changed fundamentally, the bureaucracy continued to behave in patterns engrained in the 1960s and 1970s, in the course of Saudi distributive state building.

Common Causes

How has segmented clientelism produced these outcomes? The vertical fragmentation of the Saudi political economy has affected policy on macro-, meso- and micro-levels.

THE MACRO-PICTURE: PATERNAL OVERCENTRALIZATION AROUND THE AL SAUD
Policy-making in the Saudi state has historically been a paternal process in which the Al Saud towered above the system as the only macro-level actor, who as monopolistic distributors of rent kept their clients separate and avoided substantive delegation of policy powers to any collective. The regime did not allow larger independent interest groups to emerge and fragmented society through royal and bureaucratic largesse. Although the leadership is sensitive to diffuse sentiments of social discontent over issues of equity and service delivery, most actual politics happens within the state.

While this system might have functioned relatively swiftly (if not smoothly) when it was much smaller, nowadays excessive centralization weighs down policy-making on complex questions. The occasional royally driven jolts of policy activism in all three case studies show that only very senior figures can spur the system into action. Sustained attention to technical policy issues seems much rarer as senior players often lack detailed knowledge, and centralization takes its toll on royal time budgets. Activism tends to be a reaction to immediate demands and crises, while follow-up capacity is limited.

THE MESO-PICTURE: A FRAGMENTED BUREAUCRACY

The centralization of powers in the royal family has historically combined with a fragmentation of client groups and institutions below the Al Saud, including the bureaucracy, where individual organizations tend to communicate upwards with the political patrons rather than laterally with other bodies. Oil surpluses have allowed the Saudi state to sustain a wide array of very different, parallel client bodies that have not been functionally integrated and in some cases constituted veritable worlds unto themselves, including the judiciary, the Ministry of Interior, and the Ministry of Defense and Aviation.

Whether the purpose was cooptation of social elites or the tackling of specific technocratic problems through insulated elite institutions, the hub-and-spoke system has grown more pronounced in the course of bureaucratic modernization. Early political and bureaucratic contingencies have been frozen in time.

Although reasonably qualified, the Saudi technocracy has never been integrated by a cross-cutting player such as the Ministry of International Trade and Industry (MITI) in Japan or the Economic Planning Board in South Korea.[1] The powerful Ministry of Finance was able to coordinate allocations broadly during the infrastructure development phase of the oil boom, but it is in no position to coordinate regulatory policies across different agencies. Bureaucratic fragmentation has become especially salient in regulatory reform efforts since 1999. This has proven true no matter whether the leadership tasked established or new bureaucratic players with a policy. The old reflex of adding a new and privileged administrative island to solve new problems—SAGIA—did not help integrate policies in the FIA case, as the relevant regulatory turf had already been carved up. Agency interests and hierarchies trumped the liberal interests that technocrats shared in principle, and SAGIA's internal efficiency was of limited help in the broader policy game.

1. Johnson 1982; Chibber 2002; Haggard et al. 1998.

The leadership can locally overcome bureaucratic fragmentation and policy incoherence through the reorganization of jurisdictions, as seen in the Saudization case. There are few precedents of such politically and administratively difficult consolidation, however, and reorganization is unlikely to help with policy issues that are essentially cross-cutting, such as foreign investment promotion. SAGIA's attempts to avoid the national bureaucracy by creating separate FDI enclaves along the Dubai model have also run into obstacles. Similarly, the consolidation of WTO-related reforms through strict external conditionality might have been a one-off phenomenon in which international negotiation partners acted in lieu of an overriding bureaucratic lead agency.

THE MICRO-PICTURE: BUREAUCRATIC CLIENTELISM

While bits of institutions can be shifted around, internal change of the bureaucracy on lower levels is arguably even harder to engineer in the Saudi system. The Saudi state has grown continuously from the 1950s on, reaching out to ever larger parts of society through employment and distributional policies. Most of the time, however, the bureaucratic growth has been much quicker than the availability of qualified and motivated entrants into the civil service would warrant. By co-opting a significant chunk of its populace into various state organizations, the regime has managed to distribute its externally derived wealth and prevent independent socioeconomic interest groups from emerging. In a personalized fashion, bureaucratic clientelism was present already at the inception of the modern Saudi state. From the 1960s on, the regime expanded and formalized it on a grand scale.

The large-scale bureaucratic entitlements have created a civil service that is difficult to discipline, however. Cooptation has its administrative price. Despite rigid organizational centralization, senior bureaucrats have had limited control over the day-to-day dealings of low-level administrators, who play a particularly crucial role when it comes to the nitty-gritty of regulation—the focus of our case studies. Centralization can either stifle initiative or contribute to occasional bouts of activism. It hardly provides steady and consistent incentives to bureaucrats who have little chance of upward mobility, a strong sense of entitlement, and feel safest if they remain inactive. Centralization's most pronounced effect might be the undermining of horizontal cooperation, weak to start with in a clientelist system.

Despite the generally static picture, the role of the Saudi bureaucracy in the economy has seen a "subterranean" change since the boom. With less internal mobility, fewer resources, and less managerial capacity relative to the growing private sector, it has in many regards ceased to drive economic development as it once did through its use of material inducements. With a much more mature local business class, the original rationale for many of the bureaucracy's hands-on interventions has become obsolete, and it has become a conservative force—a burden for business rather than a means of prodding an inchoate business class to expand and diversify.

THE MICRO-PICTURE: BROKERAGE NETWORKS AROUND THE ADMINISTRATION
The tradition of heavy regulation inherited from a phase of state-led growth is alive and well, with numerous intrusive regulatory tools at the disposal of firmly ensconced mid-level bureaucrats. These tend either to be slow and unresponsive or, in the more malign cases, to pursue their own interests, especially when they can exploit ambiguous or overlapping rules, as the Saudization case has highlighted in particular.

Either case furthers the emergence of administrative brokers of various kinds. Brokers who make the services of the hierarchical, unresponsive, and fragmented administration available to less connected Saudis and foreigners are as old as the Saudi state. They have multiplied in manifold ways and on several levels during the boom years, when the state became omnipresent and inordinately rich relative to society, but remained opaque. From the 1960s on, formal state policies deliberately nurtured brokers. Bureaucratization and formalization of the state has, if anything, boosted them, as it made the administration more complex, demanding, and rigid. Growth of the system has moreover made brokerage more anonymous and monetized.

If society is largely absent on the macro- and meso-levels of policy-making, its informal structures often penetrate the state deeply on a micro-level in the course of policy implementation. It is a testament to the pervasive importance of personal intercession that even SAGIA itself has become a weighty informal "broker" for foreign investors—chaperoning large foreign investors through the Saudi bureaucracy on an individual basis without streamlining bureaucratic behavior in general.

At boom times, brokerage was usually a means to gain access to the state's various material resources. This is often still the case. But as was clear in the case studies, nowadays businesses increasingly have to resort to brokerage with different organizations simply to avoid administrative hassle and, in the worst cases, to "grease the wheels." It is doubtful that the reform efforts since 1999 have changed the fabric of the state on lower levels. A 2005 survey by a senior Riyadh Chamber representative showed that 77 percent of businessmen felt they had to "bypass" the law to conduct their operations; two percentage points higher than in a survey two years earlier.[2] The comprehensive regulatory streamlining

---

2. Mandil 2005.

imposed by the WTO will be an important test case for the resilience of brokers in coming years.

MICRO-STRUCTURES AND THEIR IMPACT ON STATE CAPACITY

In several instances in the FIA and Saudization cases, regulatory change has increased pressures on businesses, which have in turn often taken recourse to informal avoidance mechanisms, in which they seemed to invest much more creativity than in collectively negotiated solutions.

The actual regulatory power of the bureaucracy tends to remain low. In all cases, information gathered about businesses was unreliable, and however heavy-handed the rules were, businesses tended to find ways to avoid them or counteract their intent. Low regulatory power is a result of traditionally weak performance incentives for bureaucrats, but also, relatedly, of low formal trust between the rigid, inward-looking, clientelist bureaucracy and claimants in society. In a stultifying equilibrium, low trust in turn is a result of low expectations of regulatory fairness or predictability—in other words, regulatory capacity.

Caution should therefore be taken not to ascribe all problems of the Saudi bureaucracy to corruption or rent-seeking. Venality has been present in our case studies to different degrees (much more in Saudization than in the FIA or WTO cases, for example) and does not explain much as such. It rather is a symptom of underlying authority and incentive structures—rigidly hierarchical, vertically divided, and clientelist—which lead to bureaucratic rigidities, and various techniques to avoid them, corruption only being one.

ECONOMIC POLICY AND VETO PLAYERS ON VARIOUS LEVELS

It might appear counterintuitive that the highly centralized Saudi system contains so many veto players. However, in some ways it is exactly overcentralization that makes vertical links predominate and weakens horizontal cooperation. The veto power of senior fiefdoms like the Ministry of Interior are relatively obvious. Absent a clear, top-down push to implement a policy, however, many less powerful actors have also been able to stymie policies or parts thereof, whether deliberately or not. Numerous agencies managed to water down the foreign investment reform or prevented it from being smoothly implemented (or both), and a wide selection of government organizations played a role in thwarting the Ministry of Commerce's WTO adjustment drive.

Dilatory vetoes in the Saudi system seem to be relatively easy to cast, requiring little more than inactivity. Even active veto lobbying is easier in a fragmented system, needing less coordination between different players than active support in favor of a specific policy. Parts of Saudi business, not otherwise much involved in deliberating policy details, have repeatedly been able to achieve the postponement of policies.[3] Both bureaucratic agencies and significant parts of business reflexively defend established stakes. A public forum to settle policy debates and build broader horizontal alliances for change—whether within the state or between state and business—is sorely lacking.

In addition to meso-level vetoes through state organizations or business groups, individual bureaucrats and their networks can also cast micro-level vetoes against the implementation of new policies. As politically powerless and fragmented as many bureaucrats are, on a small scale they possess considerable power to prevent things from happening or change rules against their original intention—perhaps not when a senior prince's attention is temporarily drawn to a specific policy, but in most other instances of day-to-day administration.

The different types of veto player act on different levels and stages of the policy process. Princes and senior ministers are most important when policies are negotiated, senior level bureaucrats count most when the accompanying regulations are spelled out, while low-level players act in the arena of practical implementation. The different actors exert different types of dilatory power: on big policy issues, on regulatory implementation, and on processual implementation, respectively. And while the impact of senior-level players is individual, low-level actors have a diffuse, collective impact. Princes are also more effective at putting up active resistance, whereas the influence of other bureaucrats tends to be passive. The latter, however, tends to be more widespread and less focused, requiring more attention from the leadership, and can in fact be harder to overcome. The further one moves down, the more structural the problem becomes. This is not to deny that high-level veto players can activate low-level subordinates or clients, though. The Ministry of Interior (MoI), for example, can block sectoral policies, and ministers can instruct their own bureaucrats to "go slow" on a given issue.

Other Reforms and the Book's Argument

This book could not cover all economic reforms that have happened since the late 1990s, and it has focused on the ones with the longest duration and the broadest range of institutions involved. Lest we incur charges of selection bias, the following section will give a brief overview of other major economic reforms since 1999—ones that were more focused on specific institutions—to complement the findings from the main cases.

---

3. Another major example not covered in this book is a plan for an income tax on foreign laborers, which was brought down by business lobbying in 2003.

Privatization

Privatization in the Kingdom was widely debated from the mid-1990s on, and a broad consensus evolved in the bureaucracy over its necessity.[4] The Supreme Economic Council (SEC) adopted an ambitious privatization strategy in June 2002, covering numerous sectors. It assigned no lead agency for the process, however, which was mostly debated in an interministerial committee under the SEC.

There, a familiar drama unfolded. Despite general agreement on the policy, individual agencies refused to yield any of their assets, reminiscent of the "negative list" debate around the FIA. Ministries proposed to privatize parts of other organizations first, in a "beggar thy neighbor"—fashion.[5] In several instances, moreover, the leadership was reluctant to impose tariff hikes that could have paved the way for selloffs, notably in the electricity and water sectors.[6] The paternal-distributive state was alive and well. Stalemate and drift ensued.

In several important instances, however, royals felt compelled to exert pressure in order not to lose credibility, and Saudi Arabia has seen four instances of privatization of relatively "easy" and profitable assets that did not require hiking consumer prices: Saudi ports services were partly privatized in the late 1990s under the leadership of the—historically efficient—Saudi Ports Authority;[7] the government corporatized Saudi Telecom and sold off 30 percent of its shares to nationals in early 2003; the Ministry of Finance sold 70 percent of the National Company for Cooperative Insurance in the winter of 2004/2005; finally, from 2007 to 2009 several ancillary services of national airline Saudia (but not the subsidized airline itself) were part-privatized. 2004, moreover, saw an international tender for the second Saudi mobile phone license, introducing private competition to Saudi Telecom. Once decided, these transactions usually happened in a clean and efficient fashion. As all of them were the responsibility of one specific agency, interagency conflicts in implementation were limited. Moreover, although all of them involved the regulation the new entities, the number of those entities was limited—unlike the broad and diffuse regulatory challenges in the three main case studies. A small number of elite technocrats could handle them. All in all, privatization, insofar as it happened, resembled the project developments of the 1970s and 1980s and was accordingly successful, certainly in regional comparison.

---

4. Supreme Economic Council 2002; Montagu 2001: 33.
5. Discussion with Western economic advisor, Riyadh, June 2004. The term "beggar-thy-neighbor" in the context of public sector reform is term taken from Waterbury 1991: 12.
6. Butter interview; Montagu 2001: 34.
7. For a detailed if very positive account, see Bakr 2001.

Capital Markets Reform

Capital markets reform was another large project that occupied the cabinet in the 2000s. After a long period of drift, the cabinet finally issued the capital markets law in 2003, creating a formally independent Capital Market Authority; previously, the stock market had been regulated by SAMA, and only national banks were allowed to act as brokers. WTO pressure seems to have accelerated the regulatory reform. Apparently due to senior conflicts over who should head the new body, it was not staffed until 2004, but then got off to a quick start.[8] The financially autonomous organization issued most of the important bylaws within a few months, after extensive informal consultation with banks and consultants.[9] Although the CMA has been fighting an uphill struggle against rampant insider trading, bankers generally recognize it as a more qualified and committed regulator than most other Middle East stock market authorities.[10] Being able to draw on the manpower and expertise of SAMA and granted authority over a "new" area of regulation, it has managed to tackle its sectorally delimited rule-making tasks quickly. The one major issue holding up the establishment of its role, characteristically, was a dispute with the Ministry of Commerce over who would be responsible for converting limited liability to joint stock companies.[11]

These cases confirm our expectations about which kinds of reforms are possible and not possible in the Kingdom's segmented political economy. There are two systemic factors which determine how difficult a reform is. One is the "depth" of a reform, that is, the degree to which it involves gathering information from and changing the behavior of a large number of actors, both in bureaucracy and in society. The second is the number of organizational players involved in government and in business, what we might call the "width" of reform. "State capacity" cannot be understood independent of these variables.[12]

Development projects and privatizations do not require much regulatory reach, and elite technocrats in one individual agency can usually conduct them. Of course, stories of individual projects differ from agency to agency, and some have been mired in corruption. Nonetheless, the biggest policy successes of the Saudi state have arguably all been individual projects—just as the large FDI successes after 2004 have been projects sponsored by individual state entities.

---

8. Supposedly Prince Sultan objected to the appointment of Jammaz Al-Suhaimy to head the CMA. Interview with Western investment banker.
9. Crucially, the CMA board determines salaries; cf. article 7 of the Capital Market Law.
10. Discussions with bankers, Riyadh, Dubai, and London, 2005–09.
11. Phone interview with Saudi investment banker, April 2005.
12. For a similar conceptualization, see Chibber 2004: 7. Chibber does not take into account the necessity for the internal mobilization of the administration on lower levels.

The story is more complex when one evaluates projects not individually but as components of a cross-cutting infrastructural strategy. Because agencies fail to coordinate, much duplication and misallocation has occurred. This is certainly true of the fragmented health and education sectors covered in the history chapter—which remain fragmented as of this writing.

Macro-coordination of projects through fiscal allocations is still easier than the coordination of diffuse bureaucratic behavior, however. The foreign investment and WTO reforms involved both many agencies and many mid-level bureaucrats, making them highly complex challenges that were more difficult to handle than the expansion strategies of the 1970s. In the Saudization case, jurisdiction was consolidated in one agency, improving policy coherence and allowing macro-control of labor markets. Its great "depth" in terms of involving mid-level bureaucracy and, in principle, all of Saudi business, however, has undermined individual regulatory policies. The contrast between the kingdom's internationally feted, project-based successes, and its broader regulatory failures is striking indeed and calls into question aggregate notions of "state capacity" for rentier states.

If this book were not about Saudi economic policy but about education policies, women's issues, the health sector, or criminal and policing issues, the story would have been quite similar, one of growth through the creation of new institutions, the parallel existence of radically different bureaucratic fiefdoms, strict centralization, clientelism, and diluted implementation of broader-based policies. As of this writing, Saudi Arabia still does not have a criminal law, but at the same time it has set up a highly efficient intelligence operation in the MoI to track down domestic militants in a surgical manner. While large parts of the university system continue to churn out unemployable graduates in Islamic studies and the humanities, the regime is using its new oil wealth to create a parallel system of well-endowed new technical universities and departments. Efforts to put women on an even keel with men in their dealings with the Saudi bureaucracy have been stymied by bureaucratic foot-dragging and deliberate sabotage by conservative administrators. Yet women will be able to study engineering at the new King Abdullah University of Sciences and Technology in the same classrooms with men. The return to large-scale project spending is in many ways a return to the 1970s—and to the strengths of the Saudi system. The costs of moving around—while at the same time maintaining—existing institutions have greatly increased, however. A 1999 government project to review administrative structures has yielded few results.[13]

---

13. Royal decree 7/b/6629 (7/5/1420).

Outlook: State and Business

Is there any hope for fundamental change? The reform of the judicial system, which King Abdallah seemed to tackle in earnest in 2009, will be one crucial and much-delayed test case. Beyond this, there have been some subterranean shifts that we might extrapolate further into the future. Although the account in the case studies was in many respects one of stasis and resistance to change, certain aspects of policy-making within the segmented Saudi system nowadays look much different from twenty years ago. For one, economic policy today is debated much more openly. In important ways, the debate is also more institutionalized—a result of Abdallah's controlled political modernization drive. In particular, business has more formal opportunities to make its voice heard through hearings in the Supreme Economic Council, through the circulation of draft laws in the Chambers of Commerce and Industry, and through discussions in the Majlis Ash-Shura—not currently a body powerful enough to integrate the workings of the Saudi state but a new and important access point nonetheless. While other socio-economic groups remain unorganized and have practically no formal access to the policy process, the inclusion of business has been to some extent institutionalized.

Several factors have limited the role of business, however. First, the state still "grants" consultation; business does not assert it as an inherent right. Even the extensively debated FIA was in the final stage of preparations shaped by interagency negotiations rather than by business demands, and in the WTO case, the regime excluded business from substantive deliberations almost all the way. The state still acts paternally, practicing top-down consultation with client groups rather than negotiation with an equal partner. Considering the importance of business for national development, the regionally unrivalled capacities of Saudi business, and the diminished resources of the bureaucracy, this seems a carryover from the time of state-led development. Business is regaining territory through its contribution to national capital formation and increased private provision of services such as education and electricity, areas largely taken over by the state in the 1960s and 1970s. It has not yet regained much political territory.

It is not only that the state dominates policy-making, but also that business has been unassertive on nondistributional issues. As much as it acts as a veto player defending its turf with respect to commercial agency privileges or employment rules, it has usually been silent on more complex policy questions. This is a result of limited policy research capacity and the relative political fragmentation of a business class that is slowly emerging from a long history of clientelism. Indirectly, it also seems to be an outcome of unpredictable government behavior on complex issues. In the Saudization case, for example, the fragmentation of the state undermined the credibility of state agencies as negotiation partners, preventing clear meso-level accountability and responsibility.

Incoherent state behavior on the micro-level additionally undermines the state's capacity for credible commitment. This increases incentives for business to seek individual ways around the rules or, in case of collective lobbying, a postponement of policies rather than a rules-oriented agreement, which might not be properly implemented anyway. In this way, micro-structures undermine political accommodation on a meso-level—a causal mechanism that the literature on state-business relations does not capture, as it engages with issues of trust, reciprocity, and predictability but does not distinguish different levels of analysis.[14] As Saudi businesses are often still tied up in individual relationships of patronage, individual techniques of coping with the state seem to dominate collective interests.

The passivity of business and the lack of interest in complex policy issues constitute no extraordinary finding; in most developing countries, the private sector's understanding of technocratic matters is limited. Considering the individual managerial sophistication of Saudi business and its high formal level of organization through the Chambers, however, one might have expected a more proactive role in policy-making. This is why in the medium term, business could move beyond its established role as passive policy-taker. It has more capacity of development than the bloated and immobile Saudi bureaucracy, and at least industrialists demonstrated in the FIA case some capacity to lobby on a specific policy issue not directly touching their distributional interests (income tax for foreign companies).

Generally speaking, the costs of dealing with a complex and unresponsive bureaucracy will gradually exceed the competitive advantage that this opacity has in the past provided for existing businesses by establishing high barriers of entry to new competitors. "Broker" and "businessman" have become quite distinct categories in many sectors, which is why a cross-cutting private sector interest in a smoother and less intrusive administration has been emerging over the years and might be the starting point for more articulate collective policy interests. To the extent that the bureaucracy shifts from distribution to predatory regulation, business will ask for increasing independence from the administration. It is this trend that might make for the most dynamic challenges to the Saudi segmented clientelism in the years to come.

---

14. Schneider and Maxfield 1997.

The Saudi Case and Comparative Politics

The account so far has been focused on the details of the Saudi case—a pivotal case for which accepted theories appear to be insufficient. Yet with this internally valid explanation established, there remains a fundamental question for any political scientist: Can anything be gleaned for comparative politics in general? Or is Saudi Arabia a freak case of oil and patrimonialism that defies generalization? The remaining pages argue against the latter proposition. The concepts this book has developed might appear unconventional, but there are good reasons to believe that they can help to throw new light on unresolved research puzzles in many other cases. I will first relate the concepts to broader debates on state-society relations, then to the growing, yet disparate, rentier state debate. The Saudi case study in this book in many ways served to disconfirm or qualify standard theoretical expectations on a "most likely" rentier case.[15] The comparative discussion of the book's new or modified explanatory tools will be conceptual and cannot engage in full-blown paired comparisons, which would be difficult without similarly detailed field research on other cases. It will, however, lay out a clear map for future research.

Whatever its individual parts look like, a striking feature of the Saudi state is its omnipresence and great importance in most people's lives. For many decades, its resources have greatly dwarfed those of any independent social formation—much more so than in Western industrial systems, which are probably the only cases with state apparatuses of comparable size.

What does the historically dominant role of the Saudi state leave of society? Not much, it seems, in the realm of formal politics, (macro-level) "social classes" and larger, organized (meso-level) "interest groups." In its politics, the Saudi rentier state seems to have conserved many of the features of traditional and less resource-intensive clientelist systems as described by Guillermo O'Donnell: "Political activation" is low, links of authority are patronage-based and multifunctional, politics passes over society, and the state acts as "patron of patrons."[16] As in pre-populist Latin American systems, there is a vertically segmented, "columnar social structure."[17]

Much more so than in traditional clientelist systems, the state has fundamentally reoriented Saudi society, making it dependent and at the same time fragmenting it.[18] The apparent weakness of society leaves us with a puzzle. If the state

---

15. Gerring 2007.
16. O'Donnell 1979: 66.
17. Malloy 1979: 7.
18. In this sense, Saudi Arabia is an extreme case of neo-patrimonialism, which reorients society towards the center but keeps it divided and demobilized; cf. Eisenstadt 1973.

has been dominating and reshaping society so thoroughly, how come its control over implementing policy vis-à-vis society has so often been patchy and the Saudi bureaucracy is so often unable to force individuals to stick to the formal rules—be it in divulging business information and applying building regulations or even more fundamental things like registering births and following traffic rules?

To understand such apparent contradictions, we have to distinguish social structures on a macro-, a meso- and a micro-scale. Unlike most other political systems, the Saudi rentier state has prevented the emergence of independent "macro" organizations such as labor unions or broad parties in Saudi Arabia and has left little space for independent meso-structures like professional associations and functional interest groups (consumer protection, sectoral business associations, regional interest groups, and so on).[19] Small-scale, informal networks, however, have not been purged by the bureaucratic state. They have arguably grown in importance in the absence of other structures and in the face of growing state resources that small-scale links—often counteracting the vertical divisions of the broader system—could make locally available.

Joel Migdal has written about how states and societies can transform each other and about how social forces can penetrate states.[20] Distinguishing different levels of analysis can lend analytical precision to his broad "state-in-society" approach. The particular Saudi constellation, combining macro- and meso-dominance of the state with micro-penetration of society, has specific implications for the shape of political negotiations, state capabilities, and policy outcomes. Unlike other systems, policies are often negotiated *within* the state, with societal groups playing a marginal role. On this level, the state is all-powerful. This can lead to internal stalemates but also to sudden policy changes, which can be decided in a knee-jerk fashion with few groups in society able to slow down or modify policy shifts.

However, when it comes to implementation, the state often lacks regulatory power and autonomy on the bureaucratic micro-level, making for a dilution of policies on the ground through small-scale social networks and interests. This helps to explain the declaratory nature of many seemingly radical reforms in Saudi Arabia. The regime is able to make decisions without great societal constraints but is all the more constrained when executing them.

Policy processes in Saudi Arabia are hence quite different from, say, cases in which social groups influence states on macro- and meso-levels but bureaucracy

---

19. The only independent and horizontally integrated networks of reasonable size seem to be those of the Islamist groups that have emerged in the 1980s. Those, however, have little to say on the economic and regulatory policies that are the focus in this book.

20. Migdal 1988, 2001.

is more autonomous on a micro-scale. In the ideal-type Western country, policy negotiations are smoother and nonstate interest groups play a much larger role, reflecting a more equal balance of social forces. Once it comes to implementation, however, bureaucracy is more independent from small social networks, acting to instantiate a policy more stringently and coherently than is the case in the Kingdom.[21] Atul Kohli's ideal-type "cohesive-capitalist" developmental dictatorship—modeled on the South Korean case—seems to incorporate yet another combination of macro, meso and micro: here, the state appears to be a capable, independent actor on all three levels.[22]

The differences across macro-, meso-, and micro-forces in the Saudi case are particularly pronounced. But there is no reason to believe that this analytical differentiation cannot be usefully applied to other cases with other kinds of constellations to give shape to the somewhat amorphous "everything influences everything" assumption of the state-in-society debate. The basic Saudi constellation, though extreme, is not fundamentally different from that of other authoritarian developing countries. In the Middle East, places as different as Egypt and the United Arab Emirates come to mind: countries with weak class and associational structures but bureaucracies that are often deeply penetrated by smaller-scale social interests. Many patterns of policy-making in the Gulf rentier monarchies may best be explained by the disjuncture of macro, meso and micro. Labor and taxation policies are often decided rashly within the state but then meet strong, if diffuse, resistance and evasion in the implementation stage.

Differentiating state-society relations by levels of analysis is also useful outside of the rentier state realm, even if the latter might contain the cases with the strongest macro-micro disjuncture. At a minimum, it allows us to understand what is special about the Saudi case—and in which respects its state functions like other developing countries. A short comparison with Brazil, a staple example in the "developmental state" debate, will illustrate this: Like many other late developers, Brazil has gone through an authoritarian phase in which the regime has used corporatist policies of cooptation to pacify and fragment business and labor interests.[23] It has thereby undermined macro-level interest groups such as national unions and peak business associations. By the same token, however, it has empowered meso-level, sectoral groups with particularist interests—for which

---

21. Of course, implementation is never perfectly smooth in any system; cf. Wildavsky and Pressman 1984. Yet there are important, real-world capacity differences between systems on different levels of analysis.
22. Kohli 2004.
23. Weyland 1996, 1998.

the Saudi rentier regime, with its individualized cooptation strategy, has never left much space.

While the Brazilian bureaucracy is shot through with micro-level clientelist interests broadly similar to those in the Saudi case, policy-making at least in the 1980s and 1990s was also bound up in meso-level bargaining. This has slowed down policy-making and, crucially, has allowed sectoral groups in business and among the middle class to extract numerous special concessions in fields such as welfare and tax policy, to the detriment of national-level policy. This has reached a level where specific state agencies engaged in open alliances with or were "captured" by particular groups,[24] which has resulted in strongly diluted and self-defeating policies. In Saudi Arabia, by contrast, societal groups have had little influence over the regime on either the macro- or the meso-level. Business could cast an occasional veto but was far from "capturing" specific agencies. Insofar as state agencies did manage their differences, therefore, the state was able to produce coherent policy packages; it certainly was more autonomous in policy-making. On the micro-level, however, the Saudi bureaucracy has hardly been more independent from social interests than the Brazilian one. This has diluted implementation in the plentiful ways described in the case studies.[25]

Social constraints on the state have a different impact on policy depending on the level of analysis. Macro-level constraints, while reducing the state's maneuverability, tend to lead to more acceptable, negotiated solutions, while micro-level constraints are always likely to compromise coherent implementation, even if they make policy more palatable. The impact of meso-level constraints depends on context, but they are likely to reduce policy coherence and work against unorganized interests. Different types of states are under different social constraints: Ideal-type Western states on the macro- and possibly meso-levels, most developing countries more on the meso- and micro-levels,[26] "capitalist-coherent" developmental states on no level, and fragmented rentier states like Saudi Arabia only on the micro-level. The Kingdom therefore incorporates features typical of other developing countries, but in an altogether unique constellation.

Differentiating levels of analysis in state-society relations can arguably help to unpack two comparative politics concepts as contentious as they are resilient: "state autonomy" and "state capacity." Much literature on state-society relations and development that draws on these terms, on rentier states as well as other

---

24. Cf. Kohli 2004: 14.
25. A similar comparison could be made with Mexico; cf. Grindle 1977.
26. This might be a useful way to conceptualize Kohli's "multiclass-fragmented" state, which in his scheme includes the developing countries of the world that have neither failed completely nor overperformed like the Asian industrializers; cf. Kohli 2004.

systems, seems to conflate macro, meso and micro, and tends to make problematic assumptions of state coherence.[27]

We have seen that the autonomy of state agents and organizations can be constrained in different parts of the policy process, and on different levels, with very different consequences for regime power and policy outcomes. The Saudi regime faces few organized macro- and meso-level constraints, yet its autonomy is compromised on the micro-level, not only in the sense that its control over low-level bureaucracy and the surrounding social networks is limited but also through numerous micro-level distributional commitments that tie down the state. Although no collective agent is acting on the regime, it has repeatedly failed to reverse its costly subsidy and employment policies.

In his study of the Saudi political economy, Rayed Krimly has attempted to show that the autonomy assumption of rentier theorists does not hold, pointing to the intense interpenetration of state and society.[28] I would argue that this is mostly true on a micro-level, less so on the meso- and macro-levels. Distinguishing levels of analysis explains why the Saudi state appears penetrated and constrained on the one hand—by diffuse clientelist expectations and networks—but frequently incommunicado in terms of systematic policy negotiation on the other—which would require larger, organized groups to communicate with. It can lack autonomy and be too autonomous at the same time. And needless to say, when political negotiations pit different parts of the state against each other rather than against societal groups, the state autonomy concept loses analytical utility.

"State capacity" seems a more cohesive concept than "autonomy": a measure of how effectively state decisions can be turned into changes on the ground. But on closer inspection, it also appears too broad an aggregate to be useful, as it varies according to the type of policy and institutions involved. The very coherence of the state can influence state capacity on the meso-level: if agencies, however efficient and goal-oriented by themselves, are not capable of coordinating their policies, implementation will suffer. Again, this is not just an ad hoc rule to accommodate the Saudi exception but applies to quite different cases. Vivek Chibber has shown how post–World War II planning authorities in India were unable to overrule other institutions in their industrialization strategy. While all bureaucratic units were internally coherent and behaved rationally, their different organizational missions undermined the coherence of national development policy.[29] Capacity can thus depend on the number of organizational players

---

27. A state can only be autonomous in a meaningful way if it acts coherently, which is often not the case in Saudi Arabia; cf. Nettl 1968.
28. Krimly 1993: 52.
29. Chibber 2002.

involved in a policy and the existence of authoritative meso-level mechanisms of concertation.

Relatedly, different parts of the state can enjoy different degrees of strength or internal efficiency.[30] The micro-level structures and capacities of bureaucracies in the Saudi case can look drastically different. The spectrum of institutions built with oil money ranges from unreconstructed patrimonialism to tightly run technocracy. Both internal efficiency and regulatory outreach can differ greatly as well. While the Saudi state has generally struggled to regulate societal actors, some sectoral regulators—SAMA for banks or the new Communications and Information Technology Commission for telecoms—have developed considerable capacity. Finally, different types of policies themselves make different micro-level demands in terms of the bureaucratic and societal mobilization they require. Some aim at changing the behavior of hundreds of thousands of actors, while others involve only a select number of technocrats.

Saudi capacity to execute large but locally delimited projects has been high—but not as a matter of course, as the fate of other oil states such as Nigeria, Libya, or Venezuela demonstrates.[31] Conversely, however, it has often been difficult to implement broader policies that involve more than a limited number of elite bureaucrats or institutions. As the case studies have shown, Saudi state capacity is best determined from case to case by the width and depth of policies. In few countries are the concomitant differences as stark as in Saudi Arabia, but there are likely to be similar distinctions in most developing countries. In many Western democracies, conversely, project implementation capacity might actually be lower than in Saudi Arabia due to the various planning restrictions and societal interests that have to be dealt with, while broader regulatory capacity is much higher. State capacity has different dimensions, which can in turn be linked back to state-society relations and the internal structure of the state on different levels.

How Does It Add to the Rentier State Debate?

Where does all of this conceptual disaggregation leave Saudi Arabia as a "rentier state"? Where does it leave rentier theory approaches in general? To start with, there is no coherent "rentier state theory."[32] Nevertheless, there is a cluster of

---

30. Huber 1995: 169–70; Shue and Kohli 1994: 322.
31. Amuzegar 1998; Karl 1997; Vandewalle 1998.
32. Mick Moore 2004: 305.

hypotheses on the impact of oil on politics and development in developing countries, postulating a variety of mostly negative effects.[33]

Common Ideas of Rentier State Theorists

TAXATION, REPRESENTATION, AND REGIME AUTONOMY

The classic political hypothesis about externally derived state income[34] is that it reduces the democratic accountability of regimes, as these do not have to tax their constituents and hence do not need to bargain with them.[35] A broader corollary of the taxation-representation argument is that of general state autonomy. Without taxation-induced political bargaining, rentier states are supposed to be *generally autonomous* from societal demands, however articulated, free to pursue policies as they please, drawing on external resources the use of which they are not held accountable for.[36]

REGULATORY DEFICIENCIES AND PERVASIVE RENT-SEEKING

Rent-induced absence of taxation is supposed to be bad not only for democracy but also for state capacity, since rentier states do not have to gather information on their constituents systematically and coerce them into contributing to the fisc.[37] States are "distributive," not extractive, which means that they lack the administrative machinery to regulate markets, as regulation requires information and, in many cases, fiscal instruments.[38] In a particularly pronounced assertion of this argument, Kiren Chaudhry has argued that the Saudi state had substantial taxation and regulation capacities, which the oil boom of the 1970s obliterated.

The rentier literature also frequently posits a predominance of individual rent-seeking behavior in bureaucracy and society.[39] With so much money around to

---

33. No work includes all of these hypotheses; they are combined in quite different ways, and some works implicitly seem to contradict hypotheses presented by other authors (for example, Karl and Chaudhry do not assume state autonomy, as the majority of rentier state theorists do). The following section merely distills the core assumptions, both implicit and explicit, that recur in the literature. To avoid further inflating a theoretically eclectic account, I have confined myself to political and institutional hypotheses. On economic ramifications (Dutch disease, and so on), see Sachs and Warner 2001; Amuzegar 1998; Gelb 1988.

34. This can be oil and mineral-related rents or aid income. The main point is that it is collected by the state and does not involve large-scale employment of domestic productive forces; cf. Jones Luong 2003.

35. Luciani 1990; Mahdavy 1970: 466; Islami and Kavoussi 1984: 10; Crystal 1995; Anderson 1987.

36. Gause 1994; Vandewalle 1998; Delacroix 1980; Ayubi 1995: 256–57.

37. Taxation is a main source of information-gathering; cf. Levi 1988: 122.

38. Vandewalle 1998; Chaudhry 1997; Jones Luong 2006. For a large-n study on the subject, see Isham et al. 2005.

39. The most pronounced proponent of the "rentier mentality" as a psychological syndrome is Beblawi (1987); cf. also Karl 1997: 57.

be distributed by the state and low capacity to regulate, slack and waste will be large. Individual strategies to get a finger into the pie are more rewarding than productive activities. Profligacy and corruption are the result.[40]

STATE DOMINANCE AND FRAGMENTATION OF SOCIETY

It is widely agreed that the size of rents tends to inflate the size and role of the state, giving it a dominant position in the economy.[41] Within the state, some authors hold, the executive tends to predominate.[42] According to other authors, an indirect outcome of state dominance is that rentier states weaken the autonomous organizational capacities of society. Social groups become dependent on the state's distributional policies and a mode of politics that discourages open, collective bargaining.[43] With their fiscal power, states can moreover remold whole parts of society and create dependent groups (what Michael Ross calls the "group formation effect").[44]

All of the arguments contain useful cues on the Saudi case, but none of them quite captures the essence of Saudi politics, applies across the board, or would explain the most striking outcomes of oil politics there. I have expounded more specific, distribution-based causal mechanisms that I believe will be transferable and adaptable to other cases of oil-rich countries, either in the form of refined, generic causal hypotheses or as more general analytical concepts which, while allowing systematic comparison, are more sensitive to causal impact of history and context.

Putting State Autonomy in Historical Context

As we have argued above, autonomy can be a problematic concept in dealing with a fragmented state like the Saudi one. Moreover, constraints on autonomy need to be broken down on different levels, and while the political use of rent can alleviate constraints on some, it can also increase them on others. When Terry Lynn Karl argues that rentier states lack connections to their citizens, she seems to conflate macro, meso and micro.[45] Similarly, when Pauline Jones Luong and Erika Weinthal propose that the rentier states can bolster their autonomy through

---

40. Karl 1999.
41. Ross 2009: 49–50.
42. Karl 1997: 59.
43. Vandewalle 1998: 25–26.
44. Ross 2001: 334; Chaudhry 1997: 140ff. For an account on Venezuela, see Karl 1997.
45. The same fallacy might underlie her contention that nationals will not be interested in how money is spent, as they pay no taxes (Karl 2007), which does not seem to be borne out in reality—it's just that these interests often are not organized.

discretionary patronage spending, this might be true on macro- and meso-levels, but it is rarely true on the micro-level.[46]

Rent distribution means incurring obligations, even if individualized, and these can reduce a regime's leeway to change institutions over time. This is perhaps the most striking feature of state autonomy in Saudi Arabia as rentier state: oil money initially gave its elites vast autonomy to create and reshape institutions. This autonomy, however, has declined precipitously.[47] Other rentier states seem to have seen similar shifts. Mechanisms of bureaucratic growth and entitlement tend to reproduce institutional trajectories once they have been decided upon.

Entitlements and fiefdoms in countries as different as the Gulf monarchies and Venezuela have proven remarkably sticky.[48] All GCC states saw institutional stagnation in the lost years of the 1980s and 1990s; the fiscal crisis did not trigger reforms as it might have in nonrentier states, and attempts to revoke entitlements by and large went nowhere. Benjamin Smith as well as Ellis Goldberg, Erik Wibbels, and Eric Mvukiyehe have argued that rents allow for the building of stable patronage networks which can even survive busts. This might bring political benefits, but it comes at an economic and institutional cost.

Before institutions and entitlements congeal, however, growing oil income seems to open vast leeway for institutional experiments and redundancy, without preordaining the form of those experiments. Saudi Arabia has seen the emergence of large and separate princely, technocratic, and religious fiefdoms. Dubai has turned itself into a quasi-corporate structure in little more than two decades, a global service hub with large state-owned enterprises staffed largely by expatriates. Qadhafi has used oil rents in an idiosyncratic experiment to turn Libya into a supposedly "stateless society" based on an ever-shifting committee system.[49] The creation and sustenance of all these structures would be hard to imagine without the sudden availability of external income and, importantly, a pre-oil political system that limits elite accountability and hence widens institutional choice.[50]

It remains to be investigated whether rapidly growing oil income generally leads to bureaucratic fragmentation. At least in the standard scenario of rent-based administrative cooptation in developing countries, one might plausibly expect

---

46. Jones Luong and Weinthal 2006.
47. Rentier state approaches have been criticized for not distinguishing the constraining and enabling features of oil income; cf. Waldner 1999: 107. To delimit those, the use of oil has to be seen in its shifting historical context.
48. Karl 1997; on Kuwaiti refusals to cut subsidies in times of crisis, see Crystal 1995: 195–96.
49. Davis 1987.
50. Weyland makes a similar case about Brazil, in that past actions of an autonomous regime constrained its maneuverability further down the road by building up clients and a fragmented bureaucracy (Weyland 1998). The initial autonomy and leeway for experimentation seems higher in rentier states, however (and the latter stasis arguably more profound).

elites to use their sudden fiscal surpluses to let important coalition partners build bureaucratic fiefdoms and, possibly, use some spare cash to set up oil-financed islands of efficiency in parallel. Fiefdom-building has certainly occurred in Venezuela, Qatar, and Kuwait.[51] Strikingly, Dubai, by many accounts the most successful rentier (now nearly post-rentier) system in the Middle East, has managed its development less through wholesale administrative reform than through creating islands of efficiency. The regime has given regulatory enclaves such as industrial cities and special technology and financial zones authority to circumvent the fragmented federal and local bureaucracy, operating under separate licensing, utility, and labor regimes. At the same time, Dubai nationals are coopted into bureaucratic sinecures, and the sheikhdom's education system suffers from serious quality problems.[52] As in Saudi Arabia, the justice system outside of the regulatory enclaves is slow-moving and opaque.[53]

Abu Dhabi and Qatar are on a similar course now, circumventing their bloated national bureaucracies with separate legal and immigration regimes for specific industrial and tourism zones and state-owned enterprises with separate governance and recruitment regimes—perhaps thereby accelerating national development, but also fragmenting their states. The structure of the main investment funds and public holdings often mirror the power structure in the senior ranks of the ruling families. All this involves degrees of redundancy and, in the phase of inception, institutional suppleness hard to imagine in nonrentier systems.

One might argue that the dynastic rather than the rentier nature of the Gulf monarchies has led to their institutional fragmentation. But while collective dynastic rule certainly abets fiefdom-building, a vastly different nondynastic case in the GCC neighborhood corroborates our assumption that rents lead to heterogeneous states in even more spectacular fashion: the Iranian Islamic republic. The revolutionary regime in Iran has bolstered its coalitions by granting the economic assets of the Shah and his cronies to various revolutionary cliques after 1979, financed through large-scale rent transfers ever since. The charitable foundations ("*bonyads*") created in the process are untaxed, are not bound by regular labor

---

51. Terry Lynn Karl (1997: 107) mentions the "top-down yet divided organization" of Venezuelan parties, with various personalized patronage structures all vying for state access, indicating equivalent patterns of fragmentation. Members of the ruling Sabah family built ministerial fiefdoms in Kuwait as early as the 1950s (Crystal 1995: 13), and the Qatari state is similarly divided among various fiefs (discussion with former ambassador to Qatar, Washington, D.C., March 2008). Several of the Libyan ministries in the 1950s and 1960s, the decades of early oil, were also run as patrimonial fiefdoms; cf. Vandewalle 1998: 48. The Kuwaiti state however also has several pockets of rationality and good management; cf. Crystal 1995: 192; Farah 1989: 111.

52. Discussions with advisors to Dubai and Abu Dhabi governments, Dubai, February 2009.

53. The United Arab Emirates ranked 145th worldwide on "enforcing contracts" in the World Bank's 2009 "Doing Business" survey.

legislation, enjoy tariff and foreign exchange privileges, and receive a vast share of the national budget. The bonyads report directly to the "supreme guide," while the regular Iranian government has no control over them and cannot audit them. Its attempts to extend control have repeatedly failed. The bonyads have their own welfare and housing programs, have internal "ministries" for various lines of service, and control large parts of the Iranian economy. The bonyad of the disinherited (*mostazafan*) is a veritable "state within a state" that reportedly employs four hundred thousand individuals.[54] It is hard to imagine that a system without large-scale external income could afford to maintain such autonomous, parallel structures completely outside of the regular government's remit.

There are many cases of fragmented nonrentier states in other world regions. It is not an issue that comparative politics has systematically addressed: the subject will require further, in-depth field research. Yet, disjointed state apparatuses are documented—often in passing—in case study literature on countries as diverse as Egypt, Ghana, Mexico, and Brazil.[55] Rent income certainly is not the only cause of fiefdom-building and intragovernmental communication failures.[56] Yet the depth and quality of fragmentation as well as its causes and consequences appear rather specific in the Saudi case and are best explained with reference to its rentier status—or more precisely, to the specific uses to which the Al Saud regime has historically put its oil money.

A nonrentier state that has struggled with particularly pronounced—and well-documented—internal fragmentation is, again, Brazil. Post–World War I Brazilian elites have modified their state by "piecemeal addition" instead of internal reform, and their cooptation strategies vis-à-vis different societal constituencies have resulted in a "baroque" sprawl of different agencies and particularist interests that has proven hard to coordinate.[57] Brazilian elites have at times also managed to create "pockets of efficiency," but those seem to have been more ephemeral than the Saudi ones.[58] The state grew more fragmented in the periods when state income grew rapidly, underlining our general point that it is *incremental* resources that

---

54. Maloney 2000; Oxford Analytica and Economist Intelligence Unit, various reports. The Islamic Revolutionary Guards Corps are another autonomous para-statal actor with large economic interests.
55. Grindle 1977; Grayson 1980; Geddes 1990: 227–28; Evans 1995: 62; Levine 1975; Kohli 2004; Henry 1994; Bianchi 1989; Handoussa 2002.
56. A "most similar" non-rentier case for a stringent comparison with Saudi Arabia is hard to find, as oil has had such a profound impact on the Kingdom's state and social structure. Suffice it to say that states with similar levels of per capita income, such as Asian industrializers, tend to be much less fragmented, while institutionally and culturally similar states such as Jordan and Morocco are also less heterogeneous.
57. Evans 1989: 578; Weyland 1996, 1998; Schneider 1991.
58. Geddes 1990; Kohli 2004: 177.

allow institutional experimentation; rentier states just happen to enjoy particularly large increments of income in certain periods.

At first glance, the meso-level fragmentation of the Brazilian bureaucracy appears quite similar to that of the Saudi one, resulting in similar policy-making problems. But meso-level fragmentation in Saudi Arabia is mostly driven by intra-state politics—both in its highly voluntarist (and, in many cases, quirky) origins and in its articulation in policy-making. Links to society exist mostly on the micro-level. In Brazil, by contrast, particularist *collective* interests in society have contributed to shaping the institutional landscape and, even more so, have had a significant say in policy-making. Atul Kohli sees Brazil as a typical "fragmented-multiclass" state that relies on plural social alliances that constrain it, with a mobilized and organized citizenry to claim particular resources and, by implication, influence institution-building. Decision-making seldom remains in the hands of apolitical technocrats.[59] In this, Saudi Arabia could hardly be more different. Similar points could be made about other "fragmented-multiclass" states such as Mexico.[60]

As important, some of the Saudi fiefdoms such as the Ministries of Defense and Interior or the religious judiciary are extreme phenomena even in comparison with the most fragmented nonrentier systems: organizations that are completely divorced from the rest of the state in their authority structures and internal rules. The same is true of the vast bonyads in Iran and arguably about some of the special zones and public enterprises in the smaller Gulf monarchies.[61] Some of these constitute institutional redundancies that probably only rentier states can afford and are more deeply ensconced than fiefdoms in many other states. Large if temporary fiscal surpluses have allowed the building of parallel states within the state, solving political conflicts and policy problems by adding new institutions. While such strategies tend to undercut the economic and tax base of the states in nonrentier cases, in the rentier realm the income is external and hence costly experiments are—at least in the short run—easier to sustain.

In addition to their level of autonomy, the very heterogeneity of Saudi (and, to a lesser degree, GCC) fiefdoms also seems unique. A judiciary that in many respects lives in the seventh century exists in parallel with highly modern technocratic

---

59. Kohli 2004: 11, 23, 400, 407.
60. Grindle 1977; Grayson 1980; Davis 2006. Social interests have been less organized in Egypt and Ghana, but these two also happen to be at least semi-rentiers.
61. One might make the case that security institutions tend to acquire fiefdom characteristics in many authoritarian systems; for Syria, see Perthes 1995; for Egypt, see Waterbury 1983, Bianchi 1989. Important Saudi, Iranian, and GCC fiefdoms exist outside of these realms, however. Another example of a vast and impenetrable economic fiefdom in a rentier state is Indonesia's oil and industry conglomerate Pertamina, which collapsed in the 1970s under massive debt (Barnes 1995).

structures, which in turn are guarded by tribally organized security forces. Unreconstructed patrimonial court structures characterize yet other sections of the state, most notably the regional governorates. It stands to reason that only rapidly growing external state income would allow for parallel institutional expansion in such different directions—often based on co-opting diametrically opposed social interests that in most other systems would have clashed. The Brazilian bureaucracy might be heterogeneous, but it does not, like the Saudi state, constitute fourteen centuries of history rolled up into one bureaucratic behemoth.

The astute political use of oil rents to fragment society also helps to explain the high level of continuity and institutional stability in Saudi Arabia—which allows the royals to shield islands of efficiency, on the one hand, but obliges them to sustain institutionalized, clientelist distribution, on the other. The various components of the Saudi state have seen few lateral challenges within the state or collective challenges from society—certainly compared to much less static Mexico or Brazil.[62] Conversely, one might argue, the inward-looking nature of Saudi bureaucracies also makes the system less supple. While bureaucratic communication in Brazil often happens through transversal networks of senior bureaucrats who have worked in different parts of the state at different phases, the static "iron pipe" structure of the Saudi state allows for less communication and lateral mobility of that kind.[63]

None of this amounts to a firm comparative conclusion. A general theory of rentier state fragmentation will require further in-depth research—not least because, as I have argued, elite agency strongly influences the use of oil funds in the early phase of state-building, and it would be ahistorical to postulate an automatic, determinist mechanism of state fragmentation. We may, however, assume a general propensity.

Differentiating the "State Capacity" and "Rent-Seeking" Assumptions

We have moved far from the original hypothesis of generalized rentier state autonomy. What about the assumptions about rentier state capacity? Dirk Vandewalle expects distributional states to have no interest in coherent economic policies.[64] This appears to be an overgeneralization. The history of the Saudi state shows that in specific areas, the major players have pursued quite coherent economic policies and draw on considerable bureaucratic capacities to implement them. Similarly,

---

62. On the deep discontinuities in Mexico from one presidential tenure to the next, see Grindle 1977: 164ff.
63. Schneider 1991. In my interviews, diplomats, consultants, and lawyers have described the "stovepiping" of bureaucratic communication in Saudi Arabia as particularly pronounced, even in comparison to other developing countries.
64. Vandewalle 1998: 162; cf. Luciani 1990.

the segmentation of administrative spheres in the hierarchical Kingdom has made it much easier to seek rents in some areas than in others. System-wide generalizations are difficult.

Nonetheless, the Saudi process of state-building has not usually generated high levels of trust between anonymous bureaucrats and citizens. The bureaucracy has been imposed rapidly and from the top down. At least on its lower levels, its purpose has often been clientelist employment at least as much as enhancement of regulatory powers—a pattern that many other rentier states share. There is reason to believe in an underlying pattern.

But is low regulatory penetration all about the absence of taxation? Is it true, as Kiren Aziz Chaudhry has argued, that the absence of taxation has removed the crucial reason for the Saudi state to penetrate Saudi society? Dirk Vandewalle's account of the bureaucratically incapacitated Libyan state seems to add force to this general argument. However, in the relatively orderly Saudi system, there have in principle been other mechanisms to gather information and extract cooperation: safety inspections, project monitoring by institutions that distribute subsidies, licensing documentation, supervision of municipal zoning, and so on. State dependence in fact would offer many avenues of collecting information.

The stark necessities of taxation can help to enforce the divulgence of information, but the more important and underlying factor that undermines regulatory penetration is likely the low level of trust in the effectiveness and fairness of the formal bureaucracy. As Margaret Levi argues, taxation does not automatically generate such trust.[65] In the Saudi case, the low level of formal trust is rooted in the absence of a long bureaucratic tradition and, more recently, the lack of accountability of the hierarchical and fragmented Saudi distributive state. For specific historical reasons little to do with taxation, trust in certain areas such as banking regulation is considerably higher compared to other countries, thanks to the quality of the specific institutions involved.

Successful taxation appears to be an imperfect proxy for regulatory power. In any case, many nonrentier developing countries have tax systems at least as uneven and weak as Saudi Arabia.[66] India in 2007 had 31.5 million taxpayers, Pakistan about 1.5 million of them.[67]

---

65. Levi 1988: 68.

66. One might argue that general domestic taxation is a necessary, but not a sufficient, condition of regulatory capacity. If "any taxation" is meant by that, then clearly this does not say too much and we have to look for additional conditions, as many states have general taxes of some kind without being good regulators. If "good and consistent taxation" is meant, then this excludes many nonrentier developing countries. However we define it, it is not a hard-and-fast criterion for explaining the regulatory weakness of rentier states in particular.

67. IANS, 29 July 2007; *Gulf News*, 2 August 2007.

In fact, the low regulatory power of the Saudi state is in many ways not worse than that of other developing countries—it is just low *relative to the available resources,* as figure 8.1 shows.[68] In fact, based on the figure, one might argue that oil income is likely to improve bureaucracy, but *not as much* as other forms of income. Judging from the bureaucracies of their nonoil neighbors in the Middle East, the Gulf monarchies would be unlikely to have governance scores as good in a counterfactual world in which they had never found oil.[69]

The pattern that is really specific to rentier states might therefore be that they combine a resource-rich and omnipresent state with *relatively* weak levels of formal societal penetration. This skewed pattern seems to shape bureaucratic environments in a specific way, notably leading to the emergence of numerous brokers who make a relatively weak and opaque but large and rich state accessible to other individuals.[70] This is a far more specific prediction than that of general "rent-seeking." Testing it will require considerable field research. The strikingly large number of "visa traders" and "cover-up" businesses in all six GCC countries—cases that differ in many other respects—gives it considerable prima facie validity.[71]

State Dominance and Fragmentation of Society in Rentier States

Saudi Arabia fully confirms the rentier debate hypothesis that states and, more specifically, executives dominate politics in rentier systems. The regime has used the bureaucracy as a tool of cooptation—again, a pattern witnessed in other rentier states. The strategy is not unique to rentier regimes, but oil rents can greatly augment its scope.

An also context-dependent but more specific hypothesis derived from the Saudi case is that top-down, oil-based distribution generates predominant vertical, clientelist links within the state and between state and society, which undermine horizontal forms of cooperation. As the hypothesis—which underlies this book—is a powerful explanans of political behavior and policy outcomes on

---

68. The figure looks similar for other governance indicators such as control of corruption, regulatory quality, and rule of law; quite in line with our macro/meso/micro account, rentiers look a good deal worse on "voice and accountability," which measures formal participatory opportunities for citizens. On the problems in comparing rentiers and nonrentiers in general, see Herb 2005.

69. Both the quality of economic information and the obstructionism of bureaucracy in Syria turned out to be considerably worse than in Saudi Arabia during a field research stint in the summer of 2008; interviews with former bureaucrats and businessmen, Damascus, July/August 2008; Institutional and Sector Modernisation Facility (an EU-funded administrative assistance program in Syria), *Conceiving and initiating public administration reform for the government of Syria: draft completion report* (Damascus 2007).

70. To be sure, opaque administrations always tend to produce brokers. The size of the state and its resources in rentier systems, however, are likely to lead to an unusual preponderance of intermediaries.

71. Cf. Hertog 2006a.

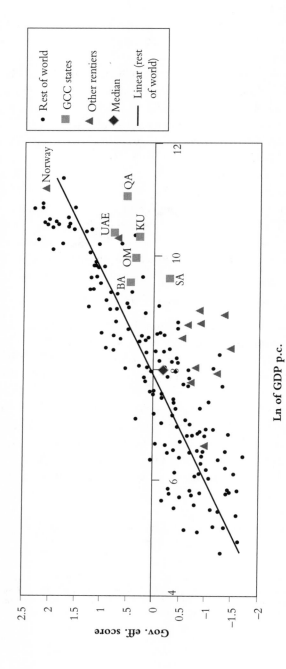

Fig. 8.1 Government effectiveness vs. GDP per capita in 2005
Source: World Bank, IMF
Note: The black line is the bivariate regression for the "rest of world" (non-rentier) cases. The rentier outlier in the upper right corner is Norway, a state that had established a solid institutional framework before large-scale rents became available.

several levels of analysis, the important task might be to delimit the conditions under which oil produces such a system of politics.[72] The idea is parsimonious, yet more tangible and sociologically grounded than global rentier theory predictions of state autonomy or nondemocratic politics.

Although he does not establish an explicit link to vertical structures of exchange, Dirk Vandewalle has also argued that rentier systems fragment societies (in his version, due to the absence of production processes that could create functionally differentiated social groups). But whatever the specific causal mechanism, it is not clear that a fragmented society on both macro- and meso-levels is an automatic outcome of rentierism.

Widespread individualized dependence on the state exists not only in Saudi Arabia but in numerous other rentier systems as well. However, once again an ahistorical view might tempt us to overextrapolate. The Saudi private sector, for one, seems to have gained more collective agency in recent years, belying an overly static view of Saudi society. Perhaps more significantly, in the case of Bahrain, as dependent on oil as Saudi Arabia, considerable parts of society are very much politically mobilized through informal but extensive sectarian networks, and public politics has a strong bottom-up component.[73] Similarly, Kuwait has a politically mobilized middle class and a merchant class that can step into the arena of politics in situations of crisis.[74] In such cases, pre-oil traditions of social organization were preserved. The state of state-society relations when oil hit these systems has crucially influenced the political uses to which the rents can be put.[75]

Kuwait, Bahrain, Venezuela, Iran, and Indonesia are oil states with stronger traditions of pre-oil national-level politics, larger independent mobilizational resources, and horizontally integrated groups in society. Even the richest regimes would have struggled to co-opt social interests to the extent that the Saudi regime has managed. Politically underdeveloped or strongly regionalized pre-oil societies like the ones in the United Arab Emirates, Qatar, Saudi Arabia, Libya, or Nigeria, by contrast, have been easier to clientelize. Levels of national integration and political mobilization at the point of "first oil" appear to be decisive variables. Before we draw general conclusions on how rentier states fragment societies, these variables, as well as the mechanisms that sustain horizontal mobilization into the oil age, demand further systematic investigation. The Saudi case can guide us toward the cases we need to look at, but it cannot answer these questions by itself.

72. The predominance of vertical links has been used in the literature to explain the absence of class politics in traditional and neo-patrimonial societies, less so to analyze state capacities and policy processes.
73. Lawson 1989.
74. Tetreault 2000; Koch 2000.
75. For a general argument along these lines, although focusing on regime survival, see Smith 2008.

References

Primary Sources

Archives

Ford Foundation archive, Institute of Public Administration, Riyadh, cited as FF/IPA
Institute of Public Administration, Riyadh, documentation center, cited as IPA
Middle East Documentation Unit, University of Durham
Mulligan Papers, Georgetown University
National Archive and Record Administration, College Park, Maryland: Record Group 59, cited as NARA
Philby papers, Middle East Center, St. Antony's College, Oxford, cited as PHILBY
Public Record Office, London: series FO 371 and FCO 8, cited as PRO
U.S. Records on Saudi Affairs, 1945–1959, Vols. 1–6 (Slough, UK: Archive Editions, 1997), cited as USRSA

Saudi Official Documents

Central Department of Statistics. *Statistical yearbook,* various issues.
Central Planning Organization. 1965. *An economic report.*
———. *Development plan.* 1970.
Interministerial Committee on Investment Obstacles. 2004. *Report,* Riyadh, English translation.
Ministry of Petroleum and Minerals. 1963. *The impact of petroleum on the economy (and social life) of Saudi Arabia* (October), Middle East Documentation Unit, University of Durham.
Ministry of Planning. *Five year development plans,* various issues.
———. *Achievements of the development plans: Facts and figures,* various issues.

Saudi Arabian General Investment Authority (SAGIA). 2003. *Izalat mu'wiqat al-bi'a al-istithmariyya fil-mamlaka al-'arabiyya as-sa'udiyya* [Removing obstacles in the investment environment in the Kingdom of Saudi Arabia], Riyadh, May.

———. 2004. Monthly investment statistics (provided as printout in May).

———. 2005. *Foreign Direct Investment Survey Report,* June.

———. 2006a. *Foreign Investment Statistics.*

———. 2006b. *At-taqrir as-sanawi li'ada' al-istithmar 1424–25* [The annual report on investment performance 2005].

———. 2007a. *At-taqrir as-sanawi lil-hayy'a al-'am lil-istithmar 1426–27* [The annual report of the Investment Authority for 2006], available at www.sagia.gov.sa/arabic/uploads1/Media/SAGIA%20Annual%20Report_2006.pdf.

———. 2007b. *Think Magazine,* issue 2.

Supreme Economic Council. 2002. *Privatization strategy for the Kingdom of Saudi Arabia,* June.

Saudi Arabian Monetary Agency (SAMA). *Annual Report,* various issues.

Um Al-Qura (official gazette of the Saudi government).

International Organizations and Foreign Governments

Economic and Social Commission for Western Asia (ESCWA). 2001. *WTO issues for acceding countries: The cases of Lebanon and Saudi Arabia* (New York: United Nations).

German embassy Riyadh. Various dispatches.

International Labor Organization (ILO). 1994. *World Labour Report* (Geneva).

International Monetary Fund. *Saudi Arabia—recent economic developments,* confidential background paper to Art. 4 report, issued after 1990/91 Gulf War.

United Nations Conference on Trade and Development (UNCTAD). 2004. *World investment report 2004* (Geneva).

U.S. Energy Information Administration. 2008 *Country analysis brief: Saudi Arabia,* available at www.eia.doe.gov/emeu/cabs/Saudi_Arabia/Electricity.html.

U.S. Foreign Commercial Service. 2004. *Country commercial guide for Saudi Arabia* (Washington, D.C.), available at http://strategis.ic.gc.ca/epic/internet/inimr-ri.nsf/en/gr122360e.html.

U.S. Foreign Commercial Service. 2003. *Agent/distributor agreement in Saudi Arabia* (Washington, D.C.), available at http://web.ita.doc.gov/ticwebsite/meweb.nsf/f41c595bf093a662852566f2004cfcf6/1aeac645b2d8ec608525682c005ba17e!OpenDocument.

U.S. Department of State. 2001. *Country reports on economic policy and trade practices: Saudi Arabia* (Washington, D.C.), available at www.state.gov/documents/organization/8190.pdf.

United National Development Programme (UNDP). 2004. *Project fact sheet: Technical assistance to Saudi Arabia to gain WTO accession* (Riyadh), available at www.undp.org.sa/Projects/008%20Accession%20to%20WTO%20-%20DA%20-%2019%20Dec%202004.doc.

World Bank. 2002. *Administrative barriers to investment in the Kingdom of Saudi Arabia* (study for SAGIA, Riyadh, August).

———. 2008. *Doing business in the Arab world, 2009* (Washington, D.C.).

World Trade Organization (WTO). 2005. *Saudi Arabia: Working party report* (Geneva: November), available at www.wto.org/english/news_e/pres05_e/wtaccsau61_e.doc.

Secondary Sources

Abir, Mordechai. 1988. *Saudi Arabia in the oil era: Regimes and elites* (London: Croon Helm).
———. 1993. *Saudi Arabia: Government, society, and the Gulf crisis* (New York: Routledge).
Albers, Henry. 1989. *Saudi Arabia: Technocrats in a traditional society* (New York: Lang).
Allen, Calvin, and W. Lynn Rigsbee. 2000. *Oman under Qaboos: From coup to constitution* (London: Frank Cass).
Al-Ammaj, Bader H. 1993. *Administration in traditional society: The case of recruitment and selection in public sector employment in Saudi Arabia* (Ph.D. dissertation, University of Southampton).
Amuzegar, Jahangir. 1998. *Managing the oil wealth: OPEC's windfalls and pitfalls* (London: I.B. Tauris).
Anderson, Lisa. 1987. "The state in the Middle East and North Africa," *Comparative Politics*, Vol. 19, No. 4, pp. 1–18.
Al-Awaji, Ibrahim. 1971. *Bureaucracy and society in Saudi Arabia* (Ph.D. dissertation, University of Virginia, 1971).
Ayubi, Nazih. 1995. *Over-Stating the Arab State: Politics and society in the Middle East* (London: I.B. Tauris).
Bakr, Mohammed A. 2001. *A model in privatization: Successful change management in the ports of Saudi Arabia* (London: Centre of Arab Studies).
Bardhan, Pranab. 1997. "Corruption and development: A review of issues," *Journal of Economic Literature*, Vol. 35, No. 3, pp. 1320–46.
Barnes, Philip. 1995. *Indonesia: The Political Economy of Energy* (Oxford: Oxford University Press).
Beblawi, Hazem. 1987. "The rentier state in the Arab world," in Hazem Beblawi and Giacomo Luciani (eds.), *The rentier state* (London: Croon Helm).
Bianchi, Robert. 1989. *Unruly corporatism: Associational life in twentieth-century Egypt* (Oxford: Oxford University Press).
Binsaleh, Abdullah Mohammed. 1982. *The civil service and its regulation in Saudi Arabia* (Ph.D. dissertation, Claremont Graduate School).
Birks, J. S., and C. A. Sinclair. 1982. "The domestic political economy of development in Saudi Arabia," in Tim Niblock (ed.), *State, society, and economy in Saudi Arabia* (London: Croom Helm).
Bligh, Alexander. 1985. "The Saudi religious elite (ulama) as participant in the political system of the Kingdom," *International Journal of Middle East Studies*, Vol. 17, No. 1, pp. 37–50.
Breton, Albert, and Ronald Wintrobe. 1986. "The bureaucracy of murder revisited," *Journal of Political Economy*, Vol. 94, No. 5 (October), pp. 905–26.
Business International. 1985. *Saudi Arabia: A reappraisal* (Geneva: September).
Chaudhry, Kiren Aziz. 1997. *The price of wealth: Economies and institutions in the Middle East* (Ithaca: Cornell University Press).
Chibber, Vivek. 2002. "Bureaucratic rationality and the developmental state," *American Journal of Sociology*, Vol. 107, No. 4, pp. 951–89.
———. 2004. *Locked in place: State-building and late industrialization in India* (Princeton: Princeton University Press).

Citino, Nathan J. 2002. *From Arab Nationalism to OPEC: Eisenhower, King Sa'ud, and the making of U.S.-Saudi relations* (Bloomington: Indiana University Press).

Coll, Steve. 2008. *The Bin Ladens: An Arabian family in the American century* (New York: Penguin Press).

Collier, David, and Steven Levitsky. 1997. "Research note: Democracy with adjectives: Conceptual innovation in comparative research," *World Politics*, Vol. 49, No. 3, pp. 430–51.

Crystal, Jill. 1995. *Oil and politics in the Gulf: Rulers and merchants in Kuwait and Qatar* (Cambridge: Cambridge University Press).

Cunningham, Robert, and Yasin Sarayrah. 1993. *Wasta: The hidden force in Middle Eastern society* (Westport, CT, and London: Praeger).

Daghistani, Abdulaziz. 2002. *Siyasat wa Ijra'at as-Sa'wada* [Policies and Measures of Saudization], Economic Studies House, Riyadh.

Davis, Diane. 2006. "Undermining the rule of law: Democratization and the dark side of police reform in Mexico," *Latin American Politics and Society*, Vol. 48, No. 1, pp. 55–86.

Davis, John. 1987. *Libyan politics: Tribe and revolution* (Berkeley: University of California Press).

Delacroix, Jacques. 1980. "The distributive state in the world system," *Studies in Comparative International Development*, Vol. 15, No. 1, pp. 3–22.

Diwan, Ishac, and Maurice Girgis. 2002. *Labour force development in Saudi Arabia*, policy paper prepared for the Symposium on the Future Vision for the Saudi Economy, Riyadh, September.

Edens, David, and William Snavely. 1970. "Planning for economic development in Saudi Arabia," *Middle East Journal*, Vol. 24, No. 1, pp. 17–31.

Eisenstadt, S. N. 1973. *Traditional patrimonialism and modern neopatrimonialism* (London: Sage Publications).

Eisenstadt, S. N., and L. Roniger. 1981. "The study of patron-client relations and recent developments in sociological theory," in S. N. Eisenstadt and Rene Lemarchand (eds.), *Political clientelism, patronage and development* (London: Sage Publications).

——. 1984. *Patrons, clients, and friends: Interpersonal relations and the structure of trust in society* (Cambridge: Cambridge University Press).

Ernst and Young. 2004. *The new Saudi Arabian tax law: A summary of certain salient corporate income tax features* (Riyadh, March).

Evans, Peter. 1989. "Predatory, developmental, and other apparatuses: A comparative political economy perspective on the Third World state," *Sociological Forum*, Vol. 4, No. 4, pp. 561–87.

——. 1995. *Embedded autonomy: States and industrial transformation* (Princeton: Princeton University Press).

Al-Fahad, Abdulaziz. 2000. "The Prince, the shaykh—and the lawyer," *Case Western Reserve Journal of International Law*, Vol. 34, No. 2, Supplemental.

——. 2004. "From exclusivism to accommodation: Doctrinal and legal evolution of Wahhabism," *NYU Law Review*, Vol. 79, No. 2, 485–591.

——. 2005. "Ornamental constitutionalism: The Saudi Basic Law of Governance," *Yale Journal of International Law*, Vol. 30, No. 2, pp. 375–96.

Fallon, Nicholas. 1976. *Winning business in Saudi Arabia* (London: Graham and Trotman).

Fandy, Mamoun. 1999. *Saudi Arabia and the politics of dissent* (Basingstoke: Macmillan).

Farah, Tawfic. 1989. "Political culture and development in a rentier state: The case of Kuwait," *Journal of Asian and African Studies*, Vol. 24, No. 1/2, pp. 106–13.

El-Farra, Taha. 1973. *The effects of detribalizing the bedouins on the internal cohesion of an emerging state: The Kingdom of Saudi Arabia* (Ph.D. dissertation, University of Pittsburgh).

Al-Fiar, Mohamed Hussein. 1977. *The Faisal Settlement Project at Haradh, Saudi Arabia: A study in nomad attitudes toward sedentarization* (Ph.D. dissertation, Michigan State University).

Field, Michael. 1984. *The merchants: The big business families of Saudi Arabia and the Gulf* (New York: Overlook Press).

Frieden, Jeffrey, and Ronald Rogowski. 1996. "The impact of the international economy on national policies: An analytical overview," in Helen Milner and Robert Keohane (eds.), *Internationalization and domestic politics* (Cambridge: Cambridge University Press).

Gambetta, Diego. 1988. "Can we trust trust?" in Diego Gambetta (ed.), *Trust: Making and breaking cooperative relations* (New York: Basil Blackwell).

Gause, Gregory. 1994. *Oil monarchies: Domestic and security challenges in the Arab Gulf states* (New York: Council on Foreign Relations Press).

Geddes, Barbara. 1990. "Building 'state' autonomy in Brazil, 1930–1964," *Comparative Politics,* Vol. 22, No. 2, pp. 217–35.

———. 1994. *Politicians' dilemma: Building state capacity in Latin America* (Berkeley: University of California Press).

Gelb, Alan (ed.). 1988. *Oil windfalls: Blessing or curse?* (New York and Oxford: Oxford University Press).

Gerring, John. 2007. "Is there a (viable) crucial-case method?" *Comparative Political Studies,* Vol. 40, No. 3, pp. 231–53.

Goldberg, Ellis, Erik Wibbels, and Eric Mvukiyehe. 2008. "Lessons from strange cases: Democracy, development, and the resource curse in the U.S. states," *Comparative Political Studies,* Vol. 41, No. 4–5, 477–514.

Grayson, George. 1980. *The politics of Mexican oil* (Pittsburgh: University of Pittsburgh Press).

Grindle, Merilee S. 1977. *Bureaucrats, politicians, and peasants in Mexico* (Berkeley: University of California Press).

Gulftalent. 2005. *Recruiting top graduates In Saudi Arabia: A survey of recent graduates of King Fahd University of Petroleum and Minerals* (Dubai), <www.gulftalent.com/home/Recruiting-Top-Graduates-in-Saudi-Arabia-English-Version-Report-3.html>.

Haggard, Stephen, Tun-jen Cheng, and David Kang. 1998. "Institutions and growth in Korea and Taiwan: The bureaucracy," *Journal of Development Studies,* Vol. 34, No. 6, pp. 66–86.

Al-Hamoud, Ahmed H. 1991. *The reform of the reform: A critical and empirical assessment of the 1977 Saudi civil service reform* (Ph.D. dissertation, University of Pittsburgh).

Handoussa, Heba. 2002. "A balance sheet of reform in two decades," in Noha El-Mikawy and Heba Handoussa (eds.), *Institutional reform and economic development in Egypt* (Cairo: American University in Cairo Press).

Hegghammer, Thomas, and Stephane Lacroix. 2007. "Rejectionist Islamism in Saudi Arabia: The story of Juhayman al-'Utaybi revisited," *International Journal of Middle East Studies,* Vol. 39, No. 1, pp. 103–22.

Henry, Clement. 1994. *Images of development: Egyptian engineers in search of industry* (Cairo: American University in Cairo Press).

Herb, Michael. 1999. *All in the family: Absolutism, revolution, and democracy in the Middle East* (Albany: SUNY Press).

———. 2005. "No representation without taxation? Rents, development and democracy," *Comparative Politics*, Vol. 37, No. 3, pp. 297–317.

Hertog, Steffen. 2006a. *Labour policy in the Gulf: Unintended consequences of regulatory ambition in rentier societies*, paper presented at Medmeeting of the European University Institute, Florence, March.

———. 2006b. "Modernizing without democratizing? The introduction of formal politics in Saudi Arabia," *International Politics and Society*, No. 3 (2006), pp. 65–78.

———. 2006c. "The new corporatism in Saudi Arabia: Limits of formal politics," in Abdulhadi Khalaf and Giacomo Luciani (eds.), *Constitutional reform and political participation in the Gulf* (Dubai: Gulf Research Center).

———. 2008. "Petromin: The slow death of statist oil development in Saudi Arabia," *Business History*, Vol. 50, No. 5, pp. 645–67.

Heydemann, Steven. 1999. *Authoritarianism in Syria: Institutions and social conflict, 1946–1970* (Ithaca: Cornell University Press).

———. 2004. "Networks of privilege: Rethinking the politics of economic reform in the Middle East," in Steven Heydemann (ed.), *Networks of privilege in the Middle East: The politics of economic reform revisited* (New York and Basingstoke: Palgrave Macmillan).

Hirschman, Albert. 1970. *Exit, voice, and loyalty: Responses to decline in firms, organizations, and states* (Cambridge: Harvard University Press).

Hoekman, Bernard M., and Michel M. Kostecki. 2001. *The political economy of the world trading system* (Oxford: Oxford University Press).

Holden, David, and Richard Johns. 1982. *The house of Saud* (London: Pan Books).

Hout, Wil. 2007. "Development under patrimonial conditions: Suriname's state oil company as a development agent," *Journal of Development Studies*, Vol. 43, No. 8, pp. 1331–50.

Huber, Evelyn. 1995. "Assessments of state strength," in Peter Smith (ed.), *Latin America in comparative perspective* (Boulder and Oxford: Westview Press).

Huyette, Summer Scott. 1985. *Political adaptation in Saudi Arabia: A study of the Council of Ministers* (Boulder and London: Westview).

Isham, Jonathan, Michael Woolcock, et al. 2005. "The varieties of resource experience: Natural resource export structures and the political economy of growth," *World Bank Economic Review*, Vol. 19, No. 2, pp. 141–74.

Islami, A. Reza S., and Rostam Mehraban Kavoussi. 1984. *The political economy of Saudi Arabia* (Seattle: University of Washington Press).

Johnson, Chalmers. 1982. *MITI and the Japanese miracle: The growth of industrial policy, 1925–1975* (Stanford: Stanford University Press).

Jones Luong, Pauline. 2003. *Rethinking the resource curse: Ownership structure and institutional capacity*, paper prepared for presentation at the International Conference on Globalization and Social Stress, co-sponsored by TIGER and Yale University, Warsaw, October.

Karl, Terry Lynn. 1997. *The paradox of plenty: Oil booms and petro-states* (Berkeley: University of California Press).

———. 1999. "The perils of the petro-state: Reflections on the paradox of plenty," *Journal of International Affairs*, Vol. 53, No. 1, pp. 32–48.

———. 2007. "Critical issues for a revenue management law," in Macartan Humphreys, Jeffrey Sachs, and Joseph Stiglitz (eds.), *Escaping the resource curse* (New York: Columbia University Press).

Knauerhase, Ramon. 1977. *The Saudi Arabian economy* (New York and London: Praeger).
Koch, Christian. 2000. *Politische Entwicklung in einem arabischen Golfstaat: Die Rolle von Interessengruppen im Emirat Kuwait* (Berlin: Klaus Schwarz).
Koelble, Thomas. 1995. "The new institutionalism in political science and sociology," *Comparative Politics*, Vol. 27, No. 2, pp. 231–43.
Kohli, Atul. 2004. *State-directed development: Political power and industrialization in the global periphery* (Cambridge: Cambridge University Press).
Koniordos, Sokratis. 2004. "Introduction," in Sokratis Koniordos (ed.), *Networks, trust, and social capital* (Aldershot: Ashgate Publishing).
Koshak, Tariq Hassan. 2001. *The Saudi budgetary process: An exploratory case study* (Ph.D. dissertation, University of Leeds Business School).
Krimly, Rayed. 1993. *The political economy of rentier states: A case study of Saudi Arabia in the oil era, 1950–1990* (Ph.D. dissertation, George Washington University).
———. 1999. "The political economy of adjusted priorities: Declining oil revenues and Saudi fiscal policies," *Middle East Journal*, Vol. 52, No. 2, pp. 254–67.
Lacey, Robert. 1981. *The kingdom* (New York and London: Harcourt Brace Jovanovich).
———. 2009. *Inside the kingdom: Saudi Arabia, 1979–2009* (London: Hutchinson).
Lackner, Helen. 1978. *A house built on sand: A political economy of Saudi Arabia* (London: Ithaca Press).
Lacroix, Stephane. 2008. "L'apport de Muhammad Nasir al-Din al-Albani au salafisme contemporain," in Bernard Rougier (ed.), *Qu'est-ce que le salafisme?* (Paris: Presses Universitaires de France).
———2010. *Awakening Islam: A history of Islamism in Saudi Arabia* (Harvard: Harvard University Press).
Lawson, Fred. 1989. *Bahrain: The modernization of autocracy* (Boulder: Westview, 1989).
Leaders in Dubai. 2006. *Business forum 2006: Doing business in the GCC survey* (Dubai).
Leff, N. H. 1964. "Economic development through bureaucratic corruption," *American Behavioral Scientist*, Vol. 8, No. 3, pp. 8–14.
Levi, Margaret. 1988. *Of rule and revenue* (Berkeley: University of California Press).
Le Vine, Victor. 1975. *Political corruption: The Ghana case* (Stanford: Hoover Institution Press).
Lippman, Thomas. 2004. *Inside the mirage: America's fragile partnership with Saudi Arabia* (Boulder: Westview).
Lipsky, George. 1959. *Saudi Arabia: Its people, its society, its culture* (New Haven: HRAF Press).
Looney, Robert. 2004. "Saudization and sound economic reforms: Are the two compatible?" *Strategic Insights*, Vol. 3, No. 2.
Luciani, Giacomo. 1990. "Allocation vs. production states: A theoretical framework," in Giacomo Luciani (ed.), *The Arab state* (London: Routledge).
———. 1995. "The oil rent, the fiscal crisis of the state, and democratization," in Ghassan Salamé (ed.), *Democracy without democrats? The renewal of politics in the Muslim world* (London: I.B. Tauris).
———. 2005. "Saudi Arabian business: From private sector to national bourgeoisie," in Paul Aarts and Gerd Nonneman (eds.), *Saudi Arabia in the balance: Political economy, society, foreign affairs* (London: Hurst).
Madi, Mohammed Abdullah F. 1975. *Development administration and the attitudes of middle management in Saudi Arabia* (Ph.D. dissertation, Southern Illinois University).

Mahdavy, Hussein. 1970. "The patterns and problems of economic development in rentier states: The case of Iran," in M.A. Cook (ed.), *Studies in economic history of the Middle East* (London: Oxford University Press).
Malik, Monica. 1999. *The private sector and the state in Saudi Arabia* (Ph.D. dissertation, University of Durham).
Malloy, James. 1979. "Authoritarianism and corporatism in Latin America: The modal pattern," in James Malloy (ed.), *Authoritarianism and corporatism in Latin America* (Pittsburgh: University of Pittsburgh Press).
Maloney, Suzanne. 2000. "Agents or obstacles? Parastatal foundations and challenges for Iranian development," in Parvin Alizadeh (ed.), *The economy of Iran: Dilemmas of an Islamic state* (London: I.B. Tauris).
Mandil, Suliman. 2003. *The investment climate in the kingdom of Saudi Arabia (facts and challenges)*, PowerPoint presentation at Riyadh Economic Forum, October, in Arabic.
Migdal, Joel. 1988. *Strong societies and weak states: State-society relations and state capabilities in the third world* (Princeton: Princeton University Press).
———. 2001. *State in society: Studying how states and societies transform and constitute one another* (Cambridge: Cambridge University Press).
Al-Mizjaji, Ahmad Dawood. 1992. *The public attitudes toward the bureaucracy in Saudi Arabia* (Ph.D. dissertation, Florida State University).
Moe, Terry M., and Michael Caldwell. 1994. "The institutional foundations of democratic government: A comparison of presidential and parliamentary systems," *Journal of Institutional and Theoretical Economics*, Vol. 150, No. 1, pp. 171–95.
Molyneux, Philip, and Munawar Iqbal. 2005. *Banking and financial systems in the Arab world* (Basingstoke: Palgrave Macmillan).
Montagu, Caroline. 1985. *Industrial development in Saudi Arabia: Opportunities for joint ventures* (London: Committee for Middle Eastern Trade).
———. 1987. *Industrial development in Saudi Arabia* (London: Committee for Middle East Trade).
———. 1994. *The private sector of Saudi Arabia* (London: Committee for Middle Eastern Trade).
———. 2001. *Saudi Arabia on the road to reform* (London).
Moon, Chung In. 1986. "Korean contractors in Saudi Arabia: Their rise and fall," *Middle East Journal*, Vol. 40, No. 4, pp. 614–33.
Moore, Mick. 2004. "Revenues, state formation, and the quality of governance in developing countries," *International Political Science Review*, Vol. 25, No. 3, pp. 297–319.
Moore, Pete. 2004. *Doing business in the Middle East: Politics and economic crisis in Jordan and Kuwait* (Cambridge: Cambridge University Press).
Nettl, J. P. 1968. "The state as conceptual variable," *World Politics*, Vol. 20, No. 4, pp. 559–92.
Niblock, Tim. 1982. "Social structure and the development of the Saudi Arabian political system," in Tim Niblock (ed.), *State, society, and economy in Saudi Arabia* (London: Croom Helm).
———. 2007. *The political economy of Saudi Arabia* (London: Routledge).
Nyrop, Richard. 1977. *Area handbook for Saudi Arabia* (Washington, DC: Government Printing Office).
O'Donnell, Guillermo. 1979. "Corporatism and the question of the state," in James A. Malloy (ed.), *Authoritarianism and corporatism in Latin America* (Pittsburgh: University of Pittsburgh Press).

Pelletière, Stephen. 2004. *America's oil wars* (Westport and London: Praeger).
Perthes, Volker. 1995. *The political economy of Syria under Asad* (New York: I.B. Tauris).
Philby, H St. J. 1955. *Saudi Arabia* (London: Benn).
Pierson, Paul, and Theda Skocpol. 2002. "Historical institutionalism in contemporary political science," in Ira Katznelson and Helen Milner (eds.), *Political science: The state of the discipline* (New York and London: W.W. Norton).
Piscatori, James. 1983. "Ideological politics in Saudi Arabia," in James Piscatori (ed.), *Islam in the politics process* (Cambridge: Cambridge University Press).
Pledge, Thomas, Ali Dialdin, and Muhammad Tahlawi. 1998. *Saudi Aramco and its people: A history of training* (Saudi Aramco).
Pressman, Jeffrey, and Aaron Wildavsky. 1984. *Implementation: How great expectations in Washington are dashed in Oakland*, 3d ed. (Berkeley: University of California Press).
Ramady, Mohammad, and Mourad Mansour. 2006. "The impact of Saudi Arabia's WTO accession on selected economic sectors and domestic economic reforms," *World Review of Entrepreneurship, Management, and Sustainable Development*, Vol. 2, No. 3, pp. 189–99.
Al-Rasheed, Madawi. 2002. *A history of Saudi Arabia* (Cambridge: Cambridge University Press).
———. 2005. "Circles of power: Royals and society in Saudi Arabia," in Paul Aarts and Gert Nonneman (eds.), *Saudi Arabia in the balance* (London: Hurst).
Al-Rasheed, Madawi, and Loulouwa Al-Rasheed. 1996. "The politics of encapsulation: Saudi policy towards tribal and religious opposition," *Middle Eastern Studies*, Vol. 32, No. 1, pp. 96–119.
Robins, Philip. 2004. "Slow, slow, quick, quick, slow: Saudi Arabia's 'gas initiative,'" *Energy policy*, Vol. 32, No. 3 (2004), pp. 321–33.
Robinson, Jeffrey. 1988. *Yamani: The inside story* (London: Simon and Schuster).
Roniger, Luis. 1990. *Hierarchy and trust in modern Mexico and Brazil* (New York and London: Praeger).
———. 2002. *Clientelism and civil society in historical perspective*, paper for "Demokratie und Sozialkapital—Die Rolle zivilgesellschaftlicher Akteure" workshop, German Political Science Association, Berlin, June.
Ross, Michael. 2001. "Does oil hinder democracy?" *World Politics*, Vol. 53, No. 3, pp. 325–61.
———. 2009. *The curse of oil wealth*, unpublished manuscript.
Al-Sabban, Aidros Abdulla. 1982. *The municipal system in the kingdom of Saudi Arabia: A case study of Makkah* (Ph.D. dissertation, Claremont Graduate School).
———. 1990. "Saudi Arabian municipalities: History, organization, and structure," in Ahmed Hassan Dahlan (ed), *Politics, administration, and development in Saudi Arabia* (Brentwood, MD: Amana Corp.).
Sachs, Jeffrey, and Andrew Warner. 2001. "Natural resources and economic development: The curse of natural resources," *European Economic Review*, Vol. 45, No. 4–6, pp. 827–38.
Al-Saleh, Mohammed Abdulaziz Abdullah. 1994. *A study of the foreign investment regulation in the kingdom of Saudi Arabia: Law and policy* (Ph.D. dissertation, University of Warwick).
Al-Salem, Abdullah Abdulkareem. 1996. *A case study of the organizational culture of the Makkah municipality in the context of the Saudi society* (Ph.D. dissertation, Templeton University).
SAMBA Financial Group. 2002. *Saudi Arabia's employment profile* (Riyadh, October).
———. 2005. *Saudi Arabia's 2006 budget, 2005 performance* (Riyadh, December).

———. 2006. *Saudi Arabia and the WTO* (Riyadh, February).
Samman, Nizar. 1990. "Saudi Arabia and the Role of the Emirates," in Ahmed Hassan Dahlan (ed). *Politics, administration and development in Saudi Arabia* (Brentwood, MD: Amana Corp.).
Samore, Gary. 1983. *Royal family politics in Saudi Arabia (1953–82)* (Ph.D. dissertation, Harvard).
Al Saud, Khalid Bin Faisal Bin Turki. 1996. *The inter-penetration of agency and structure within the Saudi bureaucracy* (Ph.D. dissertation, Swansea University).
Al Saud, Mashaal Abdullah. 1982. *Permanence and change: An analysis of the Islamic political culture of Saudi Arabia with special reference to the royal family* (Ph.D. dissertation, Claremont Graduate School).
Schneider, Ben Ross. 1991. *Politics within the state: Elite bureaucrats and industrial policy in authoritarian Brazil* (Pittsburgh: University of Pittsburgh Press).
———. 2004. *Business politics and the state in 20th-century Latin America* (Cambridge: Cambridge University Press).
Schneider, Ben Ross, and Sylvia Maxfield (eds.). 1997. *State-Business relations in developing countries* (Ithaca: Cornell University Press).
Al-Seflan, Ali Mashhour. 1980. *The essence of tribal leaders' participation, responsibilities, and decisions in some local government activities in Saudi Arabia* (Ph.D. dissertation, Claremont Graduate School).
Senany, Ahmad Mohammad. 1990. *Development planning: Public functions and private sector participation in the Kingdom of Saudi Arabia* (Ph.D. dissertation, Florida State University).
Seznec, Jean-Francois. 2002. "Stirrings in Saudi Arabia," *Journal of Democracy,* Vol. 13, pp. 33–40.
Shafer, Michael. 1994. *Winners and losers: How sectors shape the developmental prospects of states* (Ithaca: Cornell University Press).
Al-Shalan, Fahad Ahmed. 1991. *Participation in managerial decision-making in the Saudi public sector* (Ph.D. dissertation, University of Pittsburgh).
Shue, Vivienne, and Atul Kohli, 1994. "State power and social forces: On political contention and accommodation in the third world," in Migdal, Kohli and Shue (eds.), *State power and social forces: Domination and transformation in the third world* (Cambridge: Cambridge University Press).
Smith, Benjamin. 2004. "Oil wealth and regime survival in the developing world, 1960–1999," *American Journal of Political Science,* Vol. 48, No. 2, pp. 232–46.
———. 2007. *Hard times in the lands of plenty: Oil politics in Iran and Indonesia* (Cornell University Press).
Smith, Grant F. 2003. "Saudization: development and expectations management," *Saudi-American Forum Newsletter,* October, available at http://saudi-american-forum.org/Newsletters/SAF_Essay_23.htm.
Smith, Pamela, and Moin Siddiqi. 2000. "Saudi Arabia Special Report," *The Middle East,* May.
Solaim, Soliman. 1970. *Constitutional and judicial organization in Saudi Arabia* (Ph.D. dissertation, Johns Hopkins University Press).
Springborg, Robert. 1993. "Egypt," in Tim Niblock and Emma Murphy (eds.), *Economic and political liberalization in the Middle East* (London: British Academic Press).
Standard and Poor's. 2007. *Bank industry risk analysis: A bright future for Saudi banks, despite stock market collapse,* 31 May.

Streeck, Wolfgang, and Kathleen Thelen. 2005. "Introduction," in Wolfgang Streeck and Kathleen Thelen (eds.), *Beyond continuity: Institutional change in advanced political economies* (Oxford: Oxford University Press).

Tawati, Ahmed Mohmed. 1976. *The civil service of Saudi Arabia: Problems and prospects* (Ph.D. dissertation, West Virginia University).

Teitelbaum, Joshua. 2000. *Holier than thou: Saudi Arabia's Islamic opposition* (Washington, DC: Washington Institute for Near East Policy).

Tetreault, Mary Ann. 2000. *Stories of democracy: Politics and society in contemporary Kuwait* (New York: Columbia University Press).

Thelen, Kathleen. 1999. "Historical institutionalism in comparative politics," *Annual Review of Political Science*, Vol. 2, pp. 369–404.

———. 2003. "How institutions evolve: Insights from comparative-historical analysis," in James Mahoney und Dietrich Rueschemeyer (eds.), *Comparative historical analysis in the social sciences* (Cambridge: Cambridge University Press).

Tripp, Charles. 2002. *A history of Iraq* (Cambridge: Cambridge University Press).

Tsebelis, George. 1995. "Decision making in political systems: Veto players in presidentialism, parliamentarism, multicameralism and multipartyism," *British Journal of Political Science*, Vol. 25, No. 3, pp. 289–325.

Vandewalle, Dirk. 1998. *Libya since independence: Oil and state building* (Ithaca: Cornell University Press).

Vasiliev, Alexei. 2000. *The history of Saudi Arabia* (London: Saqi).

Vitalis, Robert. 1999. "Review of Kiren Chaudhry, Price of Wealth," *International Journal of Middle East Studies*, Vol. 31, No. 3, pp. 659–61.

———. 2007. *America's Kingdom: Mythmaking on the Saudi oil frontier* (Stanford: Stanford University Press).

Waldner, David. 1999. *State-building and late development* (Ithaca: Cornell University Press).

Waterbury, John. 1983. *The Egypt of Nasser and Sadat: The political economy of two regimes* (Princeton: Princeton University Press).

———. 1989. "The political management of economic adjustment and reform," in Joan S. Nelson (ed.), *Fragile coalitions: The politics of economic adjustment* (New Brunswick/Oxford: Transaction Books).

———. 1991. "Twilight of the state bourgeoisie?" *International Journal of Middle East Studies*, Vol. 23, No. 1, pp. 1–17.

Weaver, R. Kent, and Bert Rockman (eds.). 1993. *Do institutions matter? Government capabilities in the United States and abroad* (Washington: Brookings Institution).

Weinthal, Erika, and Pauline Jones Luong. 2006. "Combating the resource curse: An alternative solution to managing mineral wealth," *Perspectives on Politics*, Vol. 4, No. 1, pp. 35–53.

Weyland, Kurt. 1996. "Obstacles to social reform in Brazil's new democracy," *Comparative Politics*, Vol. 29, No. 1, pp. 1–22.

———. 1998. From Leviathan to Gulliver? The decline of the developmental state in Brazil," *Governance*, Vol. 11, No. 1, pp. 51–75.

Wickham, Carrie Rosefsky. 2002. *Mobilizing Islam: Religion, activism, and political change in Egypt* (New York: Columbia University Press).

Wilson, Rodney. 2004. *Economic development in Saudi Arabia* (London: RoutledgeCurzon).

Al-Yassini, Ayman. 1985. *Religion and state in the Kingdom of Saudi Arabia* (Boulder: Westview Press).

Yergin, Daniel. 1991. *The Prize: The epic quest for oil, money and power* (London: Simon and Schuster).

Yizraeli, Sarah. 1997. *The remaking of Saudi Arabia: The struggle between King Sa'ud and Crown Prince Faysal, 1953–1962* (Tel Aviv: Moshe Dayan Center for Middle Eastern and African Studies).

Young, Arthur N. 1983. *Saudi Arabia: The making of a financial giant* (New York University Press).

Index

Note: Page numbers in *italics* indicate figures; those with a *t* indicate tables.

Abalkhail, Mohammad, 91, 94–96, 102
Abdallah bin Abdulaziz, 247, 257
    Faisal and, 64, 66
    fiscal crisis and, 137–39
    National Guard under, 14, 17, 23, 89, 90
    Saudization and, 212, 217–21
    Tuwaijri family and, 195n59
    WTO and, 227, 237–45
Abdallah bin Faisal bin Turki, 44, 45
    foreign investment act and, 146–55, 174
    SAGIA and, 169–74
Abdalrahman bin Abdulaziz, 87
Abdulaziz bin Abdallah, 173n196
Abdulaziz bin Abdulrahman, 17
    clientelism under, 41–44
Abdulaziz bin Salman, 153
Adham, Kamal, 47n43, 74, 110
'Adwan, Abdallah bin, 46, 53
Agricultural Bank, 97
Ahmad bin Abdulaziz, 89
Alamy, Fawaz, 233, 235, 240–41
Ali, Anwar, 57
Alireza family, 47n43, 74, 241
Alireza, Mohammad, 49–50, 56
Aramco corporation, 14, 28, 44, 50, 134
    brokerage at, 110
    College of Petroleum and Minerals and, 70
    foreign investment act and, 178, 180, 181
    index of contracting activity of, 108
    lobbying efforts of, 60
    nationalization of, 128
    Petromin and, 100–101, 128–29
    Saudization and, 190, 195
arbitration law, 130*t*
Armitage, J.B., 67n33
autonomy. *See* state autonomy
Al-Awad, Awad, 174n201
Al-Awaji, Ibrahim, 83n166, 99

Badr, Fayez, 101
Bahrain, 139, 219, 275
bakhil, 121
bakshish. *See* bribery
banking laws, 127, 130*t*, 179n227
    SAMA and, 255, 264
    WTO and, 235
Barillka, William, 162n120
Bin Laden family, 74
bonyads, 268–70
Brazil, 261–62, 269–71
    clientelism in, 22n30
    patronage in, 29
bribery, 70
    "consultancy fees" and, 107
    foreign investment and, 161–62
    See also venality

brokers, 26–27, 30, 38
  administrators as, 54
  banks as, 255
  businessmen versus, 258
  competition among, 48
  consultancy fees of, 107
  foreign investment act and, 161–62
  as gatekeepers, 30, 48
  of labor import licenses, 198–99
  for military procurements, 88
  networks of, 19, 48–49, 74, 110–15, 220, 247, 251–52
  paper, 161, 193n43
  princes as, 110–11
  society of, 109–15
  for worker visas, 27, 198–203, 212–20, 217t, 273
  WTO effect on, 251–52
  See also wasta
budgetary retrenchment, 118–25, *119–23*
bureaucracy, 5, 24, 246–53
  under Abdallah, 137–41
  Brazilian, 262
  designing of, 45–47, 131–36
  early forms of, 41–42
  extent of, 85, 187
  under Faisal, 61–83
  fragmentation of, 4–5, 18–21, 31–34, 97–98, 129, 158–60, 189–91, 226–28, 244–45, 246, 249–50, 263–64, 267–71
  hierarchies in, 23–24, 84, 133–34, 247
  job classification system of, 71, 73, 104
  moonlighting in, 112
  during oil crash, 118–31, *119–23*
  Ottoman, 54
  patronage system and, 5, 16–18
  religious, 92
  social networks and, 4, 24–27, 33, 188
  wages in, 188
  Weberian, 15, 31
  See also Civil Service Bureau; clientelism
business groups, 250–53, 258
  foreign investment act and, 147–48
  regulatory demands of, 167
  Saudization and, 191–92, 202–19, 217t
  worker visas and, 214–18
  WTO and, 230–34
  See also Chambers of Commerce and Industry; entrepreneurship
business registration. *See* commercial registration
Butter, David, 155n84

Capital Market Authority (CMA), 239, 242, 255
capital market reforms, 1, 33–34, 242, 255–56
CCIs. *See* Chambers of Commerce and Industry
census, 108
Central Bank, 17, 28, 70
Central Department of Statistics (CDS), 187
centralization, 33, 248–51
  under Faisal, 62, 66–67
  interagency coordination and, 97, 102
Central Personnel Bureau, 104n116
Central Planning Organization, 67, 86
  five-year plans of, 84–85, 93, 96, 186
Chambers of Commerce and Industry (CCIs), 34, 59, 140, 257, 258
  Council of, 206, 212
  foreign investment act and, 147–48, 156, 174
  Saudization and, 202–3, 209
  WTO and, 231, 233, 236–37
  See also business groups
charitable foundations *(bonyads)*, 268–70
Chaudhry, Kiren, 15, 39, 49, 76, 265, 272
  on Fahd's economic policies, 119–21, 125
Chibber, Vivek, 255n12, 263
CITC. *See* Communications and Information Technology Commission
Civil Service Bureau, 71–73, 90, 104
  See also bureaucracy
clientelism, 21–29, 135–36, 246–53, 258
  under Abdulaziz, 42–44, 52–58
  in Brazil, 22n30
  bureaucratic jobs and, 23–24
  cooptation and, 133
  definitions of, 5, 12, 21–22
  under Faisal, 76–77, 83
  formal versus informal, 22–27, 88
  local mobility and, 52–53
  during 1970s boom, 84–103
  policy-making under, 29–35
  Saudization and, 220
  society atomized by, 115–18
  See also bureaucracy
CMA. *See* Capital Market Authority
College of Petroleum and Minerals. *See* University of Petroleum and Minerals
College of Sharia, 92
commercial agency laws, 236–37
commercial registration (CR), 163–64, 175, 200
Communications and Information Technology Commission (CITC), 238, 264
"consultancy fees," 107
consumerism, 29, 168, 234, 260
copyright laws, 130t, 238, 243

corruption, 5, 25–26, 47–49
  brokerage and, 112–13
  under Faisal, 72n90, 74–76
  foreign investment and, 164–67
  kickbacks and, 54
  nepotism and, 31
  during oil crash, 125
  regulatory reform and, 139–40, 161–62
  worker visas and, 219–21
  See also nepotism
Council of Ministers, 62, 63, 65–68
  creation of, 44–45
  Saudization and, 194
Council of Senior Ulama, 79
court system. *See* sharia courts
"cover-up" businesses, 164–67, 273
CR. *See* commercial registration
Craig, James, 71n82

Dabbagh, Amr, 172–80
Dabbagh family, 43
Daghistani, Abdulaziz, 191n28
Daley, William M., 235n62
Dammam, 205
Dar Al-Ifta, 79
Davis, John, 3n1
Decree 50, of MPC, 193–94, 196
diwan, 52, 63
domestic workers, 100, 199, 213
Dubai, 250, 267, 268

education, 49n53, 71–72, 256
  budgets for, 47n44
  Fahd's policies on, 121, 129
  female, 92
  foreign investment in, 155, 156
  religious, 92, 126, 188
  study abroad programs of, 105, 121n198
  technocratic, 99
  See also Ministry of Education
Egypt, 261, 269, 270n59
  exports of, *226*
  foreign landownership in, 144n2
  Qualified Industrial Zones in, 224n2
Eisenstadt, S. N., 43n10, 60n120
electricity policies, 119–20
  foreign investment and, 155, 178
  regulation of, 179n227
elites, 16, 22
  Brazilian, 269
  low-level vetoes and, 33
  religious, 39

state building and, 3, 58–60, 69
  See also patronage
Emaar corporation, 173, 179
entitlements, 18–19, 23, 83–84, 118–25, *119–23*, 250–51
entrepreneurship, 152, 250–52, 258
  worker visas and, 214–18
  WTO negotiations and, 230–34
  See also business groups
environmentalism, 29
European Union, WTO and, 227, 237n82, 238
exceptionalism, Saudi, 12–13, 259–75

Al Fahd, 87, 94, 206
Fahd bin Abdulaziz, 85–103
  economic policies of, 118–31, *119–23*
  education policies of, 121, 129
  entourage of, 48
  Faisal and, 64, 90–91, 110
  foreign investment act and, 148
  illness of, 138
Faisal bin Abdulaziz, 18, 61–83
  assassination of, 85
  centralization by, 62, 66–67
  civil service law of, 104
  entourage of, 48
  Fahd and, 64, 90–91, 110
  institutionalization by, 44
  religious organizations and, 92
  Saud and, 46–47, 59n110, 61–64
Faisal Foundation, 98
Faqeeh, Adel, 232
Al-Faqih, Asad, 46
Faqih, Usama, 159
  WTO and, 227–29, 234–37
FDI. *See* foreign direct investment
FIA. *See* foreign investment act
fiefdoms, 5, 14–16, 31–32, 134, 252, 268
  commoner, 92–98
  consolidation of, 64–66
  creation of, 6
  of Gulf Cooperation Council countries, 270–71
  institutions as, 47–49
  during 1970s boom, 86–98
  parallel, 95–98
  politics between, 90–92
  WTO and, 244
fishing rights, 48
Folk, Harold, 69
Ford Foundation consultancy mission, 7, 69–73, 76
foreign capital investment law, 130*t*

foreign direct investment (FDI), 178–79
  SAGIA and, 143–46, 151–84, *174*
  WTO and, 238, 241
foreign investment act (FIA), 143–84, 245, 248–50, 258
  minimum levels in, 164–67, 220
  negotiating details of, 152–56
  patterns of consultation over, 150–51
  regulatory fragmentation and, 158–60
  restricted sectors of, 153–55
  Saudization and, 147, 171, 173, 203, 220–21
  World Bank and, 146, 154–57, 169, 173, 176–77
Al-Fursan travel agency, 206

GCC. *See* Gulf Cooperation Council
Geddes, Barbara, 29
General Agreement on Tariffs and Trade (GATT), 226
General Investment Authority, 20, 32
General Organization for Technical and Vocational Training (GOTEVOT), 205, 207
Ghana, 269, 270n59
Gillibrand, Michael, 85n4
globalization, 138, 149, 231
  See also World Trade Organization
Goldberg, Ellis, 267
Al-Gosaibi family, 43
Al-Gosaibi, Ghazi, 99, 100
  as minister of labor, 212–21, 217*t*
  Zamil and, 199n87
gross domestic product (GDP), 74–75, *75*, 85, 140
  business ranking and, 177n222
  civil service wages and, *123,* 124
  Gulf Cooperation Council and, *274*
  public debt and, 137
group formation effect, 17n15
Gulf Cooperation Council (GCC), 75, 140, 178, 215n206
  customs union and, 237
  fiefdoms in other countries of, 270–71
  GDPs in, *274*
  rentier system and other countries of, 261, 267–68, *274*
  worker visas and other countries of, 186, 218, 273
  WTO and, 225, 233n48, 242n108
Gulf War (1990–91), 129–30, 137, 140

Hariri, Rafiq, 117
Hashemite dynasty, 42
health care, 103, 109, 129
  budgets for, *122*
  demographics and, 108
  foreign investment in, 155
  fragmentation of, 97–98
  See also Ministry of Health
Herb, Michael, 15, 133
Hermann, Klaus, 162n121
Higher Council of Ulama, 165
  See also ulama
Hijazi families, 48, 55, 172
Hijazi vice-regency, 63
historical institutionalism, 15, 39, 245n116
Human Resources Development Forum, 192n38, 231n29
Human Resources Development Fund, 210–11
Hungary, 182

Ibn Saud. *See* Abdulaziz bin Abdulrahman
illiteracy, 15, 49, 106
Imam Muhammad bin Saud University, 92, 126
IMF. *See* International Monetary Fund
immigration, 90, 115–16, 268
  illegal, 190
  worker visas and, 211–13
India, 263, 272
Indonesia, 270n60, 275
Industrial Bank, 67
Industrial Studies and Development Center, 96
informal markets, 201–2
Institute of International Education, 121n198
Institute of Public Administration (IPA), 70, 71, 73, 99, 104
institutionalism, historical, 15, 39, 245n116
insurance business, 9, 154, 155, 178, 235, 239, 254
Intel corporation, 173
intellectual property rights. *See* copyright laws
International Monetary Fund (IMF), 57, 138, 223
  Faisal and, 62
  foreign investment act and, 146
investor service centers, 168, 182
IPA. *See* Institute of Public Administration
iqama, 194, 199, 202
Iran, 268–69, 275
  military budgets of, 87
Islamic University (Medina), 92
islands of efficiency, 5, 28–29, 134
  foreign investment and, 178–79
  during 1970s boom, 98–101
  during oil crash, 127–28
  SAMA as, 56–58, 99
  sector-specific projects and, 32
  WTO and, 233

Jameel, Abdullatif, 236–37
Japan, 20n23, 249
Al-Jeraisy, Abdulrahman, 236
Jiluwi family, 62
Jizan Economic City, 180
joint ventures, 26, 111–14, 128
    foreign investment act and, 149, 163, 178–79
Jomaih family, 53, 241
Jones Luong, Pauline, 266–67
Jordan, 19n17, 27n49, 269n55
    Qualified Industrial Zones in, 224n2
Jordan, Robert, 237n79
Jubail, development up, 99–100
judicial system. *See* legal reforms
Juffali, Khalid, 53, 231

Kaaki family, 55n93
Kader agency, 180
Kamel, Abdallah Saleh, 173n196
Al-Kamel, Saleh, 231
Kanoo travel company, 204–8
Karl, Terry Lynn, 16n12, 265n32, 266, 268n50
Khalid bin Abdulaziz, 64, 85–103
Khashoggi, Adnan, 47n43, 65, 110
    Ministry of Defense and, 88
    National Guard and, 89
Khashoggi, Mohammad, 53
kickbacks, 54
    See also corruption
King Abdulaziz City for Science and Technology, 242–43
King Abdullah Economic City (KAEC), 173, 178n226, 179–80
    website for, 173n198
King Abdullah University of Science and Technology, 180
King Fahd Holy Koran Printing Complex, 126
King Saud University, 126
Kohli, Atul, 261, 262n25, 270
Korea, South, 20n23, 249, 261
Krimly, Rayed, 83n166, 263
Kuwait, 19, 268, 275
    Gulf War and, 129–30, 137, 140

labor markets, 185–92, 201–2, 215
    domestic workers in, 100, 199, 213
    regulation of, 130*t*, 138
    segmentation of, 187–89
    See also Saudization; visas
legal reforms, 127, 130–31, 130*t*, 256
    foreign investment act and, 158–59
    WTO and, 229, 239–41

    See also sharia courts
Lehner, Norbert, 158n97, 196n66
Levi, Margaret, 272
Libya, 3n1, 264, 267, 272, 275
line agencies, 31, 94–95, 97
    foreign investments and, 152, 238
    during oil crash, 129
    project-based development and, 102

Majlis Ash-Shura, 51–52, 63, 139–40, 257
    foreign investment act and, 146–49
    Saudization and, 209–10, 214
    WTO and, 228, 234, 236–37
Malaysia, 182
Mandil, Sulaiman, 153
Manpower Council (MPC), 190, 193, 203
    Decree 50 of, 193–94, 196
    Ministry of Labor and, 212
Mansour, Mourad, 206–7
Marafiq (utility company), 178
Mas'ud, Mohammad Ibrahim, 53
McGregor, R., 67n33
Mecca, 127
Medina, 127
Mexico, 269–71
Migdal, Joel, 7, 33, 260
migration. *See* immigration
minah, 43
Ministry of Agriculture, 44, 97, 235
Ministry of Air Force, 45
Ministry of Civil Service, 187
Ministry of Commerce (MoC), 32, 49–50, 93, 248
    commercial registration by, 164, 175
    cover-up businesses and, 165
    creation of, 56
    data collection by, 77
    "expediter" offices in, 161
    foreign investment act and, 159–61, 170–71, 175–76
    Saudization and, 190–91, 195
    sharia and, 127
    WTO and, 224, 227–29, 233–36, 243, 244, 252
Ministry of Commerce and Industry. *See* Ministry of Commerce
Ministry of Communications, 45–46, 48, 97
    budgets for, *122*
    under Khalid, 86
    transport and, 102
Ministry of Defense and Aviation (MoDA), 31, 46–47, 242, 249
    budgets of, 88, *122*, 125
    under Khalid, 87–88

Ministry of Defense and Aviation (MoDA) (*continued*)
  Saudization and, 191
  under Sultan, 14, 23, 64, 65, 87–91
Ministry of Economy, 46, 51, 170
Ministry of Education, 79, 85–86
  budgets for, 92
  creation of, 44
  health care and, 98
Ministry of Finance (MoF), 54, 97, 132
  under Abalkhail, 91
  under Abdulaziz, 44
  under Faisal, 67
  foreign investment act and, 146, 149, 151
  Ministry of Communications and, 46
  under Musa'd, 102
  under Sulaiman, 54–56
  WTO and, 229
Ministry of Foreign Affairs (MoFA), 42, 46, 172n186
Ministry of Health (MoH), 77n129, 98
  Saudization and, 190, 195
Ministry of Higher Education, 85–86, 98
  See also Ministry of Education
Ministry of Industry and Electricity (MoIE), 85–86, 96
  foreign investment and, 144–46, 160
  SAGIA and, 168
Ministry of Information, 176
Ministry of Interior (MoI), 14, 31, 89–91, 102, 249
  under Abdulaziz, 44
  under Fahd, 90, 129
  foreign investment act and, 152n62, 159, 163, 170, 176, 181
  Municipal and Rural Affairs and, 86, 90
  under Naif, 139
  Recruitment Affairs Department of, 190
  Saudization and, 190, 191, 193, 198
  sharia and, 127
  venality charges against, 202
  worker visas from, 165, 211
  WTO and, 229, 242
Ministry of Justice (MoJ), 31, 66, 79, 144
  foreign investment act and, 170
  regulatory reform and, 159
Ministry of Labor (MoL), 32, 248
  Manpower Council and, 212
  Saudization and, 186, 189–90, 193–94, 203n118, 211–22, 217t
Ministry of Labor and Social Affairs, 93, 98, 189, 211–13
Ministry of Municipal and Rural Affairs (MoMRA), 86–87, 90, 97, 129

Ministry of Petroleum, 64, 100–101, 125–26, 153, 168
Ministry of Planning, 20, 86, 97, 170
Ministry of Post, Telegraph, and Telephone (PTT), 86, 129, 235
Ministry of Public Works and Housing, 85–86, 98, 107
Ministry of Social Affairs, 106, 211–13
Ministry of Transport, 102
Mish'al bin Abdulaziz, 45–49
Mit'eb bin Abdulaziz, 45, 48, 86–87, 163
Mohammad bin Abdulaziz, 132
Montagu, Caroline, 183n243
moonlighting, by bureaucrats, 112
Moore, Pete, 14n6
Moosli, Ahmad, 46
morality police, 78, 79, 98, 126, 200n97
Morocco, 224n2, 269n55
mortgages, legal issues with, 127
mostazafan, 269
MPC. *See* Manpower Council
mu'aqqibs, 26, 27
  See also brokers
Mubarak, Hosni, 12
mudiriyat, 42
"mujahedeen" (i.e., police), 204n119
Musa'd bin 'Abdulrahman, 50
Musa'd bin Saud, 46
mutawwa, 200n97
Mvukiyehe, Eric, 267

Naif bin Abdulaziz, 14, 89, 139
  entrepreneurship and, 152
  foreign investment act and, 148
  Saudization and, 190, 193–94, 203, 205–9
  WTO and, 227
Naif, Mohammad bin, 175, 176
Najdi families, 42, 48, 53, 89, 94, 101, 113
  loyalty of, 88n23
Najdi Tuwaijri family, 89
Al-Namlah, Ali, 193, 194
National Commercial Bank, 55n93
National Company for Cooperative Insurance, 254
National Guard
  under Abdallah, 14, 17, 23, 89–91
  under Faisal, 65–66
  and Royal Guard, 132
  under Saud, 47
National Industries Ordinance, 114
National Society for Human Rights, 219
Nazer, Hisham, 93–96, 100

nepotism, 31, 70–76, 112–13, 137
  See also corruption
Nigeria, 264, 275
notary services, 127, 164, 176

"obstacle report," 169–72
O'Donnell, Guillermo, 259
oil crash, 118–28
  policy stagnation after, 128–31, 130t
oil rents, use of, 2, 11, 14–16, 28, 43, 60, 75, 85, 132, 267, 271
Oman, 19
one-stop shop. *See* investor service centers
Organization of Petroleum Exporting Countries (OPEC), 128
  Asian economic crisis and, 137
Ottoman bureaucracy, 54
"overstayers," 199

Pakistan, 272
patents, 51n74, 243
  See also copyright laws
path dependency, 16, 55, 132–33
patrimonialism, 16, 43–45, 47, 59–64, 259
patronage, 3, 20–21, 54–58, 248–49, 259
  in Brazil, 29
  bureaucratic jobs and, 5, 16–18
  definitions of, 21–22
  effective administrators and, 99
  intellectuals and, 22
  See also elites
Persian Gulf War, 129–30, 137, 140
Pertamina conglomerate, 270n60
Petroleum College, 70n67, 99
Petromin corporation, 100–101, 125–26, 128–29
Pharaon, Ghaith, 74, 110
Pharaon, Rashad, 53
piracy, software. *See* copyright laws
policy-making, 11
  under Faisal, 66–67
  after oil crash, 128–31, 130t
  parallel fiefs and, 95–98
  under Saud, 45
  under segmented clientelism, 29–35
Ports Authority, 101, 173, 180, 254
Portugal, 182
postal service, 54, 86
Presidency of Civil Aviation (PCA), 195, 206, 207
Presidency of Sports and Youth Affairs, 87
prime minister post, 64, 132
private sector. *See* business groups
privatization, 1–2, 9, 31
  Aramco and, 139
  capital market reform and, 255
  Ports Authority and, 101, 254
  Saudi Telecom and, 155, 254
  WTO and, 240
Public Works Department, 68

Al-Qadhafi, Muammar, 267
Qatar, 268, 275
Qualified Industrial Zones, 224n2
Al-Quraishi, Abdulaziz, 104

Al-Rasheed, Madawi, 27n52, 195n59
Real Estate Development Fund, 120
religious organizations, 79–80, 92, 126–27, 130, 170, 243, 260n18
  cooperative insurance and, 239, 254
  economic globalization and, 231
  education and, 92, 126, 188
  missionaries and, 126
  Zakat and, 77n126, 171, 232
  See also morality police; sharia courts
rentier state debate, 9–10, 15, 37–38, 44, 48, 78, 133, 138
  critique of, 2–6, 263–75
  history and, 55, 60
  state autonomy and, 18–20, 57, 60, 265–71
  state capacity and, 33–34, 49, 51, 57, 76, 271–73
  state fragmentation and, 5, 80, 157, 267–71
rentier system, 15, 38, 246, 259–75, *274*
  mentality of, 265n38
  providing jobs in, 187n3, 188
  sociology of, 2–4
Riyadh Development Authority, 32
Al-Riyadh Development Authority, 32, 101
Riyadh Economic Forum, 167
Riyadh governorships, 47, 62, 129
Roniger, Luis, 21, 22, 25n42, 26n47
Ross, Michael, 17n15
Royal Commission for Jubail and Yanbu (RCJY), 151, 163, 168, 178, 180
Royal Guard, 46–47, 132

Sabban family, 43
SABIC. *See* Saudi Arabian Basic Industries Company
SAGIA. *See* Saudi Arabian General Investment Authority
Salem, 'Id bin, 47n43, 53
Salha, Najib, 55n93
Salman bin Abdulaziz, 62, 99n92, 101, 195n57, 205, 210

SAMA. *See* Saudi Arabian Monetary Agency
Saqqaf, Omar, 67
Sattam bin Abdulaziz, 204–5
Saud bin Abdulaziz, 45
  entourage of, 48
  Faisal and, 46–47, 59n110, 61–64
Saudi Arabian Basic Industries Company (SABIC), 28, 100–101
  foreign investment act and, 145, 178, 181
  during oil crash, 128
  WTO and, 225, *226*, 233, 244
Saudi Arabian General Investment Authority (SAGIA), 20, 32, 247–48, 251
  under Dabbagh, 172–80
  fate of, 182–84
  foreign investment act and, 143–46, 151–84, 249–50
  Intel and, 173
  investor service centers of, 168
  "obstacle report" of, 169–72
  Saudization and, 185–86
  as super-*wasta*, 173–78
  WTO negotiations and, 224
Saudi Arabian Monetary Agency (SAMA), 28, 255, 264
  under Faisal, 67, 69–70
  foreign investment and, 178–79, 238
  as island of efficiency, 56–58, 99, 134
  Saudization and, 190–91, 195
  sharia and, 127
  WTO and, 233, 235, 238, 239, 242, 244
Saudi Arabian Standards Organization (SASO), 107, 229
Saudi Aramco. *See* Aramco
Saudi Electricity Company, 178
Saudi Gas Initiative, 139
Saudi Industrial Development Fund (SIDF), 96, 108, 120
Saudi Oger, 117
Saudi Ports Authority, 28
Saudi Telecom, 9
Saudization, 31, 139, 185–222, 230–31, 245, 248, 256
  avoidance of, 192–203
  certificates of, 171, 189, 202
  evaluation of, 215–22
  foreign investment act and, 147, 171, 173, 203, 220–21
  institutional consolidation and, 210–21
  private sector and, 191–92, 202–10
Savola Group, 232
SEC. *See* Supreme Economic Council

segmented clientelism. *See* clientelism
September 11th attacks, 172n191
Shakir, Ibrahim, 47n43
Sharbatli, Hasan, 54–55
sharia courts, 79, 127, 129, 243, 249, 257
  See also legal reforms
Al Sheikh family, 101
Shiite Muslims, 115n182
SIDF. *See* Saudi Industrial Development Fund
sinecures, 23–24, 72
Smith, Benjamin, 267
South Korea, 20n23, 249, 261
Spiekerman, Ernest T., 68n41
sports, 87
state autonomy, 19–20, 133–34, 247, 265–71
  rentier theory and, 3–6, 15, 38, 265
state capacity, 33, 252, 255–56, 263–64, 271–73
state-society relations, 3, 58–60, 69, 108–18, 246
Al-Suhaimy, Jammaz, 255
Sulaiman, Abdallah, 42, 53–57
Sulaiman, Abdulaziz, 54
Sulaiman Al-Hamdan family, 53
Sulaiman, Hamad, 54, 55
Suleim, Soliman, 99
Sultan bin Abdulaziz, 187
  Ministry of Communications and, 46n31
  Ministry of Defense and, 14, 23, 64, 65, 87–91
  Saudization and, 195, 205, 206
Sultan bin Salman, 205–7, 210
Supreme Commission for Tourism, 205–7
Supreme Economic Council (SEC), 139–40, 257
  foreign investment act and, 146, 148, 151, 153–55, 170
  privatization strategy of, 254
Supreme Petroleum Council, 128
Supreme Planning Board, 69
Suroor, Mohammad, 46, 54n93
Suwayyel, Ibrahim, 46n34
Syria, 270n59

Taher, Abdalhadi, 100
Talal bin Abdulaziz, 45–46, 48
Tariqi, Abdallah, 46, 53
tasattur. *See* "cover-up" businesses
taxation, 5, 15, 77, 130*t*, 265
  foreign investment act and, 149–50
  of foreign workers, 253n3
  imports and, 225
  regulatory power and, 272
  Zakat, 77n126, 171, 232
Al-Tayyar, Nasser, 204n119, 205–7
Tayyar travel agency, 205, 206

telecommunications industry, 163, 264
　foreign investment in, 154–55, 238
　regulation of, 179n227
Thelen, Kathleen, 16n10, 19, 245n116
transparency, 9, 28, 89, 139, 238
　foreign investment act, 148, 151, 152, 170–71
　World Bank and, 163n63, 238, 242
travel agents, 203–8
Turki bin Abdulaziz, 87
Tuwaijri family, 89, 195n59

ulama, 78–80, 101, 133
　on "cover-up" businesses, 165
　tenancy of, 126–28
umara (regional governors), 41
Umm al-Qura University, 92
unions, 29, 81–82
United Arab Emirates, 138, 139, 157, 215n206, 261, 268, 275
　Emaar corporation of, 173, 179
　foreign investment in, 172n190
United States–Saudi Joint Economic Commission, 108
University of Petroleum and Minerals, 70, 128
University of Sciences and Technology, 256
U.S. Army Corps of Engineers, 87
utility tariffs, 119–20

Vandewalle, Dirk, 271, 272, 275
venality, 161n119, 202
　See also bribery
Venezuela, 16n12, 264, 267, 268, 275
vertical links, 5, 18, 24, 246
　development of, 16, 20–21
　patrimonialism and, 275n71
veto players, 31, 240, 252–53
　ad hoc, 208–10
　low-level, 32–33, 107
　worker visas and, 216
"virtual" capital, 166
visas, 165
　brokers in, 27, 198–203, 212–20, 217t, 273
　Saudization and, 191, 200, 211–19, 217t
　visitor, 172n186, 199n
Vitalis, Robert, 63

wakeels, 110
Washington Consensus, 138, 149
wasta, 27, 116
　SAGIA and, 173–78
　Saudization and, 188, 201, 208, 213
　See also brokers
Waterbury, John, 254n5
water policies, 97, 102–3, 119–20, 163
Weberian bureaucracy, 15, 31
Weinthal, Erika, 266–67
Wibbels, Erik, 267
World Bank, 69, 138, 139, 223
　SAGIA and, 146, 154–57, 169, 173, 176–77
　on transparency, 163n63
World Bank Governance Indicators, 24
World Trade Organization (WTO), 9, 146, 223–45, 248–52
　Chambers of Commerce and, 236–37
　domestic Saudi interests and, 226–29
　Faqih's negotiations with, 234–36
　implementation of, 224–25, 241–45
　regulatory changes and, 238–40
　Saudi private sector and, 230–33
　Saudi reforms unrelated to, 240–41
　state-business dealings and, 233–34
　Yamani's negotiations with, 237–38

Yamamah armament project, 126n221
Yamani, Hashim, 146, 148
　WTO and, 237–39, 242n106
Yamani, Zaki, 94–96, 100–101
Yanbu, 99–100
Yemen, 43n11
Young, Arthur, 43, 56–57
Youth Welfare, 14

Zahran, Siraj, 54n93
Zaidan, Mohammad, 54n93
Zaini, Omar, 72
Zakat, 77n126, 171, 232
Al-Zamil, Abdalrahman, 99, 150n45
　Gosaibi and, 199n87
　on Human Resources Development Forum, 192n38, 231n29
　Saudization and, 230–31